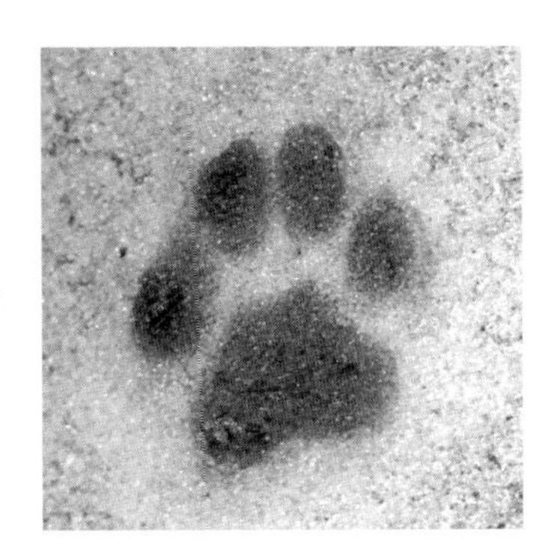

TRACKING & THE ART OF SEEING

TRACKING & THE ART OF SEEING

How to Read Animal Tracks & Sign

SECOND EDITION

By Paul Rezendes

Quill
A HarperResource Book
An Imprint of HarperCollins *Publishers*

TITLE PAGE: *Mt. Katahdin in Baxter State Park, Maine, "on fire" in the early light of a fall morning.*

HarperCollins books may be purchased for educational, business, or sales promotional use. For information please write: Special Markets Department, HarperCollins Publishers, Inc., 10 East 53rd Street, New York, NY 10022.

SECOND EDITION

Designed by Jill Shaffer
Photography © Paul Rezendes, except for the following: Page 100 © M. A. Chappell/Animals Animals; page 110 © Robert Maier/Animals Animals; page 117 © Robin Redfern/Animals Animals/OSF; page 144 © Leonard Lee Rue III/Animals Animals; page 145 © Bill MacDonald; page 161 © Mark Picard/New England Stock Photo; page 163 © William Fournier; page 209 © Daryl Pederson/AlaskaStock Images; page 219 © Clark Mishler/AlaskaStock Images; page 231 © Bradford Glass/New England Stock Photo; page 233 © Susan C. Morse; page 239 © Robert A. Lubeck/Animals Animals; page 284 © John Hyde/AlaskaStock Images; page 300 © Douglas B. Elliott; page 312 (top, bottom right) © Mark Elbroch

Black-and-white drawings © Heather K. Lenz
Quick Reference Charts designed by Robert Metcalfe

Second edition produced by Julie Stillman, Shelburne, Vermont, and Verve Editions, Burlington, Vermont.

Library of Congress Cataloging-in-Publication Data

Rezendes, Paul.
Tracking & the art of seeing: how to read animal tracks & sign/by Paul Rezendes. — 2nd ed.
p. cm.
Includes index.
ISBN 0-06-273524-1
1. Animal tracks. 2. Animal tracks — Identification. I.Title.
II. Title: Tracking and the art of seeing.
QL768.R49 1999
599—dc21 98-45867
CIP

---

04 10 9 8 7 6

To Paulette
and
To the wild within

CONTENTS

Preface to the Second Edition

In the six years since the publication of the first edition of this book, tracking has become somewhat more sophisticated. On the scientific side, several efforts are underway to establish standardized terminology and methodologies, to develop a network of trackers nationwide, to institute courses at the university level, and to legitimize tracking as a reputable scientific tool in wildlife studies, management, and conservation. On the recreational side, tracking is probably not destined to become the next craze in adventure sports, but serious interest in tracking is definitely on the rise. Part of this interest is fueled by people's desire to connect more intimately with nature and to better understand themselves. I hope that *Tracking & the Art of Seeing* will continue to accompany many enthusiasts on their tracking expeditions.

In this revised edition, I've expanded the book to include a chapter on bird tracks and sign, with more than 35 new photos and illustrations. A quick reference guide to track and trail measurements, summarized in one easy-to-use section at the end of the book, should be a welcome addition for all who want to verify what they're looking at in the field. With the increased numbers of skilled trackers collecting and analyzing data, there have been some new and exciting discoveries among my own staff and students. In this edition, I've included new information about black bears and their marking trees, tracking bobcats without tracks by locating and following their scent posts using your sense of smell, new documentation on distinguishing northern and southern flying squirrels through their tracks, and new discoveries about fishers stashing food in trees. Wherever possible, I've updated studies regarding the status of wildlife species across North America, including some exciting information on mountain lions in the northeast. Finally, all measurements have been updated and refined to reflect new data that my staff, my apprentices, and I have collected since the publication of the first edition.

Although much of the tracking information in this book is from my own observations in the field, I've consulted with many people and used various research materials to complete this project. Many thanks are due to the people who were instrumental in the publication of the first edition: Regan Eberhart, formerly of *Harrowsmith Country Life*, Wayne Grady, editor and writer, Sandy Taylor and Howard White, former editor and former director, respectively, at Camden House Publishing, and Jill Shaffer, designer and art director. A special thank-you to my friend Heather K. Lenz for her extraordinary illustrations. For their review of the original manuscript or parts thereof, I would like to thank James Cardoza, Paul Lyons, David Brown, and Susan Morse.

Thank-you's for assistance in collecting field data and in obtaining photographs of animals, their feet and their tracks for the first edition go to Bruce Wilson, John McCarter, Paul Wanta, Doug Elliott, Dave Taylor, Sharon Young, Todd Fuller, Bill Fournier, Mark Picard, Bill Byrne (and his chipmunk "Chuckles"), and Tom and Cheryl Hoenig. Also thanks to Tom French, Tom Decker, Bob Arini, Lyle Jensen, Diane Davis, Dr. Mark Pokras, Pam Landry, John Spaulding, Adrienne Miller, and John Croxton. A special thank-you to my brother Ken Rezendes and his wife, Carol, for helping us out during our research trip to Alaska.

I would also like to thank the following people for their scientific consultations on wildlife matters while doing research for the first edition: Dr. Ken Elowe, Maine Division of Fisheries and Wildlife; Daniel Harrison, professor of wildlife, University of Maine; Herman Griese, wildlife biologist, Alaska; Gordon Bachelor, supervising wildlife biologist, New York Fish and Wildlife Department of Environmental Conservation; Mark Brown, wildlife biologist, New York; Dr. Illar Muul, president of Integrated Conservation Research in Harpers Ferry, West Virginia; Dennis Voigt, Ministry of Natural Resources in Ontario, Canada; and Dr. John Theberge, University of Waterloo, Ontario, Canada.

For this second edition of *Tracking & the Art of Seeing*, I'd like to first thank my staff instructors, John McCarter and Mark Elbroch, for their integrity and dedication to the art and science of tracking, their careful review and thoughtful suggestions for improvements to the revisions, and for the too-numerous-to-mention refinements to track and trail measurements from their hours of collecting data in the field. A special acknowledgment to John for his contribution to the understanding of bobcat sign, and to Mark for his understanding of track patterns and animal movements. I am deeply indebted to them both.

An additional thank-you to Heather K. Lenz for the illustrations in the bird chapter; to Jill Shaffer for the book design; Bob Metcalfe for his excellent design of the quick reference charts; Gary Chassman of Verve Editions for helping to find the book a new home with HarperCollins; and to Tricia Medved, our editor and her assistant, Greg Chaput, at HarperCollins. To Julie Stillman, packager, editor, friend, and advisor (to name just a few), a huge thank-you for her persistence in multiple efforts on my behalf.

Thanks also to David Small for his suggestions and review of the bird chapter; Susan Morse, Dr. James Halfpenny, the Massachusetts Division of Fisheries and Wildlife, and Kristi MacDonald and Dr. George Amato from the Wildlife Conservation Society, for their assistance in updating wildlife information and population trends; to Alcott Smith and my apprentices, past and present, for their help with various wildlife studies and data collection for this second edition: Bob Metcalfe, Lydia Rogers, Keith McCormick, Dave Kay, Eleanor Marks, Charlie Perakis, Paul Wanta, Kayla Sanford, Colin Crawford, Heather Lenz, Jonathan Sargent, Kent Hicks, Jeff Kunz and Linda Huebner. An extra thank-you to Bob, Lydia, and Keith for helping collect bird track data.

Many thanks also to my family, friends, and students for their continuing love, encouragement, and support and to my office manager, Mary Ellen Scribner, who sees what needs to get done and takes care of it.

Finally, but most of all, a resounding thank-you to my wife and partner, Paulette M. Roy, for her persistence with the details so things get done right, and for the generous donation of her many, many other talents to this project. We work as a team in each and every phase, not only of this book, but of our many business pursuits in general, and our lives together as a whole. She truly deserves at least half the credit, if not more. To her I offer my deepest gratitude and love.

INTRODUCTION

A white-tailed deer run through golden hedge-hyssop.

What Tracking Means

AT ONE TIME, being able to read tracks and sign was a matter of life and death. Knowing where the food was and what the predators were doing could mean the difference between survival and extinction. Most of us would have to go back pretty far to find ancestors for whom that was literally true. I believe, however, that in a different way, it is still true.

Many people today think tracking is simply finding a trail and following it to the animal that made it. They conjure up images from old Westerns in which the Indian scout helps the good guys find the villain. Or they imagine a hunter "tracking down" a large game animal in the deep woods. Even those with some tracking skills think that the most important aspect of tracking is finding the next track. I've been asked if I could track a grasshopper over a rock. I've seen a couple of trained trackers study a track carefully—go through a few complicated techniques designed to determine whether the animal turned or not—trying to find the next track. And when I asked them whether they knew what animal they were tracking, I found that they often did not.

Don't get me wrong. I'm not saying that finding the next track isn't important, because it is, or that following an animal's trail through the forest has no significance, because it does. But I think the true meaning of reading tracks and sign in the forest has been pushed into the background by an overemphasis on finding the next track.

As I said, I think reading tracks and sign in the forest is perhaps as much a survival skill today as it was to our ancestors, but I'm talking about a different kind of survival. What I'm *not* talking about is putting food on the table. Although I have been a vegetarian for more than thirty years, I don't have a problem with people who hunt for food. I currently devote much of my time to landscape photography, but for many years I made part of my living as a wildlife photographer. In that sense, I was an avid hunter, and all the skills involved in hunting an animal for food apply to hunting an animal for a photograph. As any hunter knows, finding the next track is one of the hardest ways to find the animal. In a survival situation (or for a wildlife photographer trying to make a living), spending a lot of time and energy finding the next track could mean starvation.

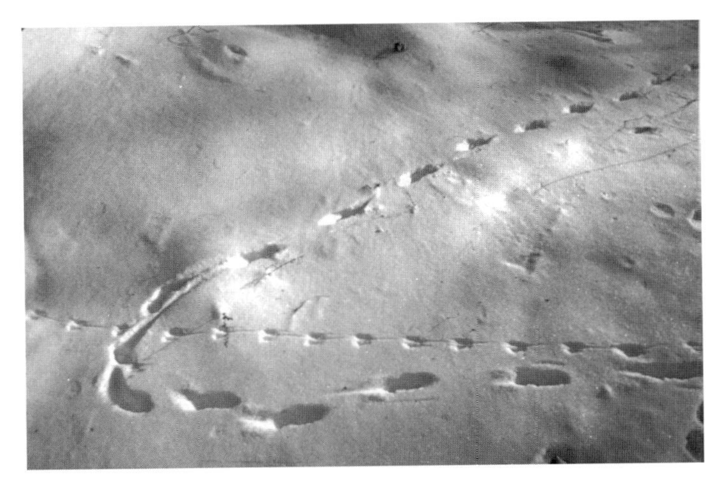

FIGURE I.1
Many interesting stories unfold when you can interpret animal tracks. A white-footed mouse left its telltale trail in the snow (moving from right to left). When a red fox happened along and picked up the scent, the fox abruptly changed its direction of travel to investigate the mouse's trail.

Tracking an animal is opening the door to the life of that animal. It is an educational process, like learning how to read. In fact, it *is* learning how to read. Following an animal's trail may bring you closer to the animal physically, but, more important, it brings you closer to it in perception. The longer you follow the animal, the deeper you enter into a perceptual relationship with its life. If you spend half an hour finding the next track, you may have learned a lot about finding the next track but not much about the animal. If you spend time learning about the animal and its ways, you may be able to find the next track without looking.

If you know an animal well, you will know where to look for it and when. That is one kind of survival knowledge. If you are to be successful in your search for an animal, you must sit and wait for it, and sitting and waiting isn't much good unless you're sitting and waiting in the right place.

You will also know yourself better. That is another kind of survival knowledge. The more intimate we become with other lives, the more aware we are of how those lives connect with and affect our own. There may be only a few obvious connections at first—two animals in the same woods, hearing the same sounds, smelling the same smells—but as we track the animal farther, we find that its trail is our own trail. As it moves, it affects its surroundings. What changes the animal changes its environment, and thus changes us. There is no separation; its fate is our fate. We are tracking ourselves in a sense.

We don't need tracks to track an animal. For much of the year, the forest is far richer in sign than it is in tracks.

Sometimes there are no tracks at all, but there is never a square yard in the forest that does not tell us something about the wildlife within it. The forest is speaking to us all the time. Some of my tracking students have looked curiously at me when I've said that, but it is most certainly true. The trees speak to us. Like the good Duke in *As You Like It,* the attentive tracker "finds tongues in trees, books in the running brooks, sermons in stones." Sometimes sign speaks in a whisper—a bent twig indicating that a deer has stopped to browse; at other times, it's a loud scream at the top of its lungs: massive hemlock dieback—porcupine! We can ask the bobcat where the snowshoe hare is, and he will leave us a trail to it. But we can just as easily ask the blueberry bush; its nibbled branches will say "snowshoe hare" just as assuredly, and be much less dependent on snow or other substrate conditions. Corn in a raccoon's scat will tell you where he has been. There is a whole novel in an owl's pellet. Unfortunately, most of us just walk by these signs, not hearing or noticing a thing.

Ultimately, tracking an animal makes us sensitive to it—a bond is formed, an intimacy develops. We begin to realize that what is happening to the animals and to the planet is actually happening to us. We all are one. Tracking and reading sign help us to learn not only about the animals that walk in the forest—what they are doing and where they are going—but also about ourselves. For me, this interconnection *is* survival knowledge and the true value of tracking an animal.

Predators and Prey

FIVE HUNDRED YEARS AGO, when Christopher Columbus first reached the shores of the New World, he brought with him a European view of nature that was diametrically opposed to that of the people he encountered here. Many of the aboriginal people who had lived on this continent for thirty-five thousand years before that perceived the forest and its creatures as extensions of themselves. They were part of the forest; they *were* its creatures. What happened to other creatures of the forest happened to them.

To the fifteenth-century European mind, nature was something that had to be controlled. Their God, after all, had given humankind dominion over the animals. The forest was something to be feared. Historian Kirkpatrick Sale, in *The Conquest of Paradise*, calls the European attitude toward nature "ecohubris." Never before in the history of civilization, he writes, "was the essential reverence for nature [so] seriously challenged, nowhere did there emerge the idea that human achievement and material betterment were to be won by *opposing* nature."

Columbus and those who came after him could have learned a lot about living harmoniously with nature from the people they encountered here, but they didn't. Instead, they imposed their European view of nature on North America. The trees in the forest were turned into lumber; the cleared land was property; the creatures of the forest were either prey or nuisance animals. Predators—mountain lions, wolves, bears, weasels, raptors—were dangerous, wanton killers often preying on the very animals the Europeans wanted to kill themselves. People shot the animals, poisoned them, trapped them, burned their habitats, dug their young out of their burrows, and felt proud of themselves for doing so.

This attitude seems prevalent even today. Some lumbermen still look at a tree and see only so many board feet. I still hear a few hunters complain about the coyote, whose role has gone from trickster to scapegoat. If the snowshoe hare or deer hunting isn't what it used to be, it's because the coyotes must be killing them off. Other people believe that if someone's dog or cat is missing, a coyote must have got it. They don't consider that *Canis latrans* was here long before *Homo sapiens;* that the snowshoe hare, deer, and coyote have coexisted since the Pleistocene without any of them becoming extinct; or that in the five hundred years since Europeans arrived on the scene, numerous native North American species of animals *have* become extinct or nearly so.

As of October 1997, a list of endangered and threatened species in the United States included 447 animals, 58 of them mammals. Some are discussed in this tracking guide: the grizzly bear, gray wolf, red wolf, and eastern cougar (mountain lion). These animals have been pushed to the edge of extinction by human activity, either because we have hunted them for fur, bounties, or trophies or because we have appropriated their natural habitats.

There is only one predator species that worries me: *Homo sapiens*. We are the most powerful predator on the planet. No other animal has been so diligent in wiping out its neighboring species. In Massachusetts, where I live, I know of an instance in which hunters removed 576 deer from a fourteen-square-mile area in nine days. We have waged war on our own environment. At the height of the wholesale slaughter of bison that occupied most of the nineteenth century on the plains, some observers described it as a war. The native bison was replaced with imported cattle. This change required that native shortgrasses be replaced with faster-growing grasses imported from Europe. When coyotes and wolves appeared to be killing the cattle, war was declared on coyotes and wolves.

We called it predator control for a while, but the underlying attitude was the same: Predators (The Big Bad Wolf) were bad; prey (Bambi) were good. If we weeded out the bad, we would have more of the good. The same philosophy has been applied to crop management. Weeds, insects, and disease are bad; corn is good. If we kill off the weeds with herbicides, eliminate the insects with insecticides, and eradicate the diseases with fungicides, the corn will become more productive. But what are the effects of all these chemicals on the environment and on human health? That question often is not even considered. Domestic corn is not a viable species. If it were left to grow all by itself in the wild, it would soon become extinct. Wild corn, though it didn't grow as high or produce as many bushels per acre, was nonetheless a viable species. It withstood every natural predator nature could throw at it. It was eaten by wild animals, infested by insects, beset by fungi, drowned by flooding rivers, parched by droughts, and frozen by frosts, and still it survived. Its predators were not bad; its predators made it what it was.

Should we manage wildlife as we manage corn, by removing its natural predators? Do we want herds of white-tailed deer to be like the cows in our fields? What makes a deer a deer? Where did it get its speed, agility, and grace? What gave the bighorn its sense of balance and the moose and bison their size and power? How did the porcupine come to be? Many of the very qualities we associate with wild animals evolved as responses to predators; the deer's speed and agility and the sheep's acrobatics are escape tactics; the moose's and bison's strength is useful for fending off attacking wolves; a porcupine's quills are its armor. If we remove their predators, we remove the very forces that made them the animals we admire.

On Isle Royale, most of our early theories about the relationship between predators and prey have been tested and found wanting. Isle Royale is a 210-square-mile National Park in Lake Superior, part of the state of Michigan but only18 miles from the Canadian shore. In the late 1940s, rumors began to circulate that a pack of wolves from Ontario had crossed the ice and taken up residence on the island. At one time, defenders of the right of moose herds to regulate themselves would have advocated exterminating the wolves on Isle Royale. But Durwood Allen, then assistant head of the U.S. Fish and Wildlife Service's Wildlife Research Division, saw the new situation on Isle Royale as an opportunity to study the relationship between predators and prey in an environment unaffected by human activities such as mining, logging, road building, farming, and hunting. He realized that Isle Royale represented a return to "the pattern of primitive times," to the ancient ways of the forest before the coming of Europeans. Allen strongly believed that the wolves, far from wiping out the fragile moose population, would actually strengthen it. He thought that although the total number of moose would drop, wolves and moose would achieve an interrelated harmony, which would be better for the environment as a whole.

FIGURE I.2
A white-tailed deer feeds on dew-covered grass.

Isle Royale became one of the best opportunities to observe predator-prey relationships. High and low fluctuations in predator and prey populations persisted. These populations were affected by each other as well as by habitat, weather, and food resources. (These fluctuations may be an important function in a larger balance that is not yet fully understood.)

Researchers watching this interplay found that a healthy moose had developed enough defense strategies so that under normal conditions, it could avoid predation either by standing and confronting the wolves or by fleeing. Only rarely are wolves able to take a moose in its prime. This means that under normal conditions, wolves must test many animals before they find one that is vulnerable. This results in wolves having a 10% or lower hunting success rate. Of course, factors such as severe winters and deep snow increase predator efficiency.

These "test" chases result in separating the old, young, sick, and lame from the healthy members of the herd. Some research studies show that about one-third of the moose taken by wolves are calves and the rest are animals that are vulnerable or more than seven years old. Of twenty adult moose killed by wolves in two winters, ten had arthritic joints and six had periodontal infections that prevented them from eating effectively. As the wolves continuously test the healthy and remove the weak, they shape the physical characteristics of their prey, leaving healthier and stronger animals to breed. In turn, the moose challenge the wolves to become efficient and successful predators.

To me, this means that the moose is not separate from the wolf. The two animals form a single organism. Nature is not made up of separate enclaves—predators in this corner, prey in that corner—but of a totality, predator and prey living together in a "dynamic equilibrium," to use wolf biologist David Mech's terminology. We have labeled and separated the moose and the wolf, and in so doing we have lost sight of their essential unity. We also have misunderstood ourselves, for the biggest separation we have imposed on the world is between ourselves and nature.

When we experience the natural world, whether it be through tracking, hiking, or just walking in the woods, we are learning about ourselves and our role in nature's process. When we encounter nature, we also encounter ourselves.

Tracking as the Art of Seeing

OUR ENCOUNTER with nature is largely a matter of seeing, and it relates to the quality of attention in our lives. This is obviously important not only to the tracker in the forest but also in every aspect of our lives. Tracking may be a very good way to learn how to pay attention to our own existence.

This quality of attention is similar to a wild animal's awareness of its surroundings. Take a white-tailed deer, for example. Once you've watched one of these animals, you realize how sensitive it is to its own immediate environment. Its eyes catch every movement in the forest; its nose

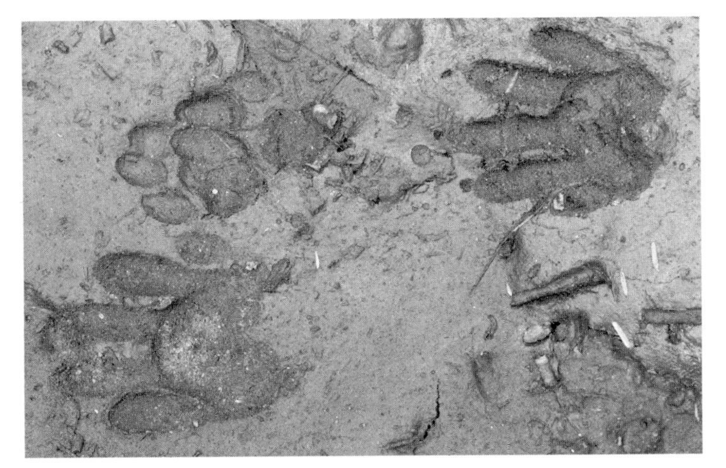

FIGURE I.3
Recorded in the mud are the detailed tracks of a raccoon and a house cat. The raccoon's front track is in the upper right, its hind in the lower left. The cat's double-registering track is in the upper left.

inspects every waft of air; its ears turn like highly sensitive antennae picking up the smallest vibrations. The deer is immersed in its senses, fully in its body; it is, as I sometimes say, living the wild within.

Most of us have lost touch with the wild within, lost our connection with nature and with ourselves. Henry David Thoreau said that we are asleep. As in sleep, our perceptions of ourselves are fragmented, narrow, and isolated, made up of half-remembered truths and out-of-context memories. In the forest, we feel threatened. Fear enters, and with it comes a need to control, to dominate. We think we are dominating the forest, but we are really trying to dominate our own fear of the forest.

Some Native Americans demonstrated a larger perspective of self. When they went into the forest to gather herbs, they did not always take the first plant they came across. Instead, they left it intact, maybe said a prayer for it, and placed an offering at its roots. They wanted the needs of the plant to be met, for they knew that if its needs were not met, their own needs would not be met either. If the animals and the plants did not survive, they would not survive.

Some Native Americans continue their traditional ways, but most of us live as if we were separate from nature. Whereas the deer is fully in its body, we have retreated into our minds. By thinking, we have set up parameters that divide the universe into things that can be categorized, and we call that understanding. This gives us a sense of power and control. We look at the forest and say, "That's a white pine. That's a white oak. Over there is a sugar maple," and we think we know the forest. But we have no real contact with those trees. We miss the details—the subtle curves of the branches, changes in the texture and color of the bark as the light fades, or the wind blowing in the dying leaves. We do not embrace the forest with our whole being; instead, we label it with our minds.

I notice this sometimes during my tracking classes. We'll be looking at a set of tracks on a muddy riverbank—some of us kneeling, others standing; some of us counting toe impressions, others looking for palm and heel pads. One of the goals is to identify the animal that made the tracks. Once we have determined the species, some of the students immediately jump up and are eager to move on. They've labeled the tracks, and now they think they know them. Other students, however, stay down on their knees, looking at the details of the tracks. They try to figure out how the animal was moving, whether it was running or walking, leaning toward the river or looking up at the trees. They are continuing to learn and to develop a sensitivity to their surroundings. This is what I encourage in my tracking classes and what I want to communicate in this book.

Thoreau called us sleepwalkers. Have you ever found yourself walking along a path in the woods and then suddenly realizing that the whole forest around you has changed? You started out among conifers, but now you are surrounded by deciduous trees. Or you realize that the birds are active and noisy, and you don't know when the change took place. You have awakened. You let the smell of fern leaf wash through you. You realize why you were asleep. You were talking to yourself, caught up in a familiar, endless dialogue. What were you talking about? You can't remember! "I never met a man who was truly awake," Thoreau said nearly 150 years ago, surrounded even then by the second-growth forest of northern Massachusetts. How much more applicable is this comment to people today. How much farther have we strayed from the wild within.

Our security does not lie in the control we have over nature, but rather in the quality of attention that we bring to our lives. If we care about our relationship with nature, or our relationship with other human beings, that caring demands our attention. Caring *is* attention. When we really care about another person, we want that person's needs to be met. We are present and attentive. That person's needs are our needs. We pay attention to them. There is then the possibility of sensitivity, intimacy, communication, and harmony. The tracker in the forest is in love with his or her surroundings. In nature, we are open to a larger perspective of self. We learn to walk carefully on this planet. We learn to *see* it.

How to Use This Book

One of the most natural (in fact, instinctual) responses to seeing the track of an animal on the ground is to ask, "What kind of animal made that?" One of the purposes of this book is to help you answer that question. Other, sometimes more interesting questions follow, and this book will help you answer those as well. But many subsequent questions depend on finding the correct answer to the first one.

It isn't as easy a question to answer as you may think. Many tracking books lead you to believe that all you have to do is compare the perfect track on the ground with the neatly drawn illustration of the perfect print in the book, and there you have it. In real life, animals don't always make perfect tracks, any more than authors always make perfect books. Tracks can change characteristics on different substrates (sand, snow, mud, clay, etc.); for example, a wolverine track in snow will look different from the same wolverine's track in mud. Tracks also can change from season to season. Sometimes an animal will spread its toes when it's walking, and sometimes it won't. No single illustration is going to cover all these possibilities. I have tried to include as many variations as space allows, and I believe that the extensive use of photographs will enhance the process of identification.

If tracks alone are complicated and changeable, sign can be even more so. The term *sign* refers to all the indications of an animal's passage through an area, or of its living in an area, that are not directly related to the imprinting of that animal's foot on the ground. These include obvious things such as droppings, or scat, remains of food, claw marks on trees or shrubs, and trails or corridors through the forest, as well as some not-so-obvious things, such as turned stones and stunted vegetation. Sign is an extremely important indication of animal activity, and getting a feel for it is an important part of becoming a competent tracker.

FIGURE I.4
The Alaska Range rises up behind the Toklat River in Denali National Park, Alaska. In the foreground, a bank of glacial silt has recorded the alternating walking gait of a caribou.

So answering the first question—What animal made this track?—is not always easy, but it can be made easier. One of the first things you should do is check the animal's range, which is given in this book. This will tell you which animals are supposed to be in a particular area. If you see what you think is a squirrel track and there are only gray squirrels in that area, you can eliminate a lot of footwork. Designated ranges for animals are not infallible, of course. They may be out of date or too large to be specific, or you may have an animal moving through or even into an area that is not native to it. Finding positive proof of the presence of an animal that is not supposed to be in an area can be an exciting experience, and there may be others who want to share it—especially the local, state, or provincial wildlife biologist.

The best way to find out what kind of animal you're dealing with is to gather a lot of supporting data. This often requires some simple equipment. Whenever I go into the woods, I carry my tracking kit, which contains a retractable tape measure at least six or eight feet long; a notebook and pencil (pens freeze); a magnifying glass; and a good, reliable tracking book.

Checking a single track may be important, but under certain conditions, it also can be misleading. If you learn anything at all from this book, I hope it is this: *Individual tracks are not always as important for determining species as are trail patterns.* A single canine print can look an awful lot like that of a coyote, a small wolf, a large fox, or a domestic dog, but wild canines walk differently than domestic canines, foxes walk differently than coyotes and wolves, and wolves have different patterns at certain gaits than coyotes. In addition, as I said before, you don't always have the luxury of a perfect track to work with. If you understand the language of trail patterns, however, you won't need perfect tracks. Substrate conditions can distort track sizes but will not alter the trail pattern to as great an extent. Most species can be identified by their walking pattern alone.

One or two measurements of a track may not be enough to identify it. You may happen upon an atypical stretch of an animal's trail, get some oddball numbers, and not be able to tally these with the averages given in most books. On the whole, I don't give mean measurements in this book because they don't mean much in specific cases. A single species can vary in size from one end of this continent to another; for instance, a gray wolf track in Alaska or Yukon Territory will have a larger mean measurement than a gray wolf track in Ontario's Algonquin Park. Instead, I give what I call track parameters. Track parameters give the smallest and largest sizes for each track or track feature.

For example, on a particular set of tracks near my home in Massachusetts, I collected the following trail widths: 1³⁄₁₆″, 1¼″, 1⅝″, 1¹⁄₁₆″, 1¼″, 1⅛″, 1″, 1⅛″, 1⅛″, 1¹⁄₁₆″, 1⅛″, 1″, 1″, 1″, 1⅝″, 1¾″, 1¾″, 1⅝″, 1½″, 1¼″, and 1″. (I discuss how to measure trail widths later in this section.) The strides for the same set of tracks were 14⅜″, 17½″, 28″, 26″, 12½″, 24½″, 10″, 9⅞″, 13¼″, 14¾″, 18½″, 17″, 15½″, 24½″, 18″, 24″, 19½″, 21½″, 21¾″, 30¼″, 24½″, and 22″. Thus, the trail width varied from a low of 1″ to a high of 1¾″, and the strides varied from 9⅞″ to 30¼″. If I have one or two numbers that are quite a bit higher or lower than the rest, I throw them out.

The above trail exhibited the 2-2 pattern typical of the weasel family—two prints side by side, one slightly ahead of the other, a space, two more prints, a space, and so on. The question was, which species was it? The weasel family includes animals ranging in size from the tiny least weasel to the river otter and wolverine. There are no least weasels in my area (according to the range information I have), so I eliminated them. The tracks were much too small to be those of a mink or any of the larger family members, such as the marten or fisher. This left two choices: the ermine and the long-tailed weasel. A glance at my notes told me that the parameters for the ermine's trail width were 1″ to 2⅛″, while those for the long-tailed weasel's were 1½″ to 3″. The tracks I had measured easily fell into the parameters of the ermine and overlapped only slightly with those of the long-tailed weasel. Since my tracks had quite a few measurements at the lower end of the ermine scale, I could also tell that they were those of a female.

Strides are not good indicators of species for the weasel family because there is a tremendous amount of overlap between members. The stride parameters given in this book are 9½" to 43" for the long-tailed weasel and 9" to 35" for the ermine. Obviously, my weasel could fit into both parameter ranges. Most measurements for my weasel were below 24", which is at the shorter end of the scale. This added more support to my ermine deduction.

If you experience difficulty squeezing your measurements into my parameter ranges, knock off your highest and lowest figures. In the previous case of the weasel, eliminating high and low scores would lower the parameters for trail width to 1⅟16" to 1⅝", which fits more neatly into the ermine slot. You also can simply take an average and see how that fits. Average trail width for my ermine was 1¼", again well within the trail width parameters.

When taking measurements, be sure that your technique is exactly the one I use. Obviously, if my measurements include claw marks (which they do) and yours stop at the end of the toes, our figures are not going to compare well. Note that authors of different tracking books use different methods of measuring, so my numbers may not always match those of other authors. Some trackers measure strides from the toe of one print to the heel of the next; some of mine are measured from toe to toe. Little things like that can make a big difference if you're talking about 12"-long Alaskan brown bear (or grizzly) tracks.

The remainder of this section focuses on the terms and methods I used to compile the track sizes in this book. Read them carefully, but remember that certain animals may require specific techniques that differ from the general rules laid out here. When that happens, I'll give you plenty of warning in the chapter on that animal.

TRACKS. When measuring tracks, get down on your knees, put your nose as close to the track as possible, and find out exactly where the track begins and ends. Make sure you are measuring the track and not the impression made by the animal's foot as it entered and exited the substrate. Tracks in snow can be a problem, as the hole at the snow's surface may be much larger or smaller

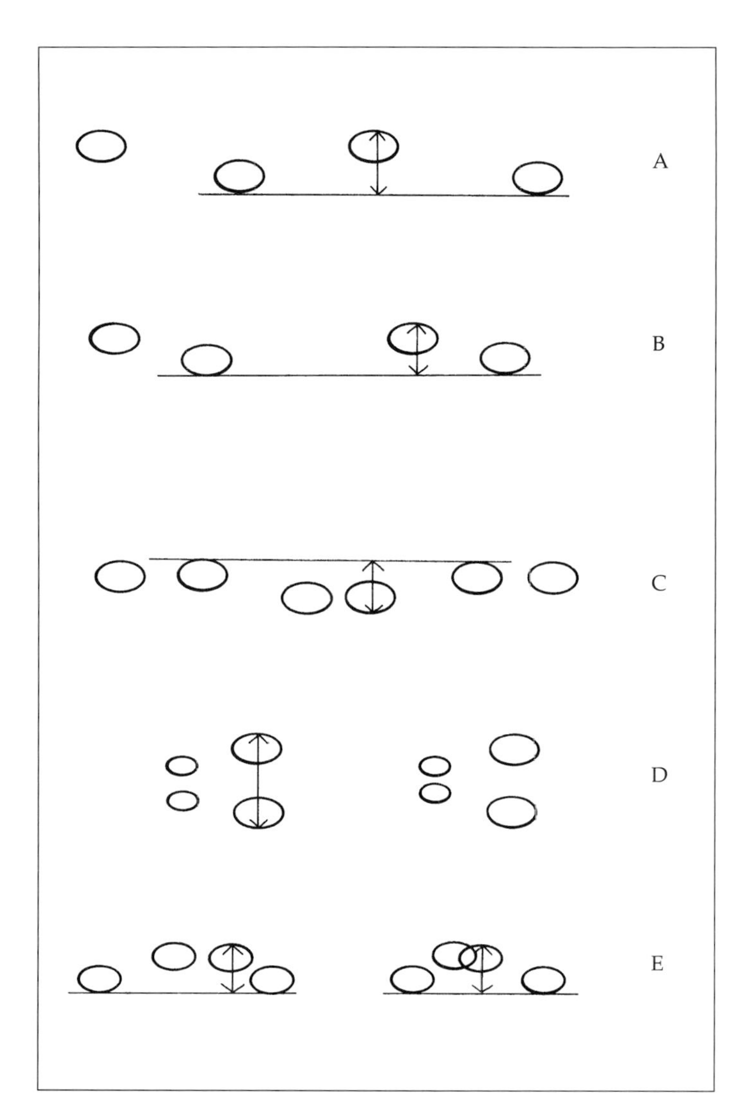

FIGURE I.5
This diagram illustrates how to take trail-width measurements of different track patterns.
A. Alternating walking or trotting pattern common to many species
B. 2-2 bounding pattern of the weasel family
C. 2-2 pattern (fast walk or trot) of bears, skunks, felines, and sometimes canines.
D. Bounding patterns of mice, chipmunks, squirrels, and many other animals
E. 3-4 and 4-4 patterns of the weasel family

than the actual track below. The hole may be larger than the track if there is foot drag at the surface or smaller if the animal's foot expanded when it put its full weight on it.

For some animals, such as bears and fishers, two sets of rear pads may show in the tracks. I call the one closest to the toes, and corresponding to the ball of the human foot, a palm pad. The rearmost pad I call the heel pad. In animals where only one pad usually shows behind the toes, such as in canine and feline tracks, I call this the heel pad as well. Technically, it is a palm pad, corresponding to the ball of the human foot, but since most people think of these pads as heel pads, I have chosen to use the familiar term.

Track width is measured across the widest part of the track, and length is measured from heel to nail, if nail and heel pad are registering. When referring to measurements given in this book, check for inclusion or exclusion of heel pads and nails. In some animals, and at some times for many animals, heel pads and nails do not show in the tracks, and so my track lengths do not include them. When this is the case, it is clearly stated.

TRAIL PATTERNS. I use different methods for measuring trail patterns, or the pattern a set of tracks makes on the substrate. Trail patterns include walking, trotting, side trotting, loping, bounding, and galloping gaits, and the methods for measuring each are illustrated in the text. For example, in alternating gaits and 2-2 patterns, I usually measure strides from toe to toe or nail to nail. With 3-4 or 4-4 patterns, strides are measured from toe to heel. In track groups, the group length is measured from heel to toe. To determine trail width, I measure from the outside of one track across the trail to the outside of the other (Figure I.5). Again, check the drawings in each chapter. In general, when measuring strides, track groups, or trail widths, try not to include foot drag. Do your best to determine where tracks begin and end. Measure from the edge of one track to the edge of the next.

SCAT. Scat, which means dung, feces, or droppings, is a very important sign. It can tell you not only what kind of animal left it but also what that animal was eating, how long ago it was there, and what other animals are in the area. Dissecting scat specimens can be very revealing, but you have to quell your natural squeamishness about examining animal feces. Measuring scat is another useful way to identify species. Concentrate on scat diameters because lengths offer little significant information. Most scat you'll find will be tubular in shape. Measure the scat at the widest part, excluding any sections that are bulbous or bunched up. I use a set of calipers to get the most accurate measurements.

A word of caution is in order here. Wild animal scat may contain parasites that can invade your own system and cause serious problems. Some people are reluctant to handle raccoon scat, for example, because (as you will read in the raccoon chapter) raccoons are hosts to a parasite whose eggs could be harmful to humans if inhaled. Also, some biologists warn that wild mammal urine can contain leptospires, bacteria that can cause infections in humans. These are serious concerns, and you should keep them in mind when working with animal scat or urine. Use a stick to break up scat samples for inspection or wear gloves.

PHOTOGRAPHY. If you have a camera, it's a good idea to build a track and sign photo library. That way, you can study the tracks at home and compare them with the photographs and illustrations in this book. Almost any camera can be used to photograph tracks, but a 35mm camera with a close-focusing or macro lens capability is preferable. I shoot positive (slide) film with an ASA of 25 or 50 for best definition, but negative (print) film with an ASA of 400 to 1000 is fine if you are not planning to publish the results. A faster ASA speed will allow you to hand-hold the camera, rather than lug a tripod around, as I do.

If possible, photograph tracks that are in the shade. Tracks with sun on them can create enormous problems. If the sun is low and the shadows fill the pad and nail marks perfectly, you have an ideal situation. Unfortunately, in my experience, perfect photographing conditions are as rare as perfect tracks. Most of the time, the shadows will not fill the tracks evenly, which results in a photograph that distorts the shape of the track. I try to solve this problem by covering the track with my own shadow while I photograph it.

If you can maintain a shutter speed faster than 1/60 of a second (slower speeds are not recommended for hand-held shots), set the lens at f8, f11, or f16. Fill as much of the viewfinder as you can with the track, or you might end up with a beautiful picture of snow or mud with a little dent somewhere in the middle. Most automatic exposure cameras can handle most mud and sand situations. Dark mud will show up lighter on the print, but the track will be easy to read. Very light sand or snow can fool an automatic camera and make your photo come out too dark (underexposed). To remedy this, try opening up one stop to let in more light. For example, if the shutter speed is 1/125, try going down to 1/60, or, if you are set at f16, go to f11. If this doesn't do it, try opening up another one-half stop.

Tracks in snow often have too little contrast to show up well on film. Pick tracks that have high contrast. Tracks made by an animal walking across black ice that has been finely dusted with snow provide plenty of contrast for a photograph. Photograph track patterns when the sun is low and casts shadows over each individual track. Usually you want the pattern, not necessarily the track details. A series of black dots will tell the story just as well. Try not to have the sun behind you; shoot across the angle of the sun. If you're an experienced photographer, try shooting into the sun. If you're not a skilled photographer, the results could be disappointing.

When you're shooting tracks, place some object of known length, such as a ruler, in the photo beside the track. You may want to measure the track when you develop the photograph, and the scale will be very helpful.

You will eventually want to shoot things other than tracks. Scat, incisor marks on acorns, and claw marks on trees are all interesting evidence of an animal's presence. The list is endless, and so is the fun.

QUICK REFERENCE CHARTS. The charts located on pages 316-321 provide an easy-to-use reference that summarizes much of the track and trail data found throughout the text. Organized first by trail patterns, then by tracks, the charts cover strides and trail widths as well as individual track lengths and widths.

Let's say I have measured eleven strides from an alternating trail pattern: 21", 17¾", 19½", 20¼", 20½", 20¼", 20½", 19½", 19¾", 20¼", and 20⅛". One measurement, 17¾", is much smaller than the others, so I discard it. The strides now range from 19½" to 21". On the chart that illustrates strides for an alternating pattern, I see that eastern coyote, bobcat, lynx, black bear, and white-tailed deer fall within the range of my measurements. From the same trail, I've also recorded trail widths from 2½" to 4". Checking the chart for alternating trail widths, three animals—opossum, eastern coyote, and domestic cat—fit the parameters, but only the eastern coyote fits both the stride and trail width criteria. So, I've determined that I'm on the trail of an eastern coyote. If you are fortunate to find some clear tracks, you might be able to further corroborate your findings.

Not every track or trail will be as easy to identify as the sample above, but using the charts will help to identify the animal more quickly, or at the very least, narrow the scope of your search.

CHAPTER 1: RODENTS

Rodentia

A beaver lodge and pond at Quabbin Reservation in Massachusetts.

White-footed Mouse
Peromyscus leucopus

Deer Mouse
Peromyscus maniculatus

Meadow Jumping Mouse
Zapus hudsonius

Woodland Jumping Mouse
Napaeozapus insignis

There are more than 120 different species of North American mice, and about half of them fall under the general rubric "white-footed mouse." The deer mouse is a type of white-footed mouse, and to me there is no perceptible difference in the tracks. There are several anatomical differences, but these change from habitat to habitat. The white-footed mouse measures up to about seven and a half inches long (including its three-and-a-half-inch tail) and weighs one-half to one ounce. Its color is gray or light brown to dull orange-brown above, with a white belly, throat, and, as its name implies, feet. The deer mouse is gray to reddish brown on its upper parts, including its tail, and white below, with longer hind feet and a tail usually longer than its body. Both animals have a bicolored tail. White-footed mice are found in deciduous, mixed, and coniferous woodlands in most of the eastern United States, ranging westward to Colorado, though barely extending into southern Canada. The deer mouse is found practically everywhere in North America, except Alaska and parts of the southeastern United States—the woodland form in deciduous or coniferous forests and the prairie form in open drylands. The woodland deer mouse is usually larger than the prairie, with a somewhat longer tail and bigger feet.

FIGURE 1.1
The white-footed mouse leaves some of the most commonly found tracks in snow-covered forests.

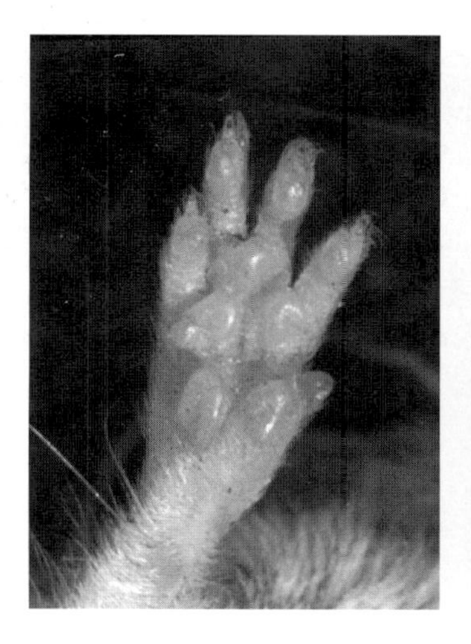

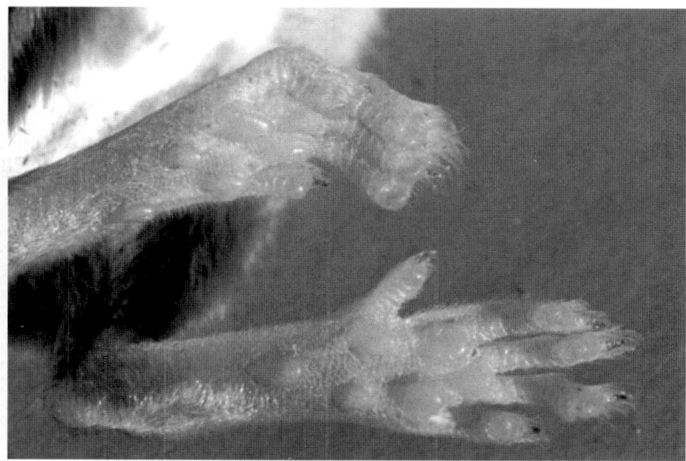

The meadow jumping mouse is in a different family altogether. It is seven to nine inches long, with the tail making up more than half of that. It appears somewhat darker in color than white-footed and deer mice, with black and buff hairs on its back, yellowish sides, and a white belly. Its range includes Alaska, all of Canada south of the tree line, and from New England south to northern Georgia and west to Montana and Wyoming.

The closely related woodland jumping mouse is slightly larger and has a white-tipped tail. It can be found in the Northeast from Newfoundland and Nova Scotia south to Maine, in Pennsylvania, and along the Appalachians to northern Georgia. Its range extends west through the Great Lakes region to Minnesota.

Both jumping mice species are profound hibernators. They begin to put on extra layers of fat about two weeks before entering their nests for approximately six months of hibernation (October or November through April or

FIGURE 1.2 *(left)*
The front foot of the white-footed mouse (right foot shown) has four toes with nails, three palm pads, and two heel pads, with a vestigial thumb located near the inner heel pad.

FIGURE 1.3 *(right)*
The hind foot of the white-footed mouse, which is larger than the front, has five toes with nails, three palm pads, and three heel pads.

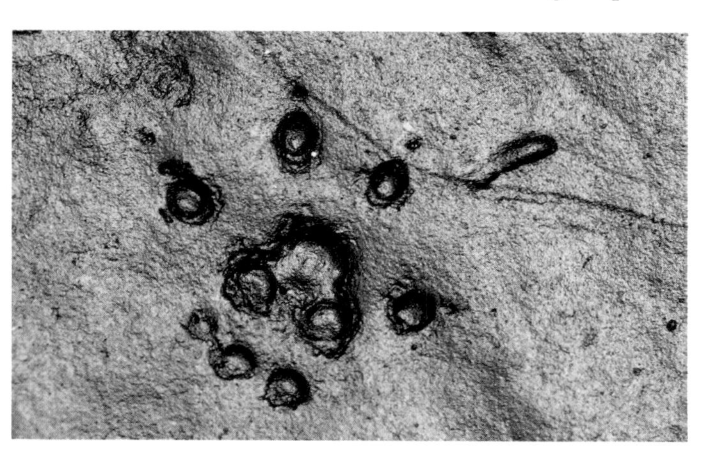

FIGURE 1.4
The right front track of a white-footed mouse has registered clearly and precisely in soft mud. These details are more likely to show up in the front track than in the hind.

FIGURE 1.5
This is an incomplete set of tracks of two mice heading in opposite directions. In the two hind tracks (on the left), only the three middle toes have registered. The two front tracks are on the right.

May). This is as long as or longer than most other mammals hibernate.

TRACKS. A mouse's forefoot (Figure 1.2) has four toes and what appears to be a vestigial thumb, three palm pads, and two heel pads, similar to a squirrel's but in miniature. The hind foot (Figure 1.3) has five toes, three palm pads, and three heel pads, although all three heel pads will rarely show in the tracks. The white-footed mouse's front track in mud (Figure 1.4) shows clearly four toes and nails, three palm pads, two heel pads, and the vestigial thumb. The front track usually measures ¼″ to ½″

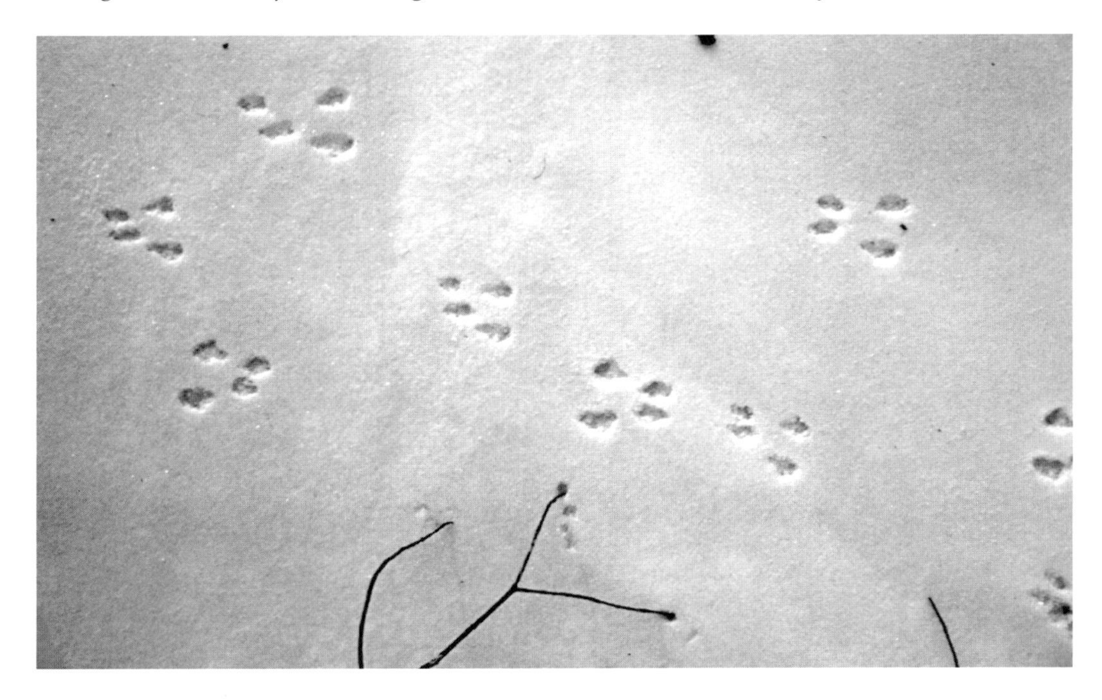

FIGURE 1.6
The common bounding pattern of the white-footed mouse in snow has a trail width between 1⅜″ and 1¾″ wide. The larger hind tracks are spread farther apart and precede the front tracks, indicating the direction of travel.

long by 3/8″ to 1/2″ wide. Hind tracks measure 1/4″ to 1/2″ long by 3/8″ to 1/2″ wide (Figure 1.5, lower center and upper left) but do not usually show all the toes and heel pads.

TRAIL PATTERNS. Mice are more easily identified by their trail patterns than by their individual tracks. Their pattern (Figure 1.6) is similar to that of the tree squirrel and chipmunk, but with a definite difference in trail width. Mice trail widths are 1 3/8″ to 1 3/4″, whereas the chipmunk's is 2 1/8″ to 3 1/3″. (Figure 1.17 later in this chapter shows chipmunk tracks on the left and mouse tracks in the lower right.) The trail widths of the jumping mice, from 1 5/8″ to 2″, may overlap those of the white-footed and deer mice. The jumping mice are not likely to leave a neat, squirrel-like pattern as does the white-footed mouse (Figure 1.7). White-footed strides are usually 3″ to 17″. Normally the meadow jumping mouse takes short bounds of 2″ to 7″. It often scurries and stays low to the ground, but when startled, it is capable of jumping 3′ to 4′. The woodland jumping mouse tends to take longer strides and is thought to be able to jump more than 10′.

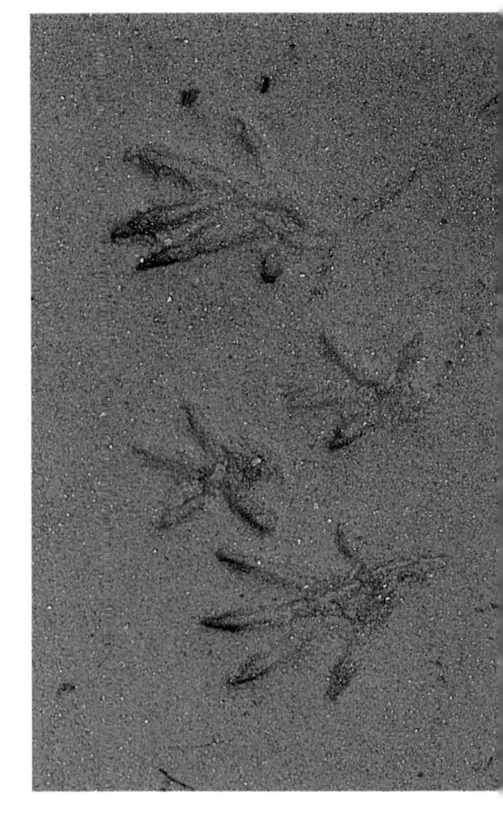

FIGURE 1.7
These jumping mouse tracks show a variation of the typical trail pattern. The larger (top and bottom) tracks are made by the hind feet, the smaller ones (in the middle) by the front feet.

SIGN: *Nests.* The mouse nest in Figure 1.8 has had the top removed to expose the inside. Mice keep their nests very clean, and unlike birds' nests, they are covered on top, with the entrance hole to the side.

Mice make their nests in ground burrows, tree cavities, wood piles, hollow logs, or abandoned birds' nests and line them with anything they can find: soft plant material, fur, or wood and bark shavings. Females have three to

FIGURE 1.8
This white-footed mouse nest was built under a panel of plywood, which was removed to expose the nest. It consists mostly of shredded bark with some strands of thread, is about 5 1/2″ in diameter, and has an entrance hole on the side.

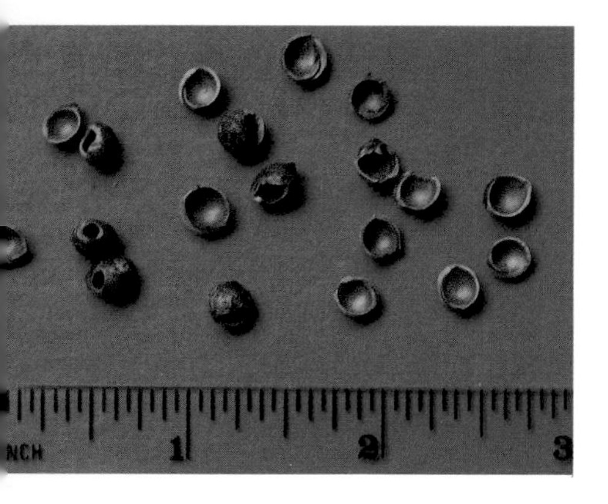

FIGURE 1.9
These wild cherry pits, opened by mice, look like they either were cracked in half or had tiny holes nibbled in them. Mice often store cherry pits for winter use.

four litters a year, and the average litter size is four. The young are altricial—born naked, blind, and helpless—but grow quickly, as the mother drives them from the nest when they are a few months old.

SIGN: *Middens.* White-footed and deer mice are good climbers and will sometimes nest in crevices of trees and store food in holes, like squirrels. Wild cherry pits (Figure 1.9) at the base of a tree are a good indication of mouse work. The pits will sometimes look as though they have been cracked opened and the seeds inside them removed, but they may simply have small nibble holes in the sides. These mice also eat acorns (Figure 1.10), which they open at the top. Sometimes they remove the top uniformly, but often the mouse will work from the top, continuing off to one side. It may make more than one entrance hole. The incisor work is very fine. With the aid of a magnifying glass, you can see tiny incisor marks along the edges of the

FIGURE 1.10 *(top)*
Usually, mice do not remove more shell than is necessary to get at the kernel, and, thus, they leave more than half the acorn intact.

FIGURE 1.11 *(bot.)*
When mice open hickory nuts, they often make more than one hole to get at the meat inside.

cuts, which appear as tiny dots. The refuse from opening the nuts will be very small fragments, much smaller than those left by chipmunks or squirrels. Mice cut away only small pieces of shell and eat as much of the exposed meat as they can; they will not cut more shell than they need to.

Mouse work on hickory nuts (Figure 1.11) is similar. Since these nuts have a harder shell than acorns, mice may do even less work on them. Sometimes they will open a hickory nut from more than one side but will still cut only enough shell to reach the meat. The middle rib of the nut is often left intact.

Flying squirrel openings on nuts can be similar. See the comparison in Figure 1.12 (mouse to the left, flying squirrel to the right). Although mice have smaller incisors, the work of flying squirrels looks smoother, and the middle rib is cut.

Mice tend to seek protection while they eat, so you will usually find their midden piles under some type of shelter, such as an overhanging rock, an exposed root system, or a tree crevice. In contrast, squirrels and chipmunks will leave middens out in the open on rocks, stumps, or stone walls or just about anywhere else on the forest floor.

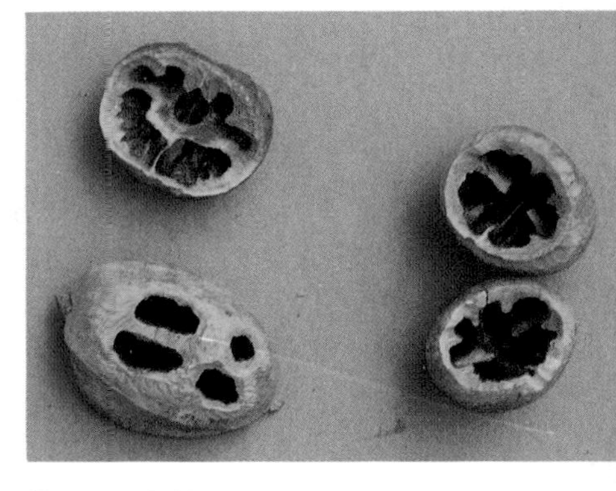

FIGURE 1.12
The hickory nuts on the left, opened by mice, have a rougher appearance than those on the right, opened by flying squirrels.

SIGN: ***Scat.*** White-footed mouse droppings (Figure 1.13), which are about the size of a grain of rice or smaller, usually have a rough, wrinkled look when seen through a magnifying glass and tend to be irregular in shape.

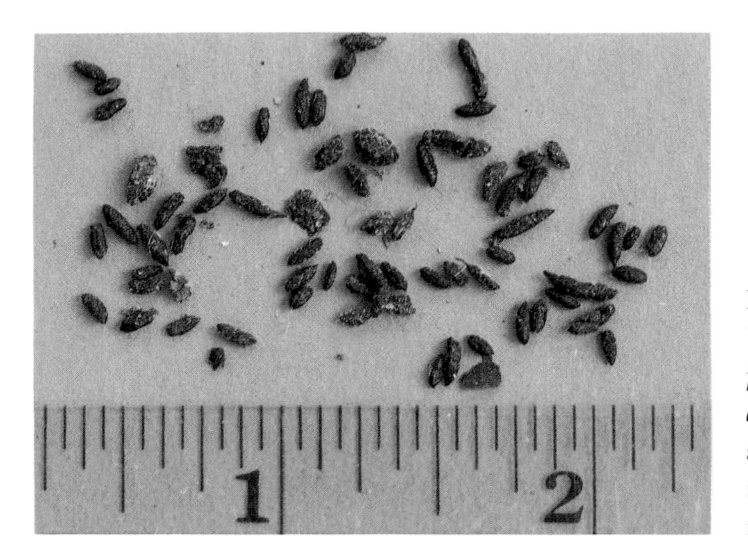

FIGURE 1.13
White-footed mouse droppings are usually irregular and rough in appearance. The scat may be randomly scattered about or concentrated in high-activity areas.

Eastern Chipmunk
Tamias striatus

Least Chipmunk
Tamias minimus

THERE ARE ABOUT nineteen species of chipmunks in North America, but by far the two most common are the eastern chipmunk, whose range is all through New England and eastern Canada as far north as James Bay and south into Georgia, and the least chipmunk, which is found primarily from the Great Lakes west to the Rockies and down into the Midwest, but not on the West Coast. The only real difference between the two is their size. The eastern chipmunk averages nine to ten inches in length, including the tail, and the least, as its name implies, averages about six and a half to seven and a half inches, including the tail. Almost everything else about them, such as markings and behavior, is interchangeable.

Both are tawny-colored members of the squirrel family, with somewhat modified squirrel tails and a series of buff and black stripes along the back (for camouflage in the forest), with the black stripes being slightly more pronounced in the western variety. Their markings are similar in color to those of a deer, and for the same reasons: They make the animals very difficult to spot against a dappled forest floor among tawny and buff leaf litter and black branch shadows. Chipmunks seek refuge and den up near rock piles, under old stumps and trees that have decayed, and in old root systems with a lot of crevices. Red squirrels often are found in the same areas, and sometimes it can be difficult to determine whether an area is being used by chipmunks, red squirrels, or both.

Although chipmunks will climb far up into trees at times, they are more comfortable on the ground and dig intricate underground denning systems rather than use tree crevices. The entrances to these dens are small (one and a half to two inches in diameter), perfectly round holes leading straight down into the ground for about seven inches, then suddenly veering off horizontally into an elaborate network of sleeping, living, and food-storage chambers. Look for these entrance holes next to rocks, old stumps, and decaying trees where there are already underground cavities.

Like squirrels, chipmunks can't store enough body fat to get them through the winter without eating, so they go into a kind of torpor that reduces their metabolism rate but allows them to wake up from time to time and stagger into a food-storage chamber for a midwinter snack before

FIGURE 1.14
The well-known eastern chipmunk is a small, lively animal often found scurrying in leaf litter on the forest floor. Its loud "chipping" call is one of the most familiar woodland sounds.

going back to sleep. They spend the summer and early fall storing vast amounts of food in their dens as well as in separate emergency caches nearby. The *Tamias* part of their Latin names means "food storer," although this may be a reference to their enormous cheek pouches. They eat pine nuts, acorns, hickory nuts, various seeds, wild cherries, and most berries: wild strawberries, blackberries, blueberries, elderberries, and Virginia creeper *(Parthenocissus quinquefolia)* berries. They'll also occasionally eat small mice and snakes, as well as insects. My wife, Paulette, and I once watched a chipmunk eating a hellgrammite, the larva of the dobsonfly *(Corydalus cornutus).* These fearsome-looking insects crawl up out of streams to pupate. They are quite big, sometimes up to three inches long, and have big "snippers" (jaws) for killing small fish. They also can inflict a painful bite on humans. We watched spellbound as the chipmunk ate the creature, chomping from head to tail while it was still wriggling around.

My friend Bob Ellis reported another unusual experience. After hearing a sharp squeal outside his window one day, he looked out and saw a chipmunk holding a mouse and gnawing at its head. According to Bob, the chipmunk ate the mouse's brains and left the rest of the carcass. Bob even supplied me with photos documenting the affair.

TRACKS AND TRAIL PATTERNS.

Chipmunks have such small feet and weigh so little (two and a half to four and a half ounces) that it's very difficult to find a clear track. But their foot structure (Figure 1.15) is very similar to that of other rodents. Each forefoot has

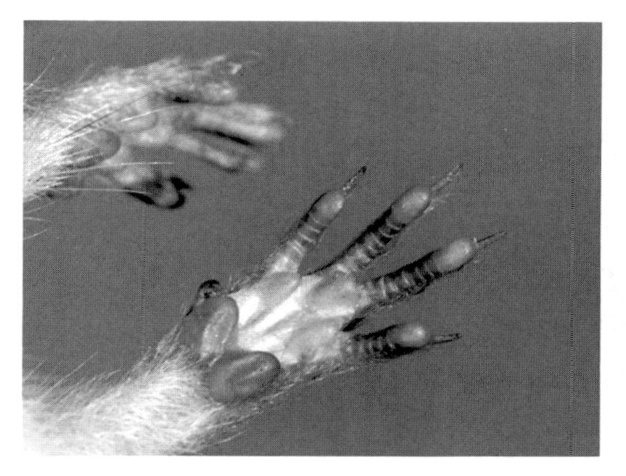

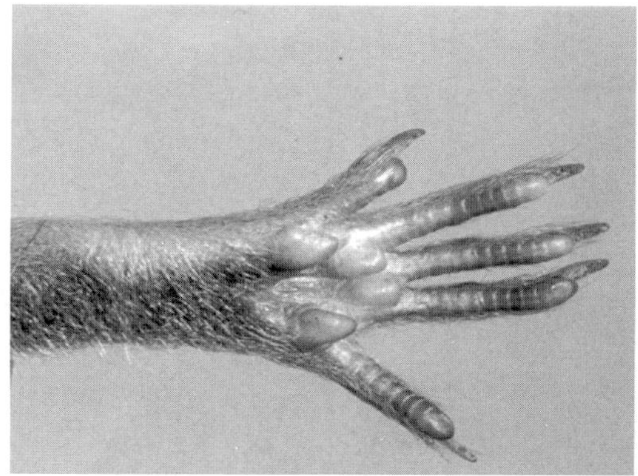

FIGURE 1.15 *(left)*
The front foot of the eastern chipmunk has four toes with long nails, three palm pads, two heel pads, and a vestigial thumb near the inside heel pad.

FIGURE 1.16 *(right)*
The hind foot of the eastern chipmunk has five toes with nails and four palm pads.

four long toes, three pads in the palm area, and two heel pads, with long nails and a little digitlike appendage, a kind of vestigial thumb, extending inward from the inside of the heel pad. Each hind foot (Figure 1.16) has five toes that are longer than those of the forefoot, four palm pads, and no heel pads. The chipmunk leaves the same trail pattern (Figure 1.17) as tree squirrels (except for southern flying squirrels), only smaller: the two front tracks almost side by side and the two hind tracks side by side and slightly ahead of the front. But the chipmunk is not consistent in its squirrel-like pattern: The hind tracks are sometimes not exactly side by side, and the front tracks are usually with one a touch ahead of the other, or are in between and sometimes ahead of the hind tracks (Figure 1.18).

The best way to tell the difference between chipmunk tracks and those of other rodents is to measure the trail width. Usually the chipmunk's trail is 2⅛″ to 3⅛″ wide, while that of the red squirrel is 3″ to 4¼″ wide and that of the gray squirrel is 3½″ to 5½″ wide. The white-footed

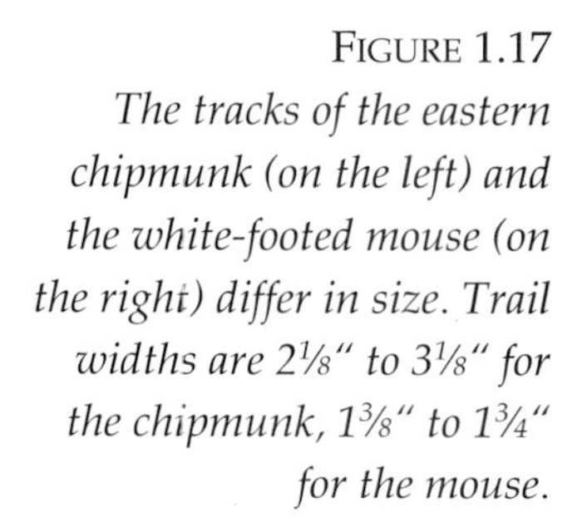

FIGURE 1.17
The tracks of the eastern chipmunk (on the left) and the white-footed mouse (on the right) differ in size. Trail widths are 2⅛″ to 3⅛″ for the chipmunk, 1⅜″ to 1¾″ for the mouse.

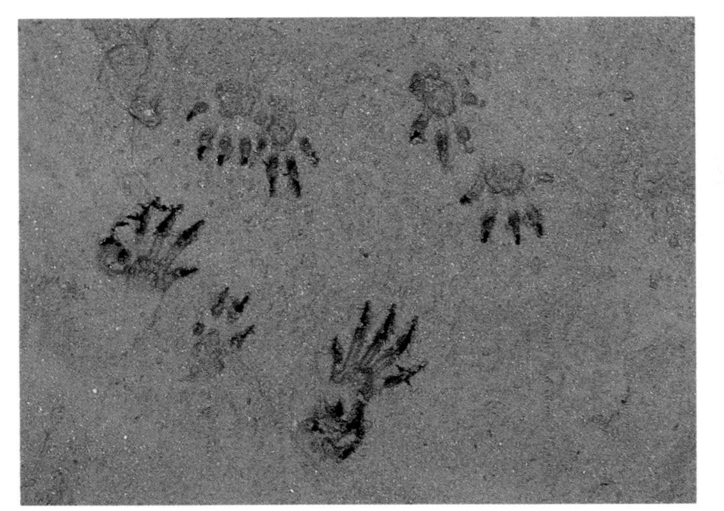

FIGURE 1.18
Here are two sets of eastern chipmunk tracks facing each other. In the upper set, the two front tracks are, more or less, in between the two hind tracks. The lower set has a similar pattern, except one front track falls just behind the right hind track.

mouse leaves a similar pattern (see Figure 1.17), but its trail is under 1¾″ wide.

The chipmunk's front track in mud measures ⅞″ to 1″ long by 7⁄16″ to 13⁄16″ wide. Its hind track is ½″ to ¾″ long by ⅝″ to 15⁄16″ wide. These measurements include nails. The full length of the hind foot often does not show in mud, which explains the longer length of the front track compared to the hind. In snow, the hind foot will register its full length, and the hind track will be longer than the front track. The chipmunk's stride is 10″ to 16″.

The dimensions given here will be slightly smaller for the least chipmunk.

SIGN: *Dens.* The chipmunk's entrance hole (Figure 1.19) is usually round, measuring one and a half to two inches in diameter. Neatness around the entrance is one of

FIGURE 1.19
The chipmunk's main entrance to its den is usually very round and neat looking, as shown here. The hole typically goes straight down for several inches before turning.

FIGURE 1.20 *(left)* *When eastern chipmunks gnaw on white oak acorns, they usually break the thin shells into small strips.*

FIGURE 1.21 *(right)* *Thick-shelled acorns opened by eastern chipmunks are often left half intact.*

the animal's identifying characteristics, as there is rarely any debris or dirt. Chipmunks are very particular about this and avoid leaving telltale mounds of dirt outside their dens by making a special work hole. It is thought that they carry the dirt in their cheek pouches and scatter it in the grass, where it doesn't show. Then they plug up the work hole and use the camouflaged entrance hole. They unplug the work hole only when expanding their tunnel systems, which they do constantly. People sometimes mistake chipmunk holes for snake holes, but snakes don't make holes—they have nothing to dig with. Snakes will, however, investigate chipmunk or vole holes looking for rodents.

You'll often hear a chipmunk before you see it; it cries out loudly before it scurries to its den. This cry can be quite startling, and I wonder if it might be a strategy de-

FIGURE 1.22 *These hickory nuts were opened by eastern chipmunks.*

FIGURE 1.23
The remains of these hemlock cones are the result of a chipmunk securing the minute seeds from within the cone. The chipmunk strips off the scales to get at the seeds, leaving the cone shafts bare.

signed to throw a potential predator off guard to enable the little animal to scurry to safety. Its high, piercing, "chipping" sound sometimes can be mistaken for that of a tree squirrel or bird. Another sound it makes, which also is loud and can carry for some distance, is not immediately recognizable as that of a chipmunk. One of my favorite forest sounds, it can best be described as a "chuck-chuck." It is thought that chipmunks use these chucking sounds to warn other animals away from their territory.

SIGN: ***Acorns and Hickory Nuts.*** Several years ago, chipmunks nearly disappeared from the area where I live in north central Massachusetts. This was after two or three years of failed acorn crops caused by infestations of gypsy moths. After a few years of restored acorn crops, however, the sounds of chipmunks once again accompanied me on my excursions into the forest.

Chipmunks have various techniques for opening acorns, depending on the thickness of the shell. Their work on white oak acorns (Figure 1.20) is similar to that of squirrels, although the strips left by squirrels may be larger. As the shell becomes thicker (Figure 1.21), the chipmunk enters through the side, gnawing away a little less than half of it. Olaus Murie, author of *A Field Guide to Animal Tracks*, reported finding basswood nuts in Minnesota opened from the side by chipmunks.

The chipmunk's entry holes into hickory nuts are about the same size as those made by flying squirrels, but they are much more ragged in appearance (Figure 1.22).

FIGURE 1.24
It is common to find shallow, circular digs in the forest floor, especially in pine needles. Chipmunks often make these, but squirrels, skunks, and small birds do also.

SIGN: ***Middens.*** When chipmunks eat, they sit on just about any object that will bring them above ground level—a stump, rock, or fallen log. Small midden piles atop these objects are common signs of chipmunk activity. They may consist of acorn shells or hemlock cones and other food materials (Figures 1.20 and 1.23). Red squirrels usually cut the scales of hemlock cones closer to the shaft than do chipmunks, leaving a smoother-looking shaft. (See Figure 1.44 later in this chapter.)

SIGN: ***Digs.*** Chipmunks will dig in the needles under white pines (Figure 1.24). I've never been able to see what they are digging for, but they leave round holes similar to those left by red squirrels.

SIGN: ***Scat.*** Chipmunk droppings (Figure 1.25) are very small and hard to find. They also vary according to the animal's diet. They can be confused with the smaller squirrels' droppings and are not easily identified.

FIGURE 1.25
Seldom found, chipmunk scat is irregular in shape and can easily be confused with that of other small rodents.

Red Squirrel
Tamiasciurus hudsonicus

Gray Squirrel
Sciurus carolinensis

Northern Flying Squirrel
Glaucomys sabrinus

Southern Flying Squirrel
Glaucomys volans

THE RED SQUIRREL (Figure 1.26) is that noisy, feisty chatterer that is always scolding intruders into its territory. It is much more territorial than the gray squirrel (Figure 1.27) and does not even tolerate other red squirrels, whereas gray squirrels seem to get along with each other rather well. There are reports of tremendous gray squirrel migrations during the early nineteenth century, with as many as a billion animals moving in a swath a mile wide to new areas. Legend has it that when the squirrels reached a river, they would float across on pieces of bark, hoisting their large tails for sails.

Paul Lyons, a wildlife biologist at the Quabbin Reservoir in central Massachusetts, feels that substantial movements of squirrels may still be possible. Some studies have documented a two-hundred-fold increase in squirrel population density after several good acorn crops. Mass movements ("migrations") of these inflated populations may result when acorn crops fail. In the fall of 1972, Bruce Spencer, a forester at Quabbin, witnessed a large movement of gray squirrels, possibly the type described by Lyons. Although most of us may never get to see a large-scale migration, the gray squirrel (which in its northern range also may be black) is now as commonly seen in city parks as it used to be in its native deciduous forests.

The red squirrel's range in North America includes most of the northern United States and Canada from North Dakota to the Atlantic Coast. In the East, it ranges south to Ohio and Pennsylvania, as well as along the Ap-

FIGURE 1.26
The red squirrel is a common sight in eastern coniferous forests of North America. It is easily identified by its rusty red back and tail, and its feisty, territorial behavior.

FIGURE 1.27
Encountered in urban areas as well as rural environments, the gray squirrel is probably the most familiar animal to people living in the eastern half of the United States.

palachian Mountains into Georgia. In the West, it ranges south from Alaska and Canada to the coniferous forests of the Rocky Mountains and into southern Arizona and New Mexico.

The gray squirrel's range extends from the Atlantic Coast (excluding northern Maine and parts of west central Florida) west to eastern North Dakota, Iowa, Kansas, Oklahoma, and Texas. It also extends slightly north into Manitoba, Canada.

Red squirrels are found in northern coniferous forests, where they live mainly on pine nuts and the winter terminal buds of evergreens, although they also eat the bark of some trees, as well as berries, nuts, insects (such as carpenter ants, which they extract from pileated woodpecker holes), birds' eggs, and nestlings. They also drink the sap of sugar maples, black birches, and other trees, and even eat mushrooms, which they hang on tree branches until they are dry enough to store.

Red squirrels live in underground dens, old stumps, and tree crevices, which they line with soft material such as moss, leaves, and grass. If natural openings are not available, they will make outside nests in conifers at an average height of about seventeen feet. These compact nests are about twelve inches in diameter, with twigs as a base and the outside formed by leaves and sometimes cones. The inside is lined with moss, grass, and leaves. Red squirrels also may renovate nests of hawks, ravens, or crows. They like to pick a nesting spot that has a thick canopy, perhaps for ease of travel and protection from weather and avian predators.

Gray squirrels, in the northern parts of their range, also use tree holes and tree crevices in winter. They insulate these holes with leaves and other soft materials. In southern areas or in summer in the North, they make large, outside leaf nests, choosing higher, older trees. When I was a young boy, I always marveled at these nests high up in deciduous trees. I could never help noticing them, especially in naked winter trees, and wondering whether anyone was home. These nests are slightly larger than red squirrel nests—twelve to nineteen inches in diameter (Figure 1.28)—and have their main entrances near the trunk of the tree, but they are more loosely made than red squirrel nests and apparently consist of a lot more leaves.

The northern flying squirrel is rarely seen, since it's nocturnal. It lives in coniferous and hardwood forests. Its most interesting characteristic is its ability to glide from tree

FIGURE 1.28
Large, outside leaf nests of gray squirrels can be seen high up in deciduous trees. They are usually 12" to 19" in diameter.

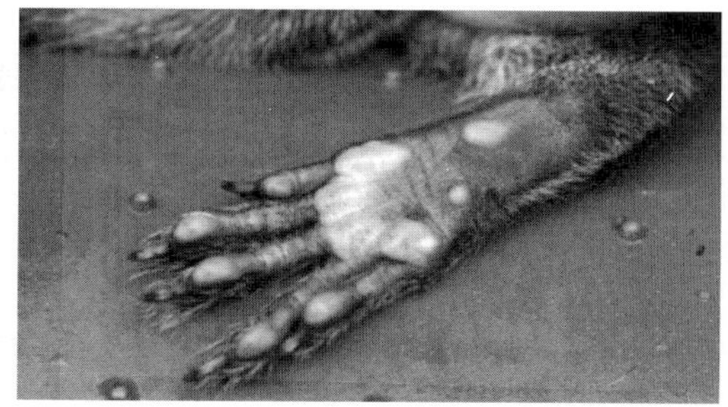

FIGURES 1.29 *and* 1.30 *The gray squirrel's front foot (left) consists of four toes with sharp nails for climbing, three palm pads, and two heel pads, with a vestigial thumb on the inside near the heel pad. Its hind foot (right) has five toes with nails, four palm pads, and two heel pads, with the area around them devoid of hair.*

FIGURES 1.31 *and* 1.32 *The hind foot of the red squirrel (left) has five toes with nails and four palm pads. The heel area is covered with hair and has no pads. Its front foot (right) has four toes with sharp nails for climbing, three palm pads, two heel pads, and a vestigial thumb on the inside near the heel pad.*

to tree, or tree to ground, by means of large flaps of skin that extend from its wrists to its ankles. It glides for distances of up to two hundred feet at an angle of thirty to forty degrees, can make right-angle turns and swoop around branches, and can pull up like a bird at the end of its glide to make a very delicate four-point landing. You'll find its tracks early in the morning, starting suddenly in the middle of a clearing and running immediately to the nearest tree trunk.

Northern and southern flying squirrels are extremely social. As many as fifteen of them have been found living in a single nest. These nests are constructed of shredded bark, leaves, moss, and grass, often in holes made by woodpeckers in conifers, with coarser materials on the outside and progressively softer materials toward the center. Sometimes the nests are lined with fur or feathers, as well as soft lichen and moss. Occasionally, outside tree nests are found. Nest diameter is about twelve inches for the southern flying squirrel and up to sixteen inches for the northern flying squirrel, which is a somewhat bigger animal.

Northern flying squirrels can be found across the southern provinces of Canada (including the Maritimes)

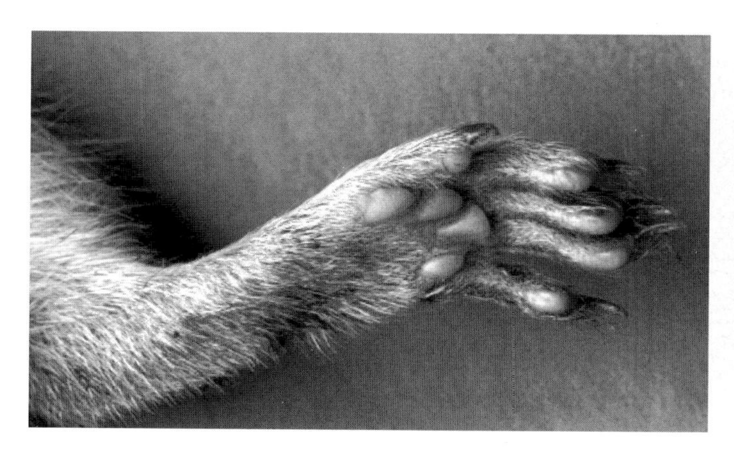
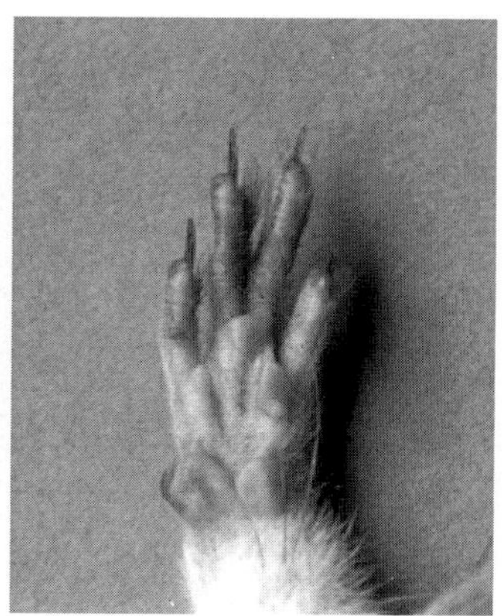

and into eastern Alaska. In the eastern United States, they range from New England west to parts of Minnesota and south through the Appalachian Mountains. They are also found in parts of the western United States, including Washington, Oregon, Montana, Idaho, California, and northern Wyoming.

The southern flying squirrel's range encompasses most of the eastern United States from the Atlantic Coast west to central Minnesota and the eastern parts of Kansas, Oklahoma, and Texas. Parts of northern New England and southern Florida are not included in this range.

Flying squirrels eat hickory nuts and acorns, fruit, seeds, fungi, buds, moths, insects, birds' eggs, and hatchlings. They hoard large winter caches in tree cavities. Southern flying squirrels also occasionally bury items under the forest ground litter.

TRACKS AND TRAIL PATTERNS. The gray squirrel's front foot (Figure 1.29) has four toes with sharp claws, three large palm pads, and two large knobs at the heel. An appendage, a kind of vestigial thumb, protrudes from one side of the heel pad. Sometimes, in a very good substrate, such as soft mud, this thumb shows up. The elongated hind foot (Figure 1.30) sometimes makes a long track but more often does not. It has five toes and four palm pads, which form a crescent behind the toes. It also has two small heel pads, with the extended area around them devoid of hair. These last two features separate it from the red squirrel, whose hind foot is covered with hair from the palm pads back (Figure 1.31); however, the gray squirrel's heel pads rarely show in its track. The red squirrel's front

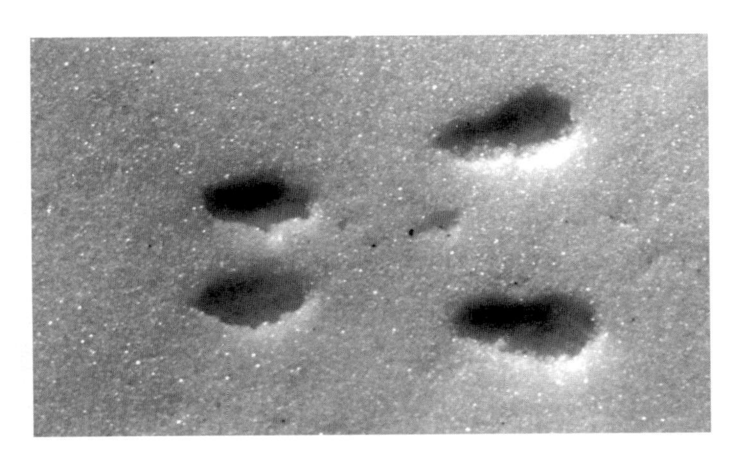

FIGURE 1.33
This is the common bounding pattern of a red squirrel in snow. The two smaller front tracks (on the left) fall close together compared to the two larger hind tracks (on the right). The trail width here is about 3⅜", and the direction of travel is to the right.

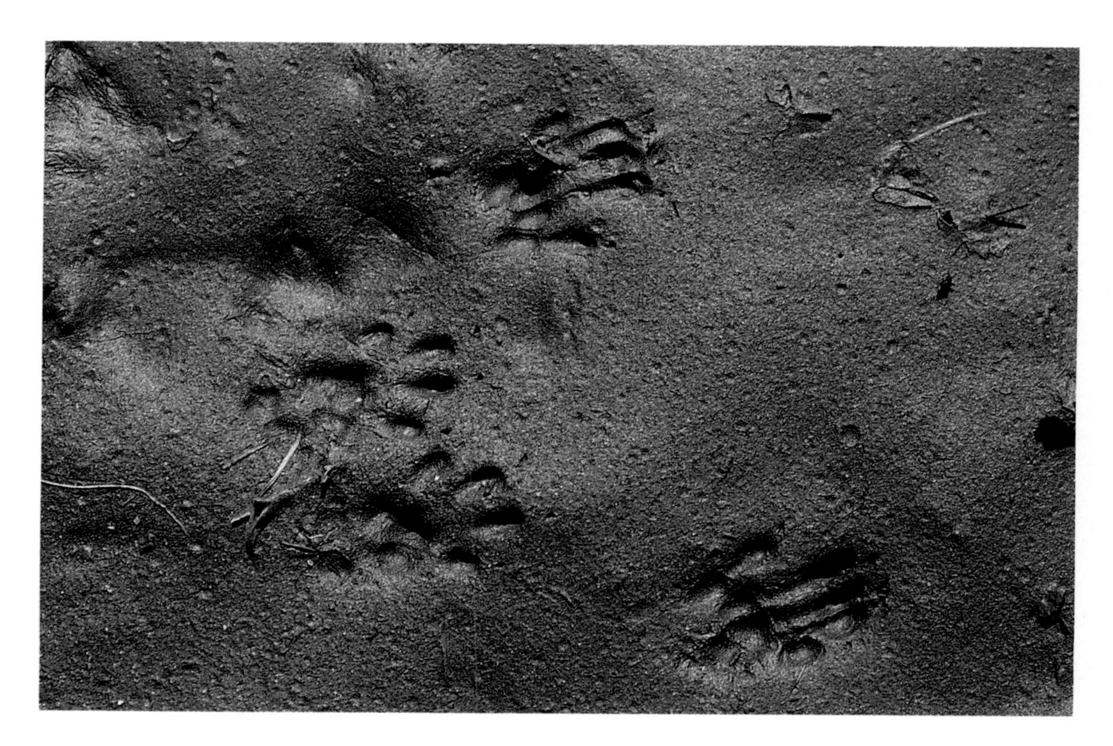

FIGURE 1.34
This is an example of the common bounding pattern of the gray squirrel: The two front tracks are on the left and the two hind tracks are on the right. Exceptionally detailed, these tracks show all toes, palm pads, and heel pads, except for the heel pads of the hind feet, which usually do not register in firm mud.

foot (Figure 1.32) has smaller, less bulbous pads, but it, too, has a vestigial fifth toe, which shows up in the track even less than does that of the gray squirrel.

There can be some overlap in track measurements of the red, gray, and flying squirrels. It is best not to rely on only one or two measurements but to take several and arrive at an average. Here are some measurements (tracks in one-quarter inch of snow) to help distinguish among them:

	Hind Track Width	*Trail Width*
Red squirrel	⅞″ to 1⅛″	3″ to 4¼″
Gray squirrel	1″ to 1½″ (occasionally to 1¾″)	3½″ to 5½″ (occasionally to 6″)
Southern flying squirrel	½″ to ¹¹⁄₁₆″	1½″ to 2⅞″
Northern flying squirrel		3″ to 4″

Since we do not often see clear individual tracks of squirrels, the squirrel's trail patterns are much better tools for determining species. When the squirrel bounds (hops), its smaller front feet land first, then the larger hind feet pass to the outside and around the front feet to land in front of them. This results in its common trail pattern. Figures 1.33 and 1.34 show two front tracks side by side and the two larger hind tracks, also side by side. The hind tracks, how-

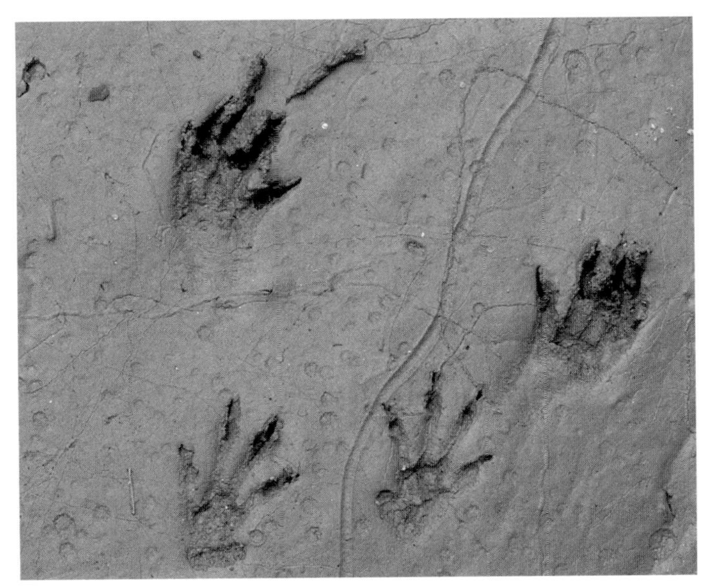

FIGURES 1.35 *and* 1.36
These two trail patterns (top and bottom) are variations of the red squirrel's pattern Grays will occasionally leave similar variations. The southern flying squirrel's trail pattern often leads with the front feet. This is a good way to distinguish the southern flying squirrel from chipmunks and the other squirrels.

ever, are farther apart and are leading the front tracks. Some squirrels (especially reds) will at times leave one front track ahead of the other (Figures 1.35 and 1.36). Shrews, many species of mice, and chipmunks also leave this familiar pattern. The easiest way to distinguish among these animals is to take measurements of the trail width and individual hind track widths. By applying the measurements given in the table, you should be able to determine which animal left the tracks.

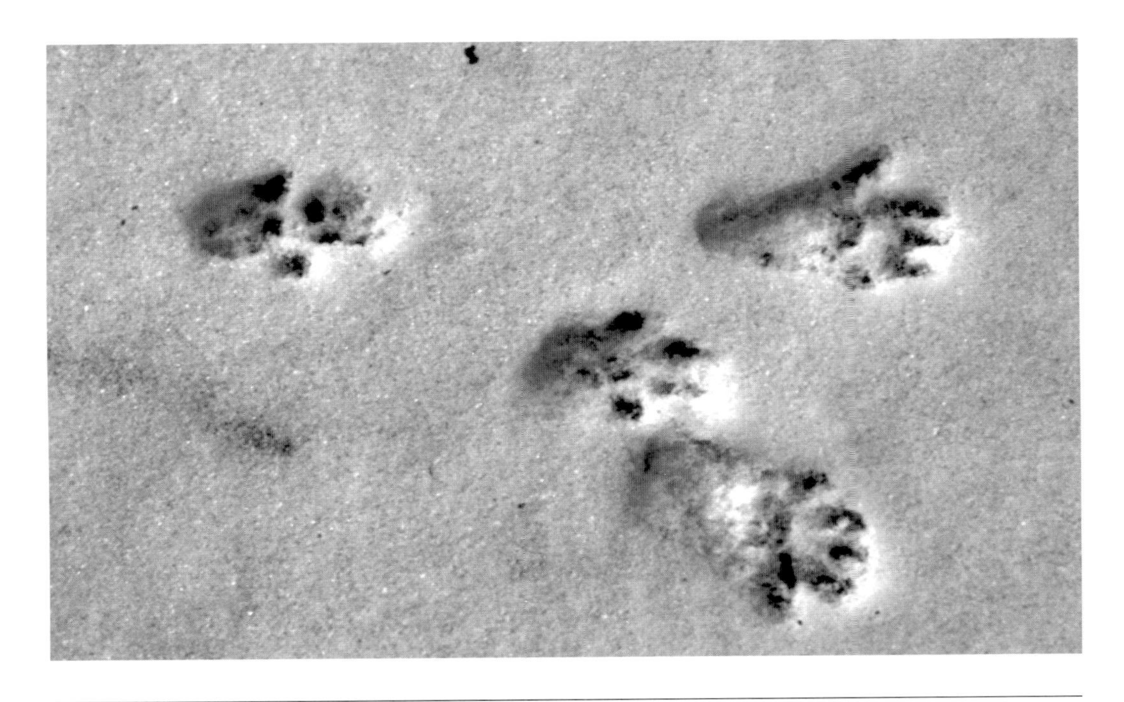

FIGURE 1.37
A trail width of 6″ confirms these as two gray squirrel trails, with their direction of travel going from top to bottom. The soft snow creates drag marks as they hop along.

Rabbit tracks are often confused with those of squirrels. With rabbits, however, the two front tracks usually appear with one well ahead of the other and almost directly in line. When cottontails or hares do place their front feet side by side, the tracks appear without any space between them. In contrast, a squirrel's front tracks normally show some space. The animals will occasionally deviate from their general patterns, so look at more than one. The more measurements you take, the more accurate your identification will be.

For example, in snow (Figure 1.33), there often are no detailed tracks, just impressions. But note the very specific pattern: two front tracks side by side with a space in between and two hind tracks side by side. That means squirrel—but which squirrel? The trail width is 3⅜″, so it must be either a red or northern flying squirrel.

In fluffy snow (Figure 1.37), there is no way you can count toes or see pad marks. All you have are indications of a squirrel's passing. Still, you can see the trail pattern: two front tracks side by side, two hind tracks side by side, and a trail width of 6″, which tells you immediately that it's a gray squirrel. Fluffy snow may exaggerate the trail width, so be sure to measure from the outside of one hind track to the outside of the other hind track.

Since there can be such variations in trail widths, make sure to take at least ten measurements of trail and hind track widths. Remember habitat considerations as well. Gray squirrels are usually found in deciduous environments, while red squirrels are mostly associated with conifers.

SIGN: ***Acorns and Hickory Nuts.*** Gray squirrels live mostly in deciduous forests and prefer acorns and other large nuts to pine or hemlock seeds. They do not store food in large underground caches, although they are known to store a number of food items in tree cavities when these openings are available. When storing items on the forest floor, the gray squirrel digs a shallow hole, places one food item in it, and covers the food. Gray squirrels have an amazing ability to locate buried nuts, misplacing only about 15% of their food supply each year. These lost nuts often grow into trees, contributing to forest regeneration.

FIGURE 1.38
These acorn strips are the work of gray squirrels. Although both red and gray squirrels peel long strips to get at the nut inside, those made by gray squirrels are sometimes wider.

Nut trees have evolved a sporadic crop pattern to ensure that gray squirrel populations never reach sufficient levels to decimate the hardwood forest. A bumper, or "mast year," crop of nuts usually occurs once every two to seven years. In lean years, gray squirrel populations decline drastically, during which time some buried nuts sprout into seedlings.

Squirrels rip their acorns open, leaving strips of shell lying around. Figure 1.38 is typical of gray squirrel work.

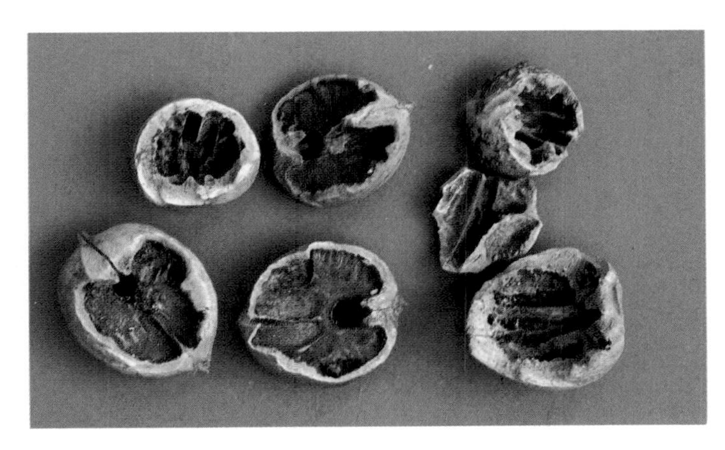

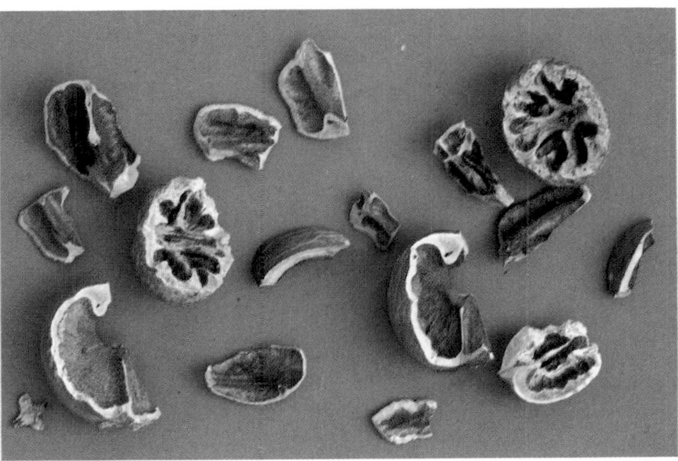

FIGURE 1.39 *(top)*
Red squirrels opened these hickory nuts, leaving large, jagged holes.

FIGURE 1.40 *(bot.)*
When gray squirrels open hickory nuts, they chip away at them, creating a ragged appearance, and often break them into small fragments. Red squirrels and flying squirrels leave the shells more intact.

FIGURE 1.41
These hickory nuts have small, round or elliptical holes with smooth edges. This indicates the work of flying squirrels, which do not chip away at the shells in the same way larger squirrels do.

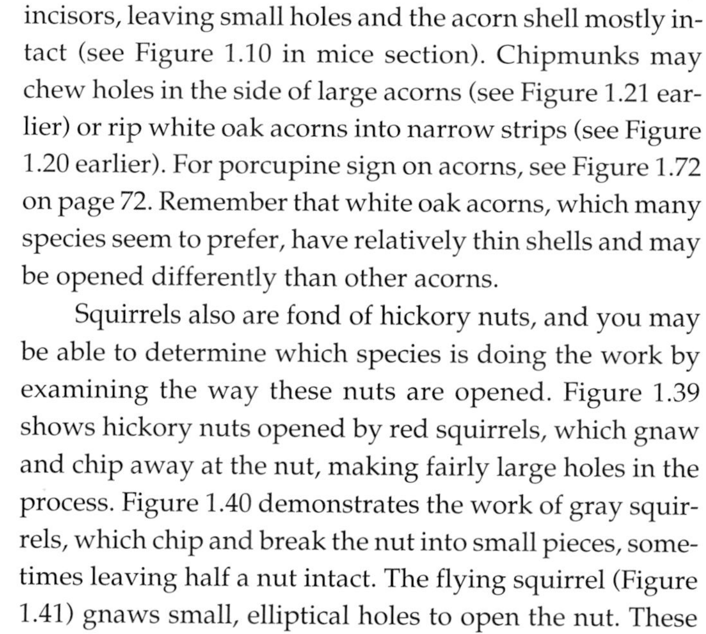

Mice gnaw away at the tops of acorns, sometimes cutting to the side or making more than one entry with their tiny incisors, leaving small holes and the acorn shell mostly intact (see Figure 1.10 in mice section). Chipmunks may chew holes in the side of large acorns (see Figure 1.21 earlier) or rip white oak acorns into narrow strips (see Figure 1.20 earlier). For porcupine sign on acorns, see Figure 1.72 on page 72. Remember that white oak acorns, which many species seem to prefer, have relatively thin shells and may be opened differently than other acorns.

Squirrels also are fond of hickory nuts, and you may be able to determine which species is doing the work by examining the way these nuts are opened. Figure 1.39 shows hickory nuts opened by red squirrels, which gnaw and chip away at the nut, making fairly large holes in the process. Figure 1.40 demonstrates the work of gray squirrels, which chip and break the nut into small pieces, sometimes leaving half a nut intact. The flying squirrel (Figure 1.41) gnaws small, elliptical holes to open the nut. These holes are very smooth in appearance. Hickory nuts opened by mice may have similar-size holes, but they will be slightly rougher in appearance (see Figure 1.12 earlier). Mice do not cut as deeply into the nut and usually leave most of the middle rib intact. Mice also often make more than one opening.

FIGURE 1.42
The ground beneath this white pine tree is covered with the scales and shafts of pine cones. An accumulation of such material is called a "midden pile" and is a definite sign of red squirrel activity.

SIGN: *Middens.* Red squirrels store huge numbers of pine cones in a single cache, stuffing them into every crevice they can find: under the roots of dead trees, in the spaces between rocks, in holes they dig themselves for this

FIGURE 1.43
These red pine cones, and the scales below them, demonstrate the various stages of the red squirrel's work of extracting pine nuts.

purpose. They'll also store hemlock cones. I once watched a red squirrel extracting seeds from hemlock cones, which are very small. The squirrel was stripping off the scales and eating the nuts at the rate of one cone every nine seconds. Hemlock scales were flying everywhere. Squirrels like to take the cones to a favorite eating spot—a tree stump or low branch—and perch there, stripping the cones down to the shaft. The piles of cone scales that accumulate over the years are called middens and can sometimes reach two feet high. Finding these middens beside a stump or at the base of a tree in a coniferous forest is a sure sign of red squirrel activity. See Figure 1.42 for an example of a midden pile. Figure 1.43 shows red pine cones stripped by red squirrels. Figure 1.44 demonstrates the difference between hemlock

FIGURE 1.44
The hemlock cone shafts on the right were cut by red squirrels, those on the left by chipmunks. Note that the red squirrel cut the scales much closer to the shaft, giving it a smoother appearance.

FIGURE 1.45
These spruce twigs indicate that a red squirrel has been feeding in the tree. The squirrel often nips the twigs to get at the terminal buds, then drops the twigs to the ground. Nip twigs may also be the result of squirrels securing cones.

cone shafts stripped by red squirrels and those stripped by chipmunks, which make small midden piles. Note that the red squirrel generally cuts the scales closer to the shaft, giving it a much smoother appearance.

SIGN: *Tunnels.* Red squirrels tunnel under the snow in winter. The tunnels lead from the home tree to a food stash or another tree. Gray squirrels usually do not make tunnels; when they do, the tunnels are not as extensive. Red squirrel tunnels may measure two to four inches in diameter, but they can squeeze through a hole as small as 1½″ across.

SIGN: *Nip Twigs.* Red and gray squirrels leave "nip twigs," as do porcupines, though the squirrels' are much smaller. These are the tips of branches, usually from hemlock, spruce, oak, or other trees, that have been nipped off and dropped to the ground (Figure 1.45.) Hemlock nip twigs are 3½″ to 11½″ long, averaging 6½″ long, with a maximum diameter of just over ⅛″. Additional nip twig diameters include: oaks up to ¼″, black locust to 5⁄16″, and pitch pine to ⅝″.

FIGURE 1.46
This is scat from both red and gray squirrels, and it is difficult to distinguish between them. Though variable, squirrel scat is usually smooth and oval in appearance.

SIGN: ***Scat.*** Red and gray squirrel scat (Figure 1.46) is quite variable but is usually oval and rather smooth. At certain times of year, with certain diets, squirrel droppings can be just small, light-brown splashes in the snow.

Flying squirrels live in the hollows of old trees. Sometimes several animals share the same hole. Over the years, the tree will fill up with flying squirrel scat and decaying wood particles. I've seen saw logs packed with these droppings stacked by the sides of logging roads. Sometimes you'll find piles of scat mixed with wood fiber around the bottom of trees as well. This material has trickled out of a hole or been dropped outside of the nesting cavity, which the squirrels are fastidious about keeping clean. It looks very similar to red and gray squirrel scat but is smaller (Figure 1.47).

FIGURE 1.47
Flying squirrel scat is most often a smaller version of the smooth and oval scat of the larger squirrels. It can also, however, be rough and irregularly shaped, as shown here.

Woodchuck, Gopher, or Groundhog
Marmota monax

Hoary Marmot
Marmota caligata

Yellow-bellied Marmot
Marmota flaviventris

DESPITE THE RHYME, the woodchuck did not get its name because it chucks wood, but because *woodchuck* is a corruption of the Algonquian word for the animal, *wejack*. The more western name, *gopher*, comes from the French *gaufre*, which means "honeycomb" and refers to the animal's complex network of tunnels, which may total forty feet from entrance to exit holes and be more than four and a half feet below the surface. The woodchuck usually makes three or four holes, as well as a plunge hole, which is hidden and drops two feet straight down. The animal may use this hole to disappear. (Its short legs are better adapted for digging than they are for running.) It also will dig a summer burrow in a field, usually on a hill with well-drained soil and close to the legumes it likes to eat, and a winter burrow (deep enough to keep the an-

FIGURE 1.48
The woodchuck keeps a watchful eye while eating. Should danger approach, an escape tunnel is usually nearby.

FIGURE 1.49
The hoary marmot is often found at high elevations. This one was in Alaska's Talkeetna Mountains.

imal from freezing), usually on a slope in a dry, forested or brushy area protected from snow. When the woodchuck hibernates, it shuts down nearly all its systems. It breathes once every five to six minutes, its heartbeat slows from 100 beats per minute to 15, and its body temperature drops from 95°F to around 45°F.

The woodchuck ranges from east central Alaska through most of southern Canada, and from North Dakota and Kansas east to the Atlantic Coast and south to Virginia and northern Alabama. The hoary marmot (Figure 1.49) is found exclusively in the western Rocky Mountains, from Alaska down to southern British Columbia and northern Washington and Idaho, where its range overlaps that of its close cousin, the yellow-bellied marmot (also known as the yellow groundhog, yellow whistler, or rockchuck). The yellow-bellied marmot's range continues south to Nevada

FIGURE 1.50 *(left)*
The woodchuck's front foot resembles that of the other marmots. It has four toes with well-developed nails, three palm pads, two heel pads, and a vestigial thumb near the inside heel pad.

FIGURE 1.51 *(right)*
Its hind foot, like other marmots', has five toes and nails, four palm pads, and two heel pads, one of which is barely noticeable.

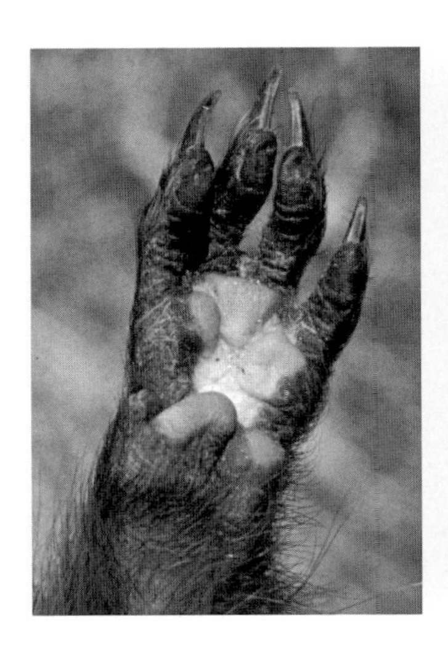

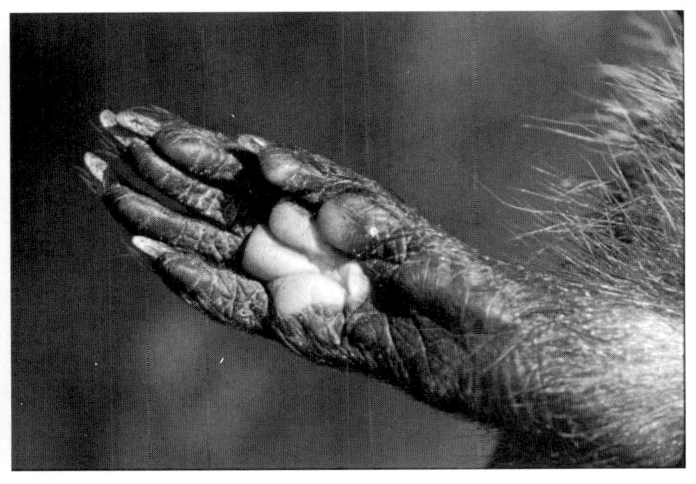

and east to New Mexico. Both animals are found at high elevations (to ten thousand feet or more). They are likely to inhabit areas with rock outcroppings or slides and nearby alpine meadows or other areas containing ample vegetation. The yellow-bellied marmot, as its name implies, is creamy yellow in color, especially on each side of its neck and across its snout, while its hoary or grizzled-looking cousin is more grayish, with a darker forehead and legs (*caligata* means "wearing boots").

TRACKS. The woodchuck's membership in the squirrel family is evident in its feet. Its front foot (Figure 1.50) has four toes, like most rodents', with two large heel pads and three palm pads, but sometimes only the toes show up in the track (Figure 1.52). Its hind foot (Figure 1.51) has five toes and four palm pads. The heel pads, one of which is barely discernible, often do not register well in the track.

Track sizes for all three marmots are more or less the same, with the hoary marmot's being at the upper end of the ranges given. I have not been able to discern any difference between the track sizes of the yellow-bellied marmot and the woodchuck. The front track (Figure 1.53),

FIGURE 1.52
The hind track (lower left) of a woodchuck almost fully registers here, except for one toe; in the front track (right), only the toes show. The tracks in the upper left and right are partial registrations of the other hind and front feet, respectively.

FIGURE 1.53
This front track (lower left) of a hoary marmot is typical of woodchuck and other marmot tracks. To the upper right is a hind track that has only partly registered.

counting the nails, is 1⅞″ to 2¾″ long by 1″ to 2″ wide. The hind track (Figure 1.52, lower left), again counting the nails, is also 1⅞″ to 2¾″ long, but it is 1⅜″ to 2″ wide.

TRAIL PATTERNS. When marmots walk, they do so with the rear foot superimposing over the front track, usually just slightly off-register, leaving an alternating walking pattern (Figure 1.54). The trail width is 3¼″ to 5½″, and strides are 5″ to 13″. I also have seen woodchucks

FIGURE 1.54 *(top)*
In the alternating walking pattern of woodchucks and marmots, the hind track usually registers close to or partly on top of the front track.

FIGURE 1.55 *(bot.)*
In the bounding pattern, the front tracks are one ahead of the other, with the hind tracks leading the front.

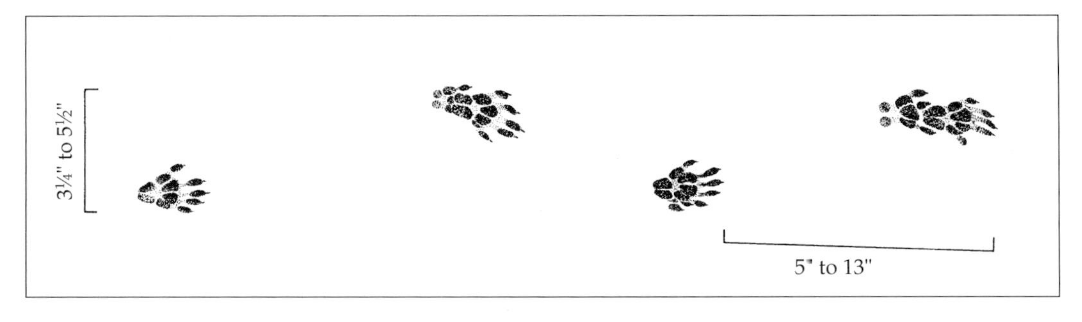

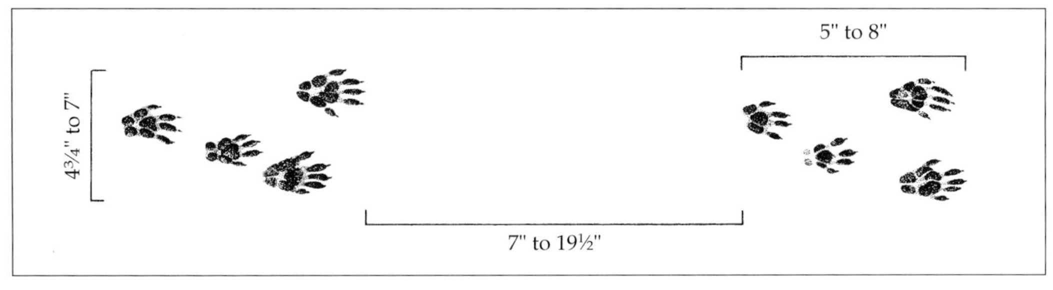

leave a bounding pattern similar to that of a raccoon's bounding pattern (Figure 1.55). Strides in this pattern are usually 7″ to 19½″, and the trail width is 4¾″ to 7″.

SIGN: *Dens.* Sometimes the mounds in front of a woodchuck's den can be one and a half to two feet high (Figure 1.56). The hole itself is usually five and a quarter to six and a quarter inches in diameter but may be larger, allowing enough space for a woodchuck to get in and out but not for a fox or coyote, which are among the woodchuck's chief predators. Foxes will use abandoned woodchuck dens, but they have to dig them out first. I've even heard of foxes living in woodchuck dens with the woodchuck still living in the same tunnel network. If the den is occupied, there should be fresh dirt at the entrance, since woodchucks and marmots are diligent diggers and often clean out their tunnels. Most main entrances will have dirt mounds; plunge holes or escape entrances, which are dug from the inside out, will not.

SIGN: *Feeding Habits.* Like all rodents, marmots may leave a forty-five-degree angle when cutting plants. Woodchucks eat a lot of succulent vegetation, as many rural gardeners can attest. Incisor marks may not always show the forty-five-degree angle, but in foods such as

FIGURE 1.56
Dirt mounds are common outside the main entrance to woodchuck dens, which have openings from 5¼″ to 6¼″ in diameter, occasionally a little larger.

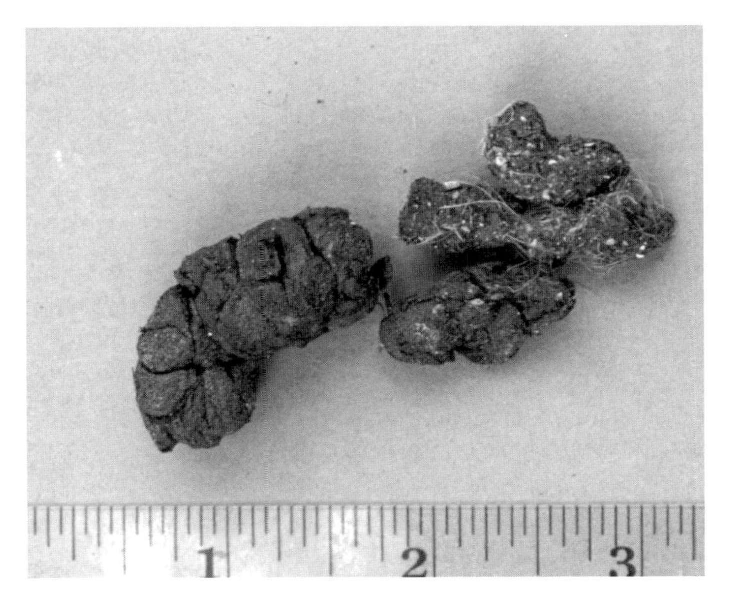

FIGURE 1.57
It is rare to find woodchuck scat because they defecate in underground chambers, or their scat is covered in the dirt mounds outside their tunnel entrances.

squashes, large teeth marks are good evidence of woodchucks versus rabbits or smaller rodents. In my experience, woodchucks seem to reach up and eat down, leaving the bottoms of the stems and some leaves intact, although this is applicable only to tall standing plants.

Woodchucks also can climb trees. Jim Cardoza, a biologist friend of mine, observed one in a peach tree, eating the peaches. They may claw or gnaw at bark, but this is uncommon. Gnaw marks will have a ragged appearance, and scratching may show nail marks.

SIGN: ***Scat.*** You will not usually find woodchuck scat (Figure 1.57), as these animals use latrines in underground chambers in their tunnel networks. They will also defecate in the dirt mounds outside their tunnel entrances, but this scat is usually covered. I remember photographing a young woodchuck digging in someone's lawn. When he finished making a small hole, he stuck his rear end in it and defecated. At that moment, he was disturbed by something (woodchucks are extremely nervous) and ran off, so I don't know whether he would have covered his scat or not.

Marmot scat is not as difficult to find. Look for it on rock outcrops and ledges. Mark Elbroch, one of my staff instructors, has observed latrines on rock outcrops that contained several hundred scats.

Porcupine
Erethizon dorsatum

THE PORCUPINE'S scientific name can be loosely translated as "the animal with the irritating back." Its common name also refers to the porcupine's most obvious feature: from the Latin *porcus*, meaning "pig," and *spina*, meaning "thorns," a reference to the thirty thousand barbed quills that are the porcupine's chief defense against predators and that have allowed it to become one of the most lackadaisical animals in the forest. It takes a lot to excite a porcupine. It doesn't move fast, and it doesn't move far. It has a very small home range and rarely moves out of it. Samuel Hearne, an eighteenth-century explorer of northern Canada, noted that porcupines are "so remarkably slow . . . that our Indians going with packets from Fort to Fort often see them in the trees, but not having occasion for them at that time, leave them till their return; and should their absence be a week or ten days, they are sure to find them within a mile of the place where they had seen them before."

The porcupine's range includes most of Alaska and Canada and a large portion of the western United States. In the eastern United States, the animal is found in New England, New York, and most of Pennsylvania, plus the northern half of Michigan and Wisconsin.

As far as I know, porcupines are primarily vegetarians, eating mostly bark (especially from conifers) in winter and more herbaceous plants in summer. In the northeastern United States, hemlocks play an important part in the porcupine's diet. They eat the foliage and to a much lesser extent the cambium. They also eat the cambium of the white pine and will feed on larch, spruce, and fir trees. In other parts of their range, where Douglas fir and ponderosa pine are found, they will eat these trees as well. The porcupine may eat hardwood bark (birch, oak, maple, poplar, and beech) in winter.

I've watched porcupines move from their winter dens in the spring as the buds start to swell on the sugar maples, and I've seen them eating these buds. One of the earliest leaves appears on wild raspberry canes, and porcupines go after them, too. In winter, they eat the canes themselves, but as spring progresses, more herbaceous material becomes available. Porcupines will be found feeding on these plants in uplands, fields, along the edges of forests, and even in the water, where they seek out aquatic plants. They

FIGURE 1.58
Endowed with a multitude of spiny quills, the porcupine is not difficult to recognize. This particular animal was feeding in a field, but porcupines frequently are found in trees.

will climb many different trees for foliage, sometimes sleeping precariously on a limb.

Porcupines venture out into fields more often than people realize. One day my wife, Paulette, and I had been doing some photography in a field near the Quabbin Reservoir, in Massachusetts. It was getting late, and the sun was about to set, which is the time porcupines start to feed. At first we saw an adult; then we noticed she had two young ones with her. Before long, we realized the field was full of porcupines coming out of the woods to gorge themselves on clover blossoms and other choice foods. It was an eerie sight, for porcupines are usually extremely solitary animals and spend a lot of time snorting and snuffing when there are other animals about; however, they all kept their distance from one another.

We got down on our hands and knees and crawled along beside some of them. They tried to hide their bellies and faces, or they turned their backs to us and displayed their quills. I got down in front of one stubborn old fellow just to see what he'd do. Most of the time, they'll just turn in the other direction, but this one had no intention of turning around. He just put his head down and kept coming. In the end, I was the one who moved aside. Sometimes a porcupine will become aggressive, whipping its tail

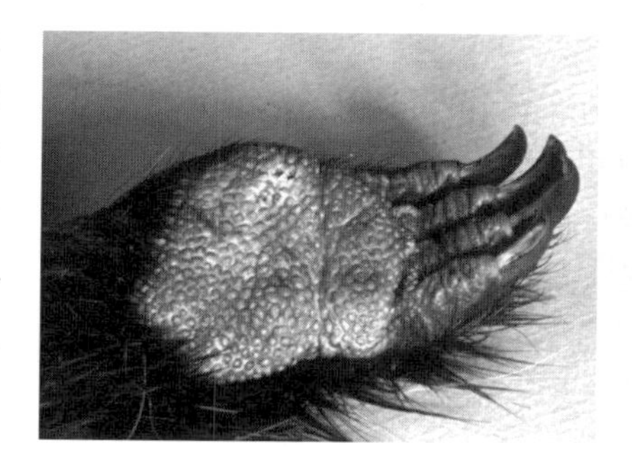

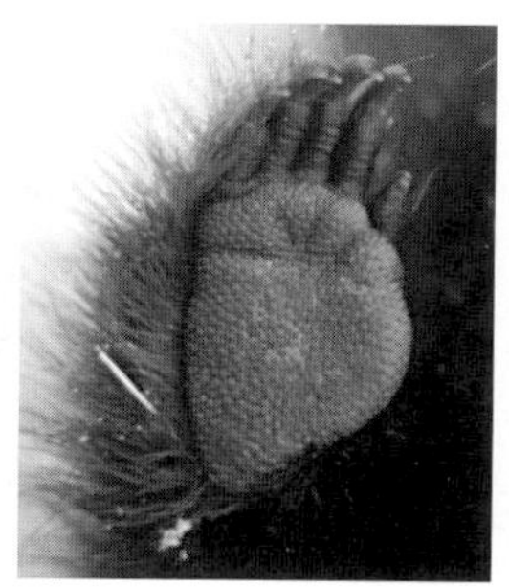

FIGURES 1.59 *(left)* and 1.60 *(right)* *The porcupine has pebbly pads and long nails for climbing on its front and hind feet. The front foot (left) has four toes and the hind (right) five.*

around and trying to slap with it. I once watched a student put his hiking stick in front of a porcupine. With unexpected speed, the porcupine slapped the stick with its tail, leaving the stick spiked with quills.

Porcupines usually weigh seven to fifteen pounds, but they can reach forty. Dan McCowan, a Canadian naturalist, has recorded a porcupine captured near Regina, Saskatchewan, that weighed just over fifty pounds. He noted that "a new-born porcupine is actually heavier than a Grizzly cub at birth."

TRACKS. The porcupine's front foot (Figure 1.59) and hind foot (Figure 1.60) both have palm and heel pads that seem to have merged together to form a single beaded or pebbly pad. This pebbled effect sometimes shows up clearly in soft mud (Figure 1.61), as do the long nails on most of the toes, which are made for climbing and digging. On harder substrates, all you may see are the round pad and nail marks. In snow, you'll often see a small hill or lump between the heel pad and the nails; this is a good indicator of a porcupine's track. Please note that the porcupine has

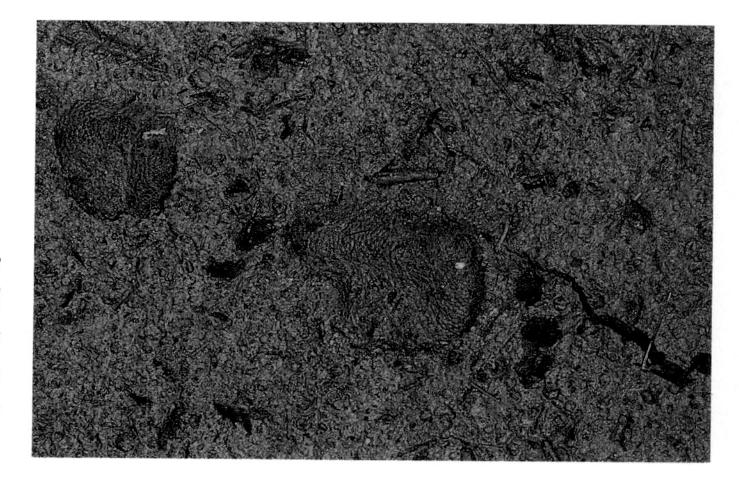

FIGURE 1.61 *The pebbly pads of both the front (on left) and the hind feet (right) show up clearly in this mud. Most of the prominent nails register well ahead of the pads.*

only four toes on the front foot. Some tracking books show front tracks with five toes, but this is a mistake. Note also that the nails don't extend out as far in the hind track (Figure 1.61, right track) as in the front (left track). Usually you will not see all five nails in the hind track.

The adult porcupine's front track measures 2¼" to 3⅜" long by 1¼" to 1⅞" wide. The hind track measures 2¾" to 3⅞" long by 1½" to 2" wide.

TRAIL PATTERNS. The porcupine is an indirect-registering animal with a diagonal walking pattern. The front foot comes down, and the rear foot steps next to, partially on top of, or, often in the case of the porcupine, slightly ahead of the front (Figure 1.62). As with many quadrupeds, the placement of the rear foot will depend on the animal's speed or gait. The faster the animal is going, the more the rear foot will overstep the front foot.

The adult porcupine's trail width may be only 5" in mud but as much as 9" in snow. Its walking stride is 6" to 10½". The porcupine's normal gait is little more than a waddle, for it has very short legs and a fat body. This makes for a very wide trail and a short stride. Figure 1.63 is a porcupine trail, about 6" to 6½" wide with a 6" to 7" stride. You can tell just by looking at this trail that it was made by a wide-bodied animal with short legs. If you look carefully at the line going down the middle of the trail—clearly a tail drag—you can see little lines made by the quills. This trail is screaming porcupine every step of the way.

Figure 1.64 is another porcupine trail, this time in deep, fluffy snow. Here the stubby-legged porcupine turns into a snowplow, making a deep trough as it shuffles back and forth from its den to its feeding area. More than one porcupine may be traveling this route. In Figure 1.65, you can see that as the porcupine walks, it brings its foot out and around and then in toward the trail, so that the tracks end up arcing in toward the center of the trail, creating a pigeon-toed appearance.

Unlike raccoons, porcupines do not have specific latrines and will urinate and defecate wherever they might be (though large accumulations of scat will pile up in and around dens). Often they will urinate at the entrance to the den, or even as they walk along. Consequently, a streak of

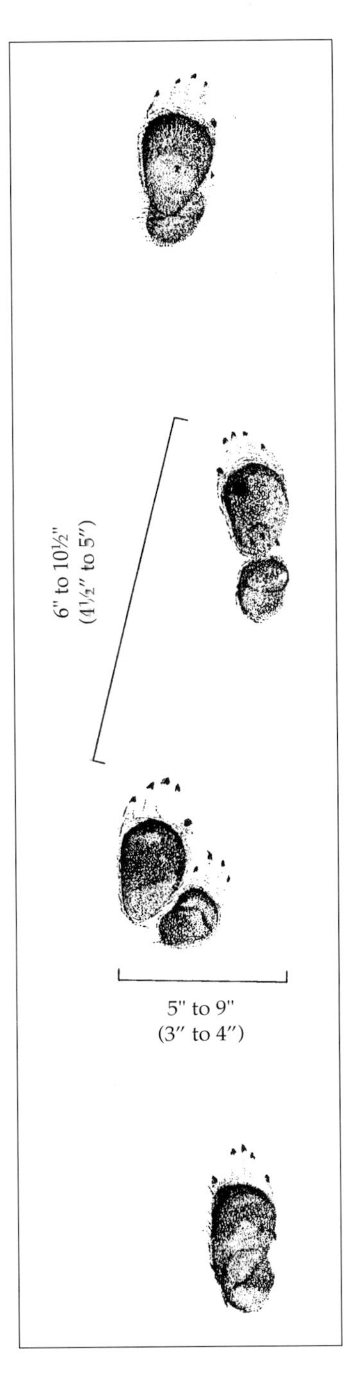

FIGURE 1.62
This trail illustrates the porcupine's alternating walking pattern. Measurements for immature porcupines (under two years) are in parentheses.

FIGURES 1.63 *(top)*, 1.64 *(bot. left)*, and 1.65 *(bot. right)* *These three trails exhibit sure signs of a porcupine's passage: tail drag with lines made by quills (top), a trough in deep snow (bot. left), and an S-shaped trail created by its pigeon-toed walk (bot. right).*

yellow in the snow is a sure sign of a porcupine. This urine has a very strong, recognizable smell reminiscent of pine or turpentine, and it stays around for quite a while. Even when there is no snow cover, you can smell the urine along a porcupine trail if it is well used.

SIGN: *Browse.* Early spring often finds the porcupine climbing to the top of a sugar maple for a feast of newly swollen buds (Figure 1.66). I also have seen porcu-

pines in red maples in spring, but they seem to prefer sugar maples. To obtain the buds, the porcupine nips off the tips of a branch, strips the tip of its buds, and then drops it to the ground. The ground under a tree being browsed by a porcupine will often be littered with these "nip twigs." You also will find porcupine scat under the tree. Check the nip twigs for the forty-five-degree-angle cut so common among rodents, and look for the coarse incisor marks of the porcupine (Figure 1.67).

In the fall, just before the trees begin to change color, I often find porcupines in oaks, after the acorns. As with sugar maples, the porcupine will litter the ground with nipped oak twigs.

Hemlock boughs scattered on the forest floor, with the tips cut at a forty-five-degree angle, also indicate porcupine activity. If you look closely at the cuts (Figure 1.67), you can see the porcupine's incisor marks on the twigs. Red squirrels will cut off hemlock boughs, but they'll usually cut off smaller pieces to get at the cones on the tips. A squirrel's incisor marks are very fine, almost imperceptible, compared to those of the porcupine.

In some areas, hemlock growth will be completely stunted by the browsing of porcupines. They are very consistent animals; once they get used to eating an individual tree, they'll return to it again and again for years. The result is often visible, even from a distance. The activity of porcupines is clearly evident in Figure 1.68. The animals will usually start at the top of the tree, eating the foliage and tender bark. This browsing can give the tree a very flat-topped appearance. Porcupines also will move along the

FIGURE 1.66
Porcupines are excellent climbers. This one was high in a sugar maple, eating the newly swollen spring buds.

FIGURE 1.67
These hemlock branches, about ½" in diameter, have been cut by porcupines at the 45-degree-angle typical of the rodent family. Notice the incisor marks across the ends. Porcupine nip twigs have been recorded up to 1⅛" in diameter.

FIGURE 1.68
The top of this hemlock shows the effects of repeated browsing by porcupines, which often begin feeding at the top of the tree, then proceed along the limbs, leaving clumps of growth on the ends.

lower limbs, cutting off the twigs and branches, eating the choice foliage, and letting the rest fall to the ground. Sometimes they stop before reaching the tips, resulting in a naked limb with clumps of undisturbed branches at the end, but they'll often eat these tips as well. Forest hemlocks may be severely stunted and flat-topped near ledges or rocky areas where porcupines are denning.

In addition to sugar maples, white oaks, and hemlocks, other common nip twigs are from apple, aspen, and black birch trees.

SIGN: *Quills.* Oddly enough, porcupines fall out of trees fairly often. Uldis Roze, a wildlife biologist who spent many years studying porcupines in the Catskills, once examined fifteen adult porcupine skeletons and found that nine of them showed evidence of broken bones. Another consequence of a porcupine falling out of a tree is the danger of its being stabbed by its own quills (Figure 1.69). Roze was surprised to find, however, that the quills actually contain a fatty acid that acts as an antibiotic, which explains why other animals rarely develop infections as a result of being stuck by porcupine quills. Although it's hard to see the barbs on the business end of the quills, they *are* there.

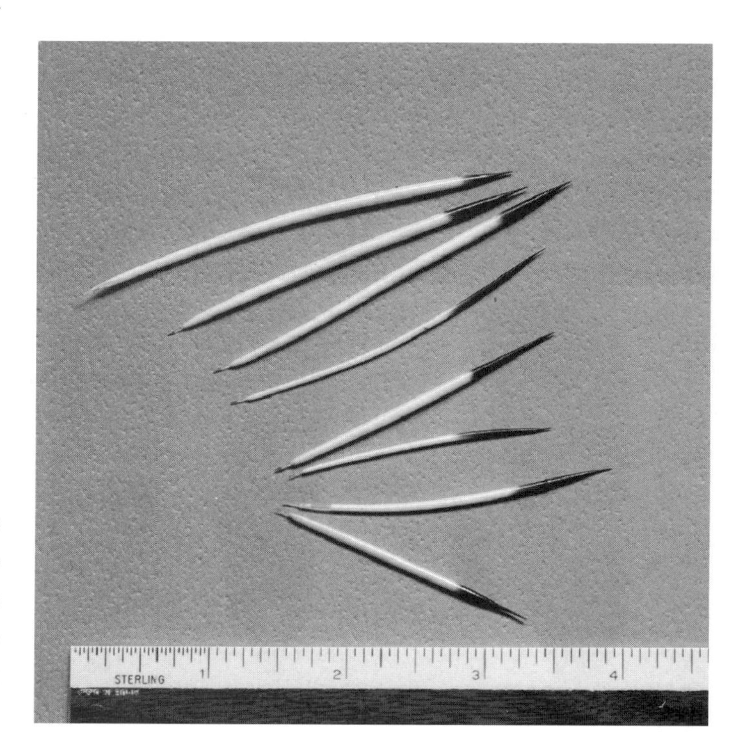

FIGURE 1.69
Porcupine quills are hollow and have microscopic barbs on their black tips. These barbs allow the quills to move in only one direction—forward.

FIGURE 1.70 *(left)*
Porcupines debark trees to feed on the inner (cambium) layer of bark. Incisor marks do not usually show when they debark white pines.

FIGURE 1.71 *(right)*
When porcupines feed on the bark of hardwoods such as oaks, incisor marks are usually evident. Teeth marks run across or, in most cases, at a slight angle to the branch or trunk

A friend of mine had a German shepherd that got into a porcupine and had quills right through its cheek, in its tongue, way back in its throat, and all over its face. We had to use pliers to get the quills out. Because of the barbs, every time the animal moved a muscle, it worked the quills in deeper. Eventually, the quills could go right out the other side of an animal. If the quills encountered an organ along the way, they could prove fatal.

Coyotes, foxes, raccoons, and bobcats rarely attack a porcupine but will eat a porcupine carcass by turning it upside down and, starting at the unquilled belly, consume the animal from the inside out, until all that's left is the hide, head, and feet. Fishers will prey on porcupines and are reputed to have a puncture-proof stomach that can accommodate the quills. A great horned owl may grab a porcupine out of a tree and drop it to the ground to kill it. Dan McCowan has described a photograph of the heart of a great horned owl that had two porcupine quills embedded in it. Nevertheless, the owl lived until it was shot while raiding a poultry farm.

Look for porcupine quills in and around dens, in porcupine trails, and at the bases of trees frequented by these animals.

SIGN: *Debarking.* I've seen white pines stripped of their bark, from ground to crown, by porcupines. The debarking shown in Figure 1.70, however, is more typical of

FIGURE 1.72
These thin-shelled acorn fragments were cut by a porcupine. Note that they are not in long strips like those made by squirrels.

the sign porcupines leave on white pines and other conifers. When feeding on hardwoods such as beeches or sugar maples, they first chew off the outer bark and let the chunks fall to the ground, then they eat the living inner bark, or cambium, which contains the most food value, lots of roughage, and a minimal amount of nitrogen. When a porcupine chews on a hardwood tree, its incisors leave a typical patchwork pattern on the inner bark (Figure 1.71).

SIGN: ***Acorns.*** Acorns have a high caloric content and are important sources of fat for animals preparing for winter. Porcupines seem to be especially fond of white oak acorns, which are lower in tannic acid and therefore lack the bitter taste of most acorns. Animals open acorns in many different ways. Smaller ones, such as mice, chew away at the shell until they make small holes from which

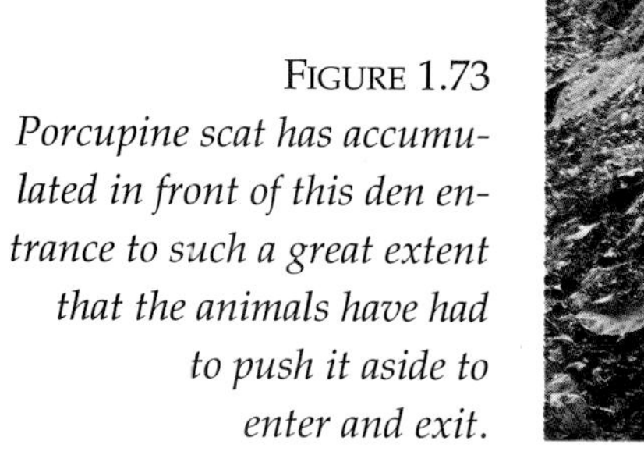

FIGURE 1.73
Porcupine scat has accumulated in front of this den entrance to such a great extent that the animals have had to push it aside to enter and exit.

FIGURE 1.74
Ledges with crevices provide good denning areas for porcupines. As is evident here, exceptionally large piles of scat may be found in and around their dens, possibly providing some insulation and deterring other animals from taking over their living quarters.

they extract the meat. Gray squirrels peel off the shell in large, elongated strips. Porcupines cut more deeply into acorns, leaving an inconsistent, ragged appearance. Note, however, that Figure 1.72 shows thin-shelled acorns. Acorns with thicker shells may have a different appearance.

SIGN: *Dens.* Porcupines gravitate toward ledges with crevices for denning and with a good food supply close by. Within the porcupine's range, a combination of hemlocks and broken ledges will almost guarantee the presence of porcupines. Look for browsed hemlocks and scat accumulations in and in front of rock crevices and caves. Another strong attractant is the white oak in early fall; look for nip twigs. White pine and poplar stands also are good indications that porcupines may be in the area.

Porcupines will den up in any hole they can find, such as hollow trees, logs, rock crevices (Figure 1.73), and even abandoned beaver lodges and bank burrows left high and dry by a retreating water level. In winter, they rarely move very far from their dens. One study showed that a porcupine's winter range was about 12% of its summer range. This sedentary existence is a form of hibernation. The porcupine's purpose is to use as little energy as possible in the cold weather. As a result, huge piles of scat will accumulate in and around the entrance to a porcupine's den (Figure 1.74). In Figure 1.73, you can see that porcupines have piled up so much scat—more than three feet of it—that they've actually had to push it aside because it was blocking the entrance to the den. Most people are disgusted to find that porcupines literally live atop their own dung, but it doesn't seem to present a problem to the porcupine. On the contrary, I believe that porcupines derive some benefits from such living arrangements. The dried dung probably adds some insulation to the den, and the scat may deter other animal species from taking over the den.

SIGN: *Digs.* Several years ago, I began to notice some curious sign in hemlock stands: smooth, round holes about the size of a quarter in the middle of a larger (six to twelve inches in diameter) dig. It looked to me as though something smooth and round had been taken out of the holes, but I couldn't figure out what. I talked to some biologist friends, and they suggested that maybe grubs had been removed. Grubs curl up in the ground, they said, and then raccoons or skunks come along, dig them up, and eat them. I was not satisfied with this explanation, however, for it did not seem to match what I was seeing. So I kept looking—for years. Every time I was near a hemlock stand, I would get down on my hands and knees and investigate the digs. There were the same scrapes, the same smooth, round holes in the ground, as if someone had taken a hickory nut and pressed it into the earth.

My friend Colin Garland noticed that a few of these digs had porcupine droppings in or near them. I'd noticed that, too, but I'd labeled it a coincidence. I thought the droppings might have fallen out of the trees. Colin's remark stuck with me, though, and I began to notice those porcupine droppings more and more—quills, too—and in places where porcupines weren't feeding in the trees. In short, I started to wonder whether porcupines were making these digs.

Eventually, I began to notice a lot of small, round puffball-like mushrooms that seem to have been cast aside and were dusty inside—they had gone dry. I wondered whether they were connected to the digs. There was nothing about them in the studies I was reading, however.

As time went on, I was pretty sure that deer were making some of the digs, because sometimes I would see a place where a deer hoof had pushed aside a lot of dirt. I also thought about raccoons, which are excellent diggers. And I thought about porcupines—what might they be digging for?

These questions kept me occupied for quite a while, until I came upon several instances in which the animal had cut some roots out of the way to get at whatever it was taking out of the holes. The roots had been cut at the familiar forty-five-degree angle of the porcupine, and the incisor marks were the size of a porcupine's. I knew for sure that no raccoon, with its canine teeth, would make a cut like that, so it had to be a porcupine. That's when I remembered the dry puffball-like mushrooms. I couldn't find any documentation about porcupines eating mushrooms, but I verified the digging with wildlife photographer Bill Byrne.

FIGURE 1.75
False truffles grow underground and are dug up by porcupines, flying squirrels, and possibly deer and other animals.

Then one day I found a fresh puffball still in its hole at the bottom of a dig, as if the animal had been frightened away in mid-meal. When I took it out, there was that nice, neat hole I'd been seeing. I took the mushroom and the spores to Doug Elliott, a naturalist friend, who sent it to a mycologist (Bill Roody) in North Carolina. The mycologist identified it, to the best of his ability, as a false truffle, of the genus *Elaphomyces* (Figure 1.75).

If we are paying attention, nature has many ways of showing us how things are connected. The truffle attaches itself to the root of the hemlock and sends out its own roots. These roots are much more efficient than hemlock roots at picking up minerals and water, and they end up feeding the hemlock, allowing it to grow more vigorously. The hemlock, in turn, gives sugars to the mushroom that the mushroom can't produce. Then the mushroom sends out this fruiting body, which emits a scent that attracts the porcupine. The porcupine comes and eats it, the spores go into the porcupine's intestinal tract, and when the porcupine defecates, it sows the spores. The porcupine feeds on the hemlock but also gives something back. Many trees, such as hemlocks, spruces, firs, pines, Douglas firs, oaks, birches, and alders, depend on the symbiotic association of the

FIGURE 1.76
These pellets, measuring $\frac{9}{16}$" to $1\frac{1}{8}$" long, are the more common variations of porcupine scat.

mycelium of different fungi, a phenomenon evident in fossils of plant rooting systems some 400 million years ago.

Squirrels and deer also dig for these mushrooms. Squirrel digs are much smaller than those of porcupines, while deer digs are as large as or larger than porcupines', and usually more triangular.

SIGN: *Scat.* Porcupine scat is variable, depending on diet (Figure 1.76). In winter, the scat is very fibrous because porcupines eat almost nothing but bark. The scat is usually in the form of little pellets, often with a curve or an asymmetrical shape, sometimes similar to that of a cashew. Deer pellets can look very similar to porcupine pellets, but deer pellets are usually symmetrical, often with a point at one end and a dimple at the other. Porcupine pellets are somewhat coarser and more fibrous than those of deer and are occasionally connected by a membrane (Figure 1.77). They also have a stronger, more pinelike scent.

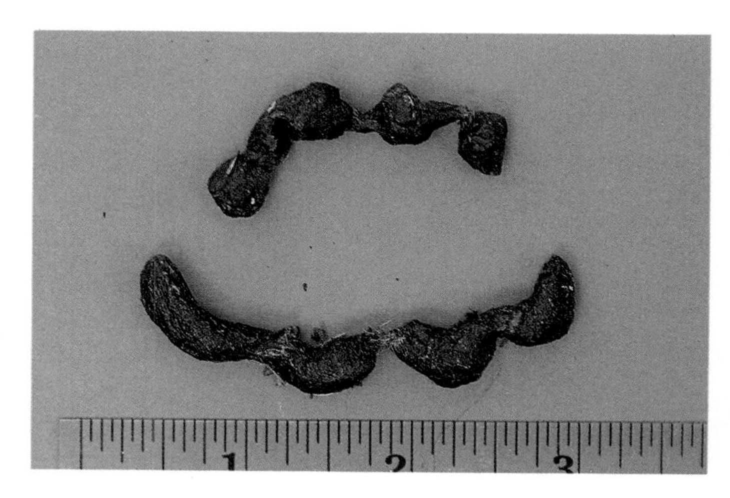

FIGURE 1.77
Occasionally, porcupine pellets are connected by a membrane, as pictured here.

Muskrat

Ondatra zibethicus

ALTHOUGH THE MUSKRAT is the largest of the North American microtines (New World mice), it's also appropriate to think of it as a small beaver. Like the beaver, the muskrat has a thick layer of waterproof underhair and long, shiny guard hairs. Its color varies from reddish to brown to almost black. It has a naked tail and naked feet. The two hind feet are partially webbed. It can stay underwater for up to seventeen minutes. The muskrat builds lodges and digs burrows, and although it doesn't build dams, its tremendous appetite for water plants creates open water in marshy areas that provide perfect habitat for migrating waterfowl. Muskrats are found in increasing abundance everywhere in North America, except Florida and parts of Texas, Arizona, California, and the British Columbia coast. The Newfoundland muskrat *(Ondatra z. obscurus)*, a subspecies, is similar to the common or eastern muskrat, except that it lives exclusively in bank burrows rather than lodges. The muskrat has long, sharp incisors outside its lips, which allow it to gnaw on plants in

FIGURE 1.78
The muskrat inhabits most of North America and is almost always found near water. Like the beaver, it builds lodges and digs bank burrows, but its long, ratlike tail helps to distinguish it from its larger cousin.

FIGURE 1.79
The muskrat's front (on the left) and hind (on the right) feet have five toes and long, well-developed nails, except for the inside toe of the front foot, which is hardly more than a small nail. The hind foot has stiff hairs surrounding the toes, which I believe help to support the animal in soft mud and aid in swimming. When the hairs show up in a track, they outline the toes, giving the appearance of a double track.

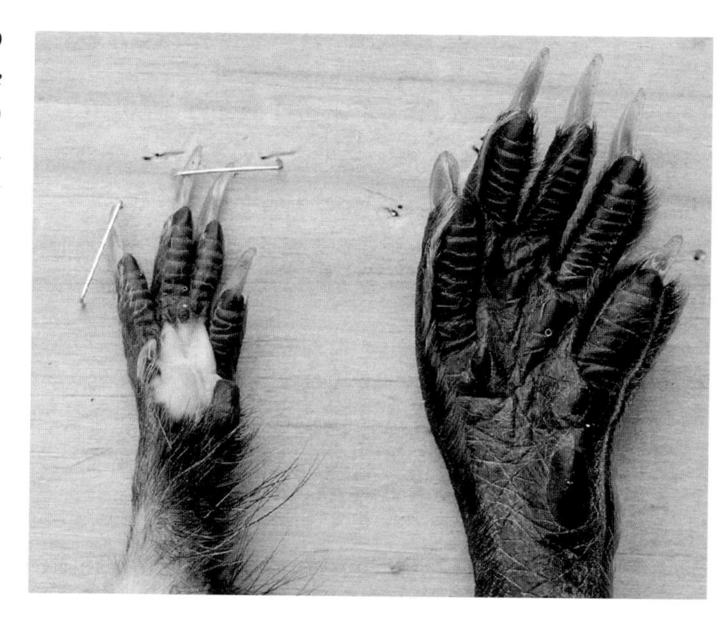

water without drowning. The muskrat will slap its tail on the surface of the water to warn neighbors of an intruder, but because its tail is narrow and only slightly flattened, the sound is not nearly as arresting as that of the beaver.

The muskrat gets its common name from the Algonquian word for the animal, *musquash*, and fits nicely with its species name, which refers to two musk glands located under the skin near the anus. During breeding season (three times a year in southern ranges), these glands swell up and secrete an oily, pungent liquid that, when deposited on a scent post (usually a twist of grass at the water's edge), announces the male's maturity and territory.

Muskrats are not always easy to see. Once, when I was visiting a pond to photograph wild turkeys, I spent two days just standing beside the water, blending into the background, waiting for turkeys to show up. After half a day and no turkeys, I began to watch the muskrats instead. I came to understand their routines and to see the places they frequented, the little corridors they traveled, and where their bank burrows were. By the second day, I knew them well enough to photograph them. I could take some pictures by panning with them across the water, but they moved so quickly that it was much better to figure out where they were going before they got there and to set up my camera beforehand.

Although several muskrats may share a single lodge, they also can be very territorial, defending their burrows and ponds to the death. Like beavers, muskrats smell and hear superbly, but they don't see very well. If you stay still and keep downwind of them, they won't detect you.

There were a lot of muskrats in the pond I was photographing, and I quickly noticed how scrappy they were with each other. Every once in a while, two muskrats would find each other, either by catching one another's scent or literally bumping into each other by accident. Then they would fight, making strange chirping noises that I found quite comical. At one point, two males fought for a while until one of them ran away. The other took off after him, chasing him around the pond, making those little chirping sounds, and constantly complaining, as though he were swearing up a storm. Sometimes, if the wind was wrong and there was no trail to follow, they would pass each other without even knowing it. This went on for quite a while, the chaser chirping vituperatively until the one being chased seemed to tire of the verbal abuse and just disappeared. The chaser continued running and scolding for some time, but in the end he, too, disappeared.

TRACKS. Muskrat front and hind feet (Figure 1.79) both have five toes, but the fifth toe of the front foot is very small—a rudimentary thumb—and hardly ever shows up as a toe in the track. You can just make out the nail. Sometimes in soft mud (Figure 1.80) you can see a mark where the fifth toe should be. Notice the tremendous difference in size between the hind and front tracks (Figure 1.81). The front track is 1⅛" to 1½" long by 1⅛" to 1½" wide—usually as wide as it is long. The hind track measures 1⅝" to 2¾" long by 1½" to 2⅛" wide.

FIGURE 1.80 *(left)*
This muskrat front track shows four elongated toes and nails, and if you look closely at the upper left of the photograph, you can see the nail imprint of the small vestigial thumb.

FIGURE 1.81 *(right)*
These muskrat tracks, recorded in fine clay, show the larger hind track (on the left) and the front track (on the right). The line through them is from tail drag.

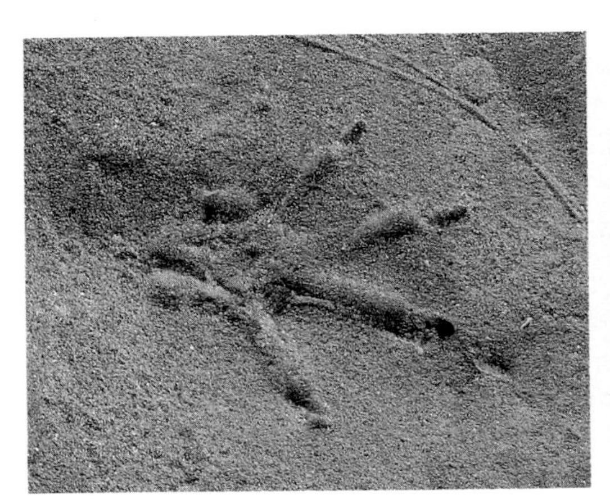

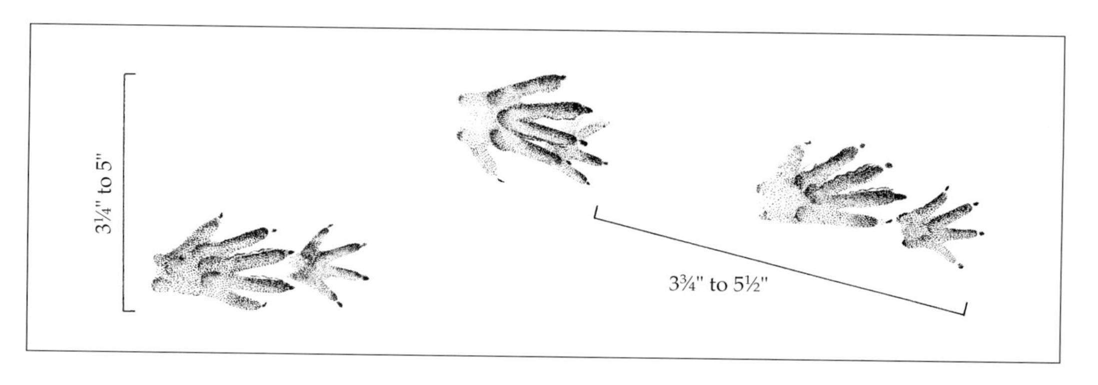

FIGURE 1.82
This trail demonstrates the muskrat's typical alternating walking pattern. The hind track registers close to or partially on top of the front track.

TRAIL PATTERNS. The muskrat's walking pattern (Figure 1.82) is similar to that of the woodchuck and many other quadrupeds: an indirect-registering pattern in which the large hind track is next to or partially on top

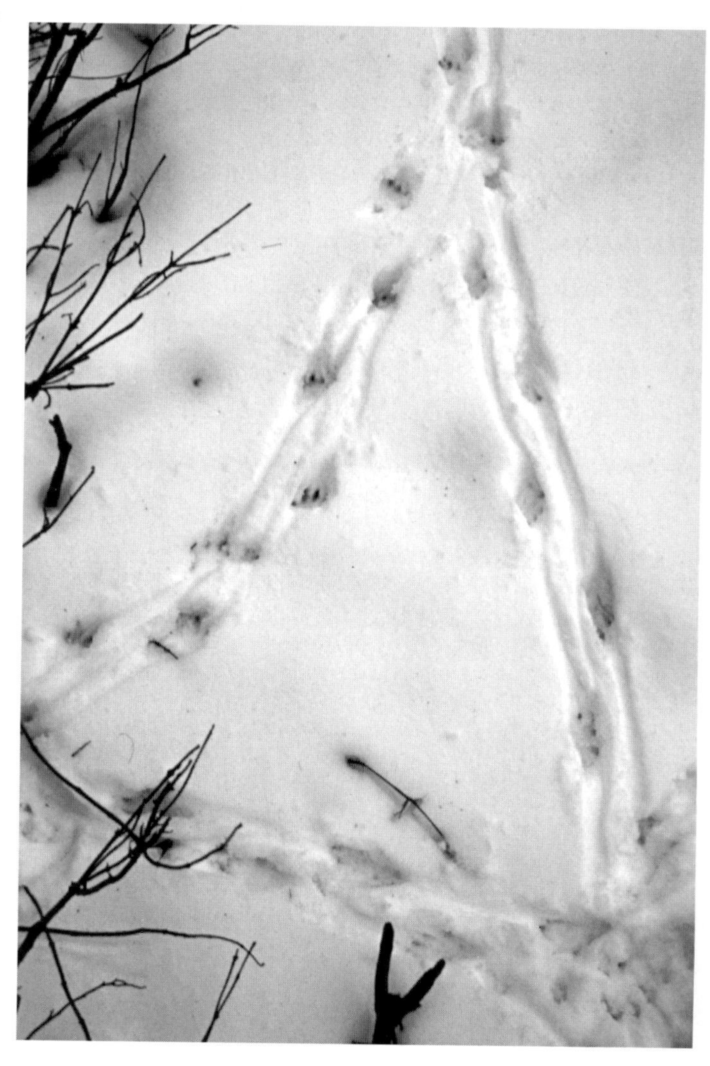

FIGURE 1.83
A muskrat traveling in shallow snow will leave lots of foot drag, often accompanied by tail drag.

of the smaller front track. In shallow snow (two inches or less), you can expect a lot of foot drag, which may be mixed with tail drag (Figure 1.83). As the snow deepens to two and a half to three inches or more, the trail becomes a plow line, with lots of tail and foot drag, although the individual tracks may be seen inside the trough. Strides are 3¾" to 5½", with a trail width of 3¼" to 5".

FIGURE 1.84
Bank burrow entrances are usually 6" below the surface of the water, but this one, which measures 5" in diameter, has been left exposed by receding water. The willows at the entrance have been partly dragged into the den and will be used for food.

SIGN: *Bank Burrows.* Muskrats dig intricate burrows (Figure 1.84) into the banks of bodies of water. The entrances begin about six inches below the surface of the water, are five to six inches in diameter, and run up to forty-five feet inland, leading up to living quarters above the water line but sometimes still four to five feet below ground level. The den itself is only slightly larger than the tunnel, about six to eight inches in diameter, and a smaller air shaft, usually plugged with loose twigs or other vegetation at the top, connects the tunnel to the surface. You may find muskrats making dens in the exterior walls of beaver lodges, even when the beavers are still in residence.

SIGN: *Lodges.* Muskrats also construct lodges (Figure 1.85) that are every bit as complex as beaver lodges. Depending on geographic location, construction can begin in August and go on until late October. The muskrat begins with a large platform made of mud piled with small aquatic

FIGURE 1.85
This muskrat lodge was constructed of aquatic plants and debris, small branches, phragmites, and mud.

FIGURE 1.86
It can be difficult to distinguish between a muskrat feeding station, shown here, and a small lodge.

plants such as cattails, phragmites, sedges, weeds, small sticks, and leaves, all mixed with more mud. When the pile is about two feet high, the muskrat excavates it from the inside, hollowing it out until a chamber large enough to house a family is formed. The domes are usually elliptical in shape and slightly lopsided, and they may be added to year after year. The lodges I have observed have been three to five feet in diameter, but some can be up to eight feet in diameter and from one and a half to four and a quarter feet above water level, with a wall thickness of three to twelve inches. Inside, each family member may construct its own sleeping chamber, so that the whole thing resembles a kind of muskrat rooming house. In other situations, only one or two chambers are made, usually by younger muskrats; the larger houses are made by adults. There will be one or more plunge holes leading to tunnels that connect the lodge to open water.

I have found predators such as red and gray foxes visiting these lodges in winter, sometimes every second day or so, and often defecating on them. You can see the story of their visits written in the snow.

SIGN: *Feeding Stations.* Muskrat feeding stations (Figure 1.86) can take the form of small lodges or huts and are sometimes hard to distinguish from their actual dwelling places. Push-ups are another type of feeding station that are generally much smaller and more numerous than lodges. Muskrats build push-ups over holes in the ice by pushing up quantities of mud and aquatic plant material and then allowing the piled debris to freeze, thus keeping the holes open and the feeding animals protected from predators and the cold. That way, muskrats can spend more time in the pond area foraging for food without having to return to the lodge. You also may find feeding platforms and other hard-to-classify shelters. Look for cut vegetation as a sign of muskrat feeding.

Although muskrats are primarily herbivores, feeding on cattails, bulrushes, sedges, arrowheads, and other aquatic plants, they also eat clams, mussels, snails, crustaceans, fish, and sometimes young birds. A pile of empty mussel shells could indicate muskrat activity.

SIGN: *Scat.* Muskrats often defecate on rocks, logs, or any object that protrudes above the water line, making its scat (Figures 1.87 and 1.88) one of the easier muskrat sign

FIGURE 1.87
Muskrat scat is often found on prominent objects in or near the water.

FIGURE 1.88 *(top)* *Muskrat pellets may stick together to form these irregular-shaped masses.*

FIGURE 1.89 *(bot.)* *Determining the age of scat can be difficult, although sometimes the distinction is clear. The muskrat scat at the top of this photograph is very fresh, whereas the scat to the left is old, weathered, and matted down.*

to find. The scat is composed of little pellets stuck together, although sometimes the pellets are separate. As the scat weathers, it becomes a flat mass and is much harder to identify (Figure 1.89), but no other aquatic animal leaves scat quite this size. The pellets are 3/8" to 9/16" long by 3/16" to 1/4" in diameter. Pellet groups (Figures 1.88 and 1.89) measure 1 5/8" long by 5/8" in diameter.

Beaver

Castor canadensis

THE BEAVER IS THE only member of the Castoridae family left in North America; an extinct form included the genus *Castoroides,* a genus of giant beavers that shared Alaska with the woolly mammoth. Today the genus has declined to a single species. (The so-called mountain beaver, *Aplodontia rufa,* is not a true beaver but a terrestrial muskratlike animal confined to the warm, wet forests of the Pacific Coast.) When explorer David Thompson crossed North America in 1784, he found that the continent "may be said to have been in the possession of two distinct races of beings, man and the beaver," with man occupying the highlands and the beaver in solid possession of the lowlands. Modern estimates place the number of beavers in precolonial North America at more than sixty million, and yet by 1930 the beaver had to be protected, having been trapped nearly to extinction. Fortunately, the beaver has been restored to most of its original range, which includes all of North America, except the extreme north and parts of southern California, Nevada, Arizona, and Florida.

Although not as big as it once was, the beaver is still North America's largest rodent (and the second largest in the world, after South America's capybara), weighing

FIGURE 1.90
Across North America, the beaver utilizes a long list of foods, including trees such as poplar, aspen, sweet gum, birch, willow, Douglas fir, pine, ash maple, and oak, as well as plants such as yellow pond lily, fragrant water lily, and ferns.

twenty-eight to seventy-five pounds. It's worth mentioning that in some cases, a beaver's growth may be indeterminate; that is, it may continue to grow as long as it lives, although at a reduced rate. Wildlife biologist Jim Cardoza has recorded a Massachusetts beaver weighing in at ninety-three pounds. On average, beavers weigh about half that and measure about forty-five inches long, including the tail.

The beaver's tail is one of the animal's remarkable features. Fifteen inches long and up to seven inches wide, flat as a shovel and scaly, it performs a variety of functions. It is used to store fat over the winter and serves as a rudder or sculling oar when the animal is swimming or maneuvering large trees in water. It acts as a temperature regulator, a rear-end balance when the animal is carrying building materials, and a prop when the beaver is cutting down trees. The beaver also has an interesting habit of slapping its tail on the water, perhaps to warn other beavers of intruders or simply to scare away intruders.

Once I was sitting by a beaver pond when some wild turkeys came down for a drink. The pond's resident beaver kept slapping its tail on the water long after any warning would have been useful information. If I had made the slightest sound, those turkeys would have been off like lightning, but they paid absolutely no attention to the beaver, which seemed to annoy it to no end. I'm convinced that it was slapping its tail as a territorial announcement: "Get out of here; that's my water you're drinking!" Ordinarily, a beaver slap has a startling effect, even when it's expected. I remember a beaver coming up to within six feet of me and just staring at me, then all of a sudden slapping its tail so loudly and suddenly I nearly jumped out of my skin.

Another notable feature of the beaver is its large incisors, which are each about one-quarter inch wide. You can often determine whether a beaver has been gnawing on a tree by measuring the width of the prominent incisor marks.

Like muskrats, beavers can remain submerged for up to fifteen minutes. In addition, they have valves that close off their ears and nostrils, lips that close behind their incisors (leaving them exposed to carry branches and other things while underwater), and a clear membrane that protects their eyes.

Beavers were assiduously sought by early trappers for their fur, which is thick, brown (the name *beaver* comes from the Anglo-Saxon word *beofor,* meaning "brown"), and watertight. The beaver keeps it that way by rubbing it with an oil secreted from its anal glands. Also located near the anus is another set of glands, called the castor glands, which secrete the castoreum that beavers use for scent marking. This substance was the instrument of the beavers' undoing. Beavers cannot help themselves from examining the scent wherever they find it, and early trappers quickly learned to bait their traps with castoreum. It was used in the manufacture of perfume, and some Native Americans used it as a medicine. It is reputed to contain the same active ingredients as aspirin. Many beavers also were taken for their meat. French settlers ate beaver on fast days, as the pope had classified this aquatic mammal as a fish because of the scales on its tail.

Beavers are extremely affectionate with one another. I have often seen them swimming in circles in one another's arms, rubbing noses, or cozily munching on a communal twig. They live in family groups, or colonies, that include a breeding pair and four or five of their immediate offspring, ranging in age from newborn to about two years old. At around age two, kits normally leave the parental lodge and establish colonies on their own. In poor habitats, where new colonization sites are limited, the kits may stay with the parents longer or recently dispersed young may return to the parental colony. This social structure allows the young to acquire valuable skills from their parents while contributing to the colony's work force, which is important because the method that beavers have evolved to ensure their survival requires a great deal of individual strength as well as interfamily cooperation.

Humans are without a doubt the beaver's worst enemies, but in some areas, wolves may take a lot of beavers. According to one study, the diet of wolves on Isle Royale in Lake Superior at times was composed of 11% beavers. In Ontario's Algonquin Park, over a nine-year period during which the deer population declined, beavers became an important food source for wolves, showing up in 55% of examined wolf scat. Other predators of the beaver include coyotes, bobcats, lynx, bears, mink, wolverines, and river otters—although I must say that I have seen otters and beavers sharing the same hole on a frozen pond, surfacing only minutes apart and seemingly uninterested in each other's presence.

Beavers are nature's engineers. Native Americans believed that the beaver was sent from above to create the world below, and at least metaphorically, that makes sense. At first there would be a stream running through a forest, flanked by tall trees and thick undergrowth. Then the beaver would build a dam, and soon the trees were down, a pond formed, and more trees died because of the high water. In a few years, the banks were cleared and there was lush, herbaceous growth on the shore and lily pads and cattails in the warm, shallow water. Great blue herons and great horned owls nested in the dead snags, while smaller birds and mammals lived in the cavities. Wood ducks and other waterfowl stopped to feed and rest on the pond, then amphibians, reptiles, otters, and muskrats moved in. Before long, a whole new ecosystem had been created. When the dam broke, the place where the pond was eventually became a meadow, and yet another of nature's cycles began. Sixty million engineers must have wrought a lot of changes in the geography of the world below, diversifying the forest and creating new and different habitats for many different species.

The beaver has been of great benefit to many wildlife species across North America, as well as to the environment itself. Beaver ponds are important water storage systems—for example, slowing and trapping runoff and releasing it gradually. The silt-laden water slows down in the pond, releasing the heavier particles so that the water is clearer downstream, and plant communities established in the sediment help to stabilize the floodplain. The stored water also can raise subterranean water tables, an asset to downslope forests and agricultural crops. Although beavers have been cited in some cases as a negative factor in trout habitats, they have surely proven their worth in the larger picture.

TRACKS. Good beaver tracks are rare because they are usually obliterated by the tail drag. If a beaver is not dragging its tail, its dragging something else. When you do find a clear track, you might notice that, unlike most rodents, whose front and hind feet have four and five toes,

respectively, the beaver has five toes on both the front and hind feet, although most of the time only three or four toes show up in the tracks. The front track (Figure 1.91) is 2⅞" to 3⅞" long by 2¾" to 3½" wide, while the hind track (Figure 1.92) is 5" to 7" long by 3¼" to 5¼" wide. Track widths can vary quite a bit depending on how many toes register. An identifiable characteristic of the beaver's track is the deep impression made by the animal's large, broad nails. Also, the two knobs at the back of the front foot show up in the heel portion of the track, which is similar to that of the woodchuck. Sometimes webs also show up in the hind track.

FIGURE 1.91 *(left)*
The front track of the beaver usually shows only three toes, sometimes four, as in this photo, but seldom all five. The heavy, broad nails are usually a good identifying feature of a beaver track.

FIGURE 1.92 *(right)*
The beaver's hind foot also has large, broad nails. In this track, you can just barely make out all five toes. Notice also the webbing between them.

TRAIL PATTERNS. Beavers use the same paths over and over as they waddle back and forth between their pond and a feeding area or another wetland, and eventually a clear trail will form on the bank (Figure 1.93) or through the undergrowth. This may be 12" wide. Check around for other beaver sign—gnawed tree stumps, wood chips, cuttings at the water's edge (Figure 1.94), or the distinct smell of castoreum along the trail. In winter, such paths may lead out onto the ice, and the snow will be marked with drag lines from branches that were hauled to

FIGURE 1.93
Beavers often drag branches, limbs, or their tails behind them, erasing most of their tracks. This beaver trail shows a series of drag marks up and down the bank.

FIGURE 1.94
Beaver cuttings, sometimes completely stripped of bark, are often found at the water's edge, usually at a trail head where the animal is traveling between the pond and woods.

a breathing hole, where they were eaten in relative safety.

The beaver has an alternating walking pattern that is most often indirect registering. Strides are 6" to 10", and trail widths are 6" to 10¾" (Figure 1.95).

SIGN: ***Dams.*** When a beaver moves into a new area, one of the first things it does is construct a dam (Figure 1.96) to raise the water level high enough to form a pond. It begins by cutting down trees with its strong, sharp incisors. A beaver can cut down a five-inch willow in three minutes. When the tree is down, the beaver cuts it into shorter, manageable lengths, which it hauls to the water and floats to an advantageous spot in the stream's course. Human engineers have attested to the fact that beavers always seem to know the best spot for building a dam. They place branches in the water so that the butt ends are facing downstream and the wide ends act as snags to catch smaller debris, which the beavers arrange across the stream. Once they have a base, they push mud and debris up onto it from the bottom of the stream. This is demonstrated in Figure 1.97, which is a side view of a broken dam. The beavers have pushed up mud

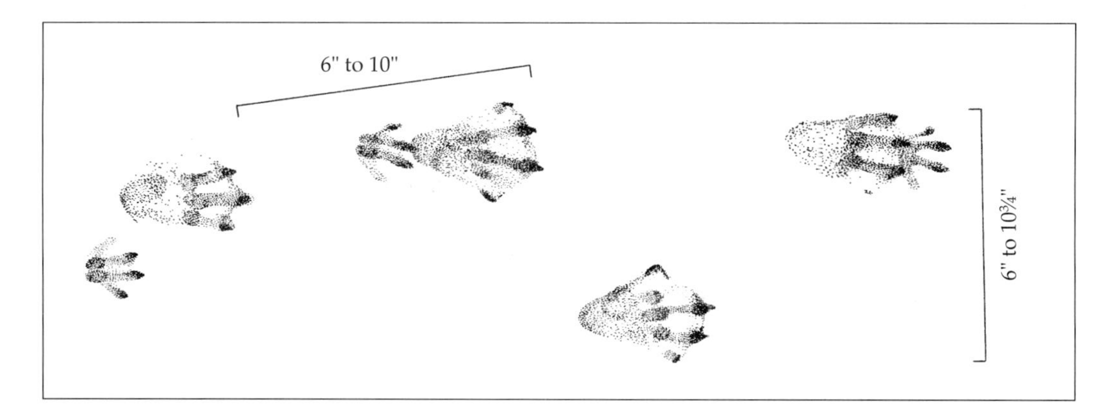

and debris on the left (upstream) side, while the back side consists mostly of small branches and limbs.

I have seen dams that were well over one hundred feet long and six feet high. David Thompson described crossing a beaver dam that was wide enough for two horses and more than a mile and a quarter long.

SIGN: *Lodges.* Beaver lodges (Figure 1.98), built up and added to by several generations, can be as high as ten feet (Figure 1.99) above the pond floor. The largest lodge ever recorded was forty feet across and sixteen feet high. Lodges are huge piles of branches excavated from within and covered with mud, debris, and aquatic plants that the beavers bring up from the pond bottom. They do not, as popular myth has it, pile the material on their flat tails and drag it up the bank and onto the lodge, but rather collect mud and debris in their forepaws, roll it into a ball, carry the ball up the side of the lodge, and pat the mud and debris mixture into place. The mixture freezes in winter to

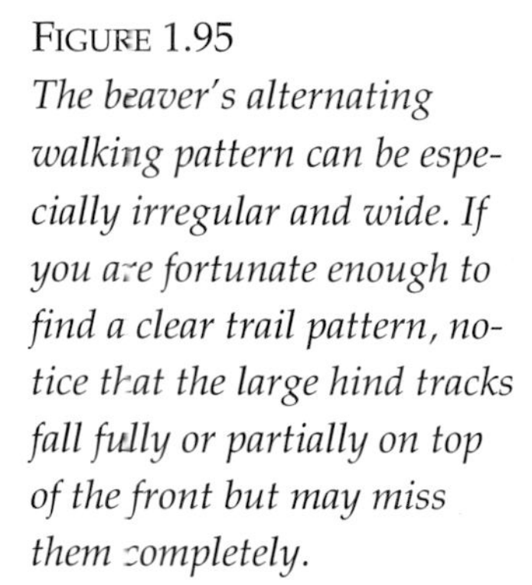

FIGURE 1.95
The beaver's alternating walking pattern can be especially irregular and wide. If you are fortunate enough to find a clear trail pattern, notice that the large hind tracks fall fully or partially on top of the front but may miss them completely.

FIGURE 1.96
Beaver dams are an engineering feat, sometimes constructed with surprisingly large logs and stones. Although not completely watertight, they can hold back a large volume of water, creating a sizable pond.

FIGURE 1.97
This cross section of a beaver dam shows how it is built. The beaver lays branches and limbs across a stream (downstream side on the right), then pushes debris and mud up onto them from the upstream side (on the left).

form a shell hard enough to withstand most predators, except perhaps a determined bear, which may be hitting the lodge before winter. The entrances are all underwater, and a loosely covered opening is left at the top of the lodge for ventilation. This is sometimes called the smoke hole because in deep winter, vapor can be seen rising from the lodge through the hole.

Not all beavers make lodges; some prefer to dig bank burrows. These long tunnel complexes, dug into the sides of slow-moving streams or ponds, have underwater openings about twelve inches in diameter and are similar to, but bigger than, those of muskrats. Other beavers dig temporary bank burrows and live in them until the water level becomes high enough to enable them to build a lodge. They may build the lodge right over the bank burrow, using the burrow as an underwater entrance.

FIGURE 1.98
Beaver lodges are often constructed in the interior of a body of water, like this one in a frozen wetland. They may also be built on the banks of a pond.

FIGURE 1.99
A receding water level left this abandoned beaver lodge high and dry. Added onto year after year, a lodge can reach an impressive size. The man standing next to this one is 6′3″ tall. The largest recorded lodge was 40′ across and 16′ high.

SIGN: *Canals.* Beavers also make canals, often as much as two feet wide and two feet deep. They use these to get closer to feeding areas without having to leave the safety of the water and as convenient ways of getting felled trees to the pond and from there to the dam.

SIGN: *Cut Trees.* Beavers eat the cambium of trees such as alders, willows, aspens, and cottonwoods, as well as their leaves, buds, and twigs. (Their forefeet are so dexterous that they can chew off the bark on twigs while rolling them with their "fingers," just as we eat corn on the cob.) Look for stout branches and even trunks completely stripped of bark, with wide incisor marks up to a quarter inch wide (Figures 1.100, and 1.101). When the beaver gathers branches inland, it will usually drag them to the water's edge and eat them there, for greater safety.

FIGURE 1.100 *(left)*
Beavers do not eat wood but instead are after the inner (cambium) layer of bark. Once they have downed a tree, they may strip it completely of outer bark to get at the more nutritious cambium.

FIGURE 1.101 *(right)*
A beaver ventured almost 100′ from the water's edge to cut this aspen, one of its preferred foods.

FIGURE 1.102
This red oak is a testament to just how large a tree a beaver can cut down.

There is almost no limit to the size of tree a beaver will tackle. I've seen them move logs five to six inches in diameter, cutting them up into three- or four-foot lengths. And in one memorable case, I came upon a red oak that had been toppled by beavers. The tree must have been well over two feet in diameter (Figure 1.102). Other large trees showing beaver work are seen in Figure 1.103.

FIGURE 1.103
These trees are well over 2′ in diameter at the base, where the beaver was cutting.

SIGN: ***Winter Caches.*** In colder areas where ice forms in the winter, beavers will cut down a large quantity of food branches and pile them up in front of their lodge before the pond freezes, anchoring them in the mud at the bottom of the pond. When the pile gets high enough, branches stick up above the water line. Later, when the pond is completely frozen over, the beavers still have a source of food below the ice. A pile of fresh branches near the lodge (Figure 1.104) is one sign that the lodge is in use.

SIGN: ***Scent Mounds.*** Many animals make sign posts—moose and deer make antler rubs, canines and felines spray trees—and beavers are no exception. A beaver's scent mound (Figure 1.105) is a territorial marker. A fresh beaver scent mound is an indication that a beaver has taken up residence in the area.

FIGURE 1.104 *(top)* *Fresh branches sticking out of the ice in front of this lodge indicate that it is active.*

FIGURE 1.105 *(bot. left)* *Beavers made this scent mound by dragging debris from underwater and mixing it with forest litter.*

FIGURE 1.106 *(bot. right)* *The yellowish substance atop this scent mound is called castoreum, a secretion from the beaver's scent gland located near the anus.*

FIGURE 1.107
Beaver scat resembles compact balls of sawdust. Look for scat in the water for beavers rarely defecate on land.

To make a scent mound, the beaver usually places material from a stream or pond onto the shore and mixes it with other debris already there. In some instances, you may find scent mounds composed exclusively of materials from the pond shore, stream bank, or forest floor. (Occasionally beavers use existing grass tussocks along the edges of streams and ponds as scent mounds.) The beaver then deposits a yellowish-orange secretion called castoreum (Figure 1.106) from its castor gland onto the mound. You can smell this scent from thirty feet away if the wind is right. Some people say it smells like a horse barn or horse leather. When you smell it, you won't need tracks to know there are active beavers in the area. These scent posts are perhaps the best indication of the presence of beavers.

SIGN: ***Scat.*** Beavers almost always defecate in the water. Look for scat on the pond bottom if the water is clear. If you find it in the winter, it'll be nothing but sawdust (Figure 1.107) because of the dryness of the beaver's winter diet.

CHAPTER 2: RABBIT FAMILY

Leporidae

Snowshoe hare trail in a high-speed run.

Eastern Cottontail Rabbit
Sylvilagus floridanus

Snowshoe Hare
Lepus americanus

White-tailed Jackrabbit
Lepus townsendii

RABBITS AND HARES belong to the order Lagomorpha, which comprises gnawing mammals having two pairs of upper incisors, one behind the other. They are famous for their prodigious reproduction rate. Some can have up to twelve young per litter, and some as many as four to seven litters per year in their southern range. This is only one of the animals' answers to an extremely high predation rate. The others are speed and camouflage. They have exceptionally long hind legs, and some of the bones are fused together to provide even greater thrust. Their skulls have cavities in them, much as birds' skulls do, to make them lighter and to assist their hearing. These animals are buff to brown colored in summer to blend in with their chosen surroundings. In winter, the snowshoe hare and the jackrabbit turn white. Despite these protective features, the cottontail is North America's primary prey species. Cottontails have been trapped extensively for their fur, although it is much less durable than that of the otter, and their flesh has provided ready food for every carnivore in the forest, including *Homo sapiens*.

Both rabbits and hares (jackrabbits are technically hares) belong to the family Leporidae, but each has adapted differently to different habitats. Cottontail young are altricial. They are born helpless, naked, and blind but grow extremely quickly. Snowshoe hare young are precocial at birth. They are more developed, their eyes are open, and they have a coating of hair. Snowshoe hares prefer open wooded or swampy areas, where they can take advantage of their great speed to elude predators. Cottontails, though

FIGURE 2.1
Cottontail rabbits usually do not venture far from cover. The lure of newly sprouting greens, however, coaxed this one into the open.

FIGURE 2.2
Snowshoe hares turn white in winter, allowing them to blend in with their snow-covered surroundings. In the absence of snow, however, the sharp contrast of their pelage has the opposite effect. That's how I found this hare. I was able to get within 12½′ of it without it bounding away.

found in wooded areas, gravitate toward farmland, pastures, hedgerows, and densely thicketed areas where they can hide. Both are about the same length—seventeen to eighteen inches, with hares sometimes up to twenty inches, including a two-inch tail—but the snowshoe hare weighs more (four and a quarter pounds, on average, compared to the cottontail's two and a half pounds). The hare's hind legs also are larger, averaging five and a half inches long, and the toes have a much wider spread than those of the cottontail.

The white-tailed jackrabbit is larger than either the cottontail or the snowshoe hare, weighing from five and a half to seven and a half pounds. Its summer color is grayish brown, with the hair on its face tipped white, to give it a grizzled look. Its ears are gray, and its tail and underparts are white. In winter, in the northern part of its range, it turns even whiter than the snowshoe hare.

Eastern cottontails are widely distributed throughout the continental United States, from the Dakotas south to parts of Arizona and New Mexico and east to the Atlantic Coast. They also extend just slightly north into southern Canada. New England cottontails are found in New England southwest through the Allegheny Mountains.

Snowshoe hares have a northern range extending from Alaska across most of Canada to the Atlantic Coast and south into much of the northeastern United States and along the Allegheny Mountains. They also inhabit the Great Lakes region and in the West are found south into the Rocky Mountains and in parts of Washington, Oregon, and northern California.

FIGURE 2.3
Snowshoe hares (this one in its summer coat) are associated with more open woods than cottontails, but they do require some dense, brushy cover.

FIGURE 2.4
Always on the alert and ready to dash off in a flash, the white-tailed jackrabbit (here in its summer pelage) is larger than both the cottontail and the snowshoe hare.

White-tailed jackrabbits are found mostly in the Midwest and prairie regions, with some ranging into the Rockies, Cascades, and Sierras.

TRACKS AND TRAIL PATTERNS. The front and rear feet of a cottontail (Figures 2.5 and 2.6) do not have large toe pads and are entirely covered by hair, which means that their tracks rarely show toe or pad marks in snow. The rear track is less than 1½" wide, and the front is less than 1¼". The rabbit's trail pattern (Figure 2.7) is probably one of the best known in the forest. The two small front tracks are one in front of the other, while the two hind tracks are side by side ahead of the front tracks. Squirrel tracks look similar, except the squirrel's two front tracks are usually side by side. That is the chief difference between squirrels and rabbits; size isn't a reliable guide.

Sometimes the rabbit's rear foot will leave a sharp-looking, pointed track (Figure 2.7), sometimes the hind tracks will be closer together (Figure 2.8), and sometimes the front tracks will be almost side by side, but the basic triangular pattern is the most common. When the rabbit

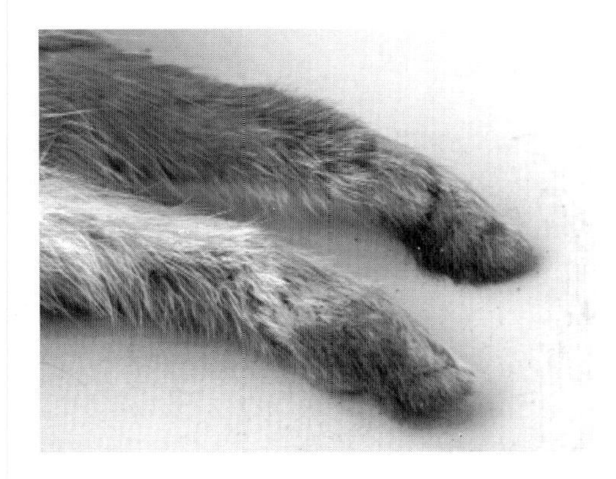

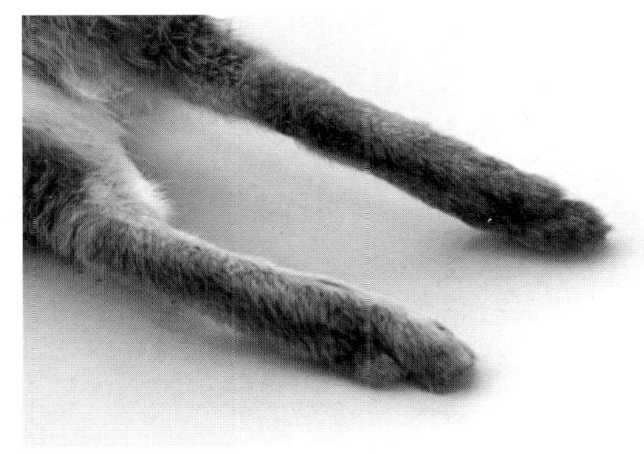

puts its front feet side by side, they will nearly touch, almost or at times actually making a single track. When the squirrel puts its front feet side by side, they hardly ever touch. Figure 2.9 is a comparison of squirrel and rabbit tracks. Usually, as the rabbit increases speed, the distance between its front and rear tracks also increases.

Rabbit tracks in sand (Figures 2.10 and 2.11) may show the toe pads. This can be very confusing. Sand also will register nail marks, which usually do not show up in snow. Pay attention to the trail pattern, which is the lagomorph's varying but readily identifiable signature.

Snowshoe hare tracks are bigger than those of cottontails. A hare's hind track is usually more than 1½″ wide; its front track is usually more than 1¼″ wide. The hare's rear toes can spread up to 4½″ across (Figures 2.12 and 2.13), which, combined with its abundant hair, enables it to run on top of snow (hence its name) while its predators (bobcats or coyotes) sink into the snow.

FIGURES 2.5 *(left) and* 2.6 *(right)*
The bottoms of the cottontail's feet are densely furred, with five toes on the front (left), four toes on the hind (right), and sharp nails on both front and hind feet.

FIGURE 2.7
This set of cottontail tracks, in a dusting of snow, demonstrates the pointy appearance that the hind feet leave under certain conditions. This is a good identifying feature of a cottontail track.

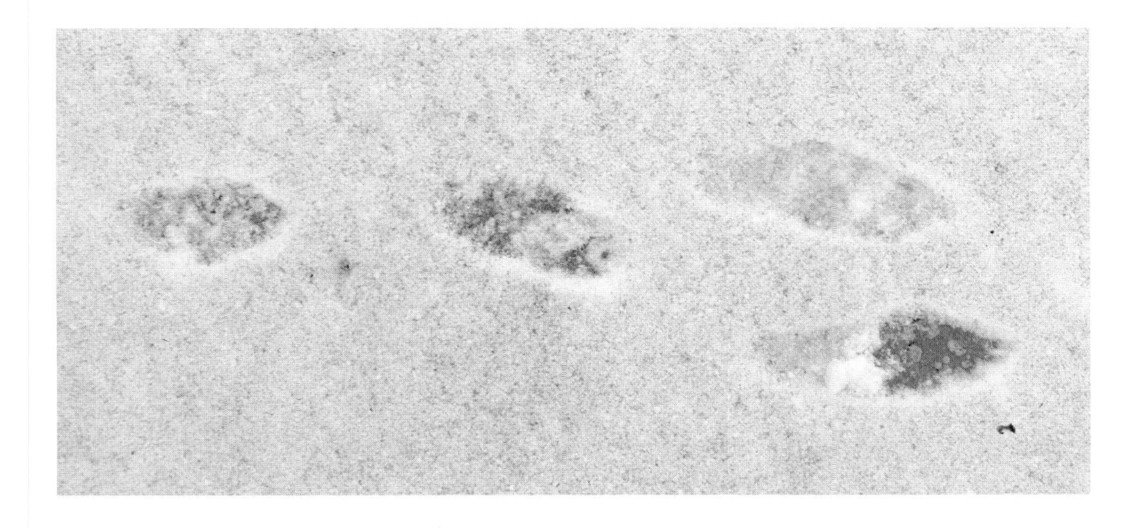

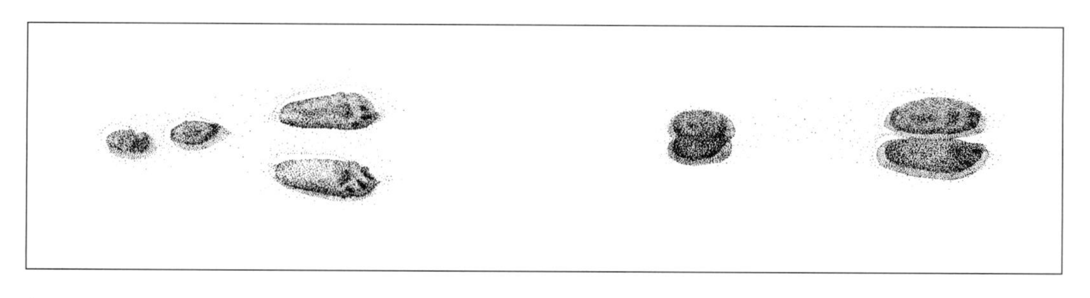

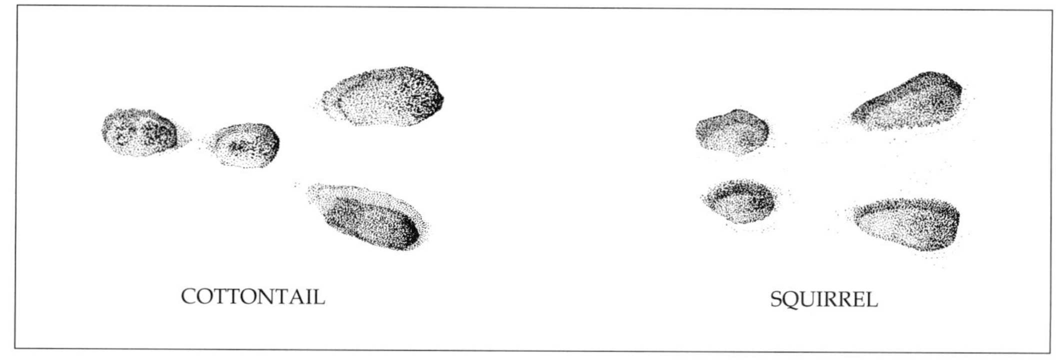

FIGURE 2.8 *(top)*
The cottontail tracks on the left show the common triangular pattern of the rabbit family; a variation is on the right.

FIGURE 2.9 *(bot.)*
This shows the difference between the common rabbit track pattern (left) and the common squirrel pattern (right). The rabbit's front feet are one in front of the other; the squirrel's are side by side.

Figure 2.13 shows a snowshoe hare's tracks with the hind toes spread out, but I find that unspread toes (Figure 2.14) are more typical. Tracks with unspread toes can be confused with those of a cottontail. The most reliable way to tell them apart is to measure the width of the hind track.

A white-tailed jackrabbit's hind track is 3½″ to 6¾″ long and usually more than 1½″ wide. Its tracks look more robust than those of a cottontail, but its hind foot does not spread as much as that of a snowshoe hare. While the snowshoe hare's hind track will "snowshoe out" to more than 4″, that of the jackrabbit will rarely exceed 2½″. Often the jackrabbit's rear heel will not register, in which

FIGURE 2.10
When cottontail tracks register in sand or mud, the nails and toe pads are prominent.

FIGURE 2.11
The cottontail's tracks on the left are of the front feet, with only four toes showing in an asymmetrical pattern; on the right, all four toes of the hind feet register.

case its track can be confused with that of a canine. (I also have seen people misidentify snowshoe hare hind tracks as those of a canine [Figure 2.15].) If the jackrabbit's rear heel is not registering, the track may be 3⅜″ long by 1½″ or more wide. The jackrabbit's stride is 1′ to 10′. Some sources claim it can jump up to 20′. I've recorded strides up to 9′8″ of a snowshoe hare being chased by a bobcat.

SIGN: *Forms.* North American rabbits do not dig holes to live in. European rabbits do, but if *Alice in Wonderland* had taken place in North America, Alice would have had to fall down a woodchuck hole. Rabbits here make forms—shallow depressions in the ground, usually under a dense thicket—and line them with fur and soft grass when having their young. Snowshoe hares do not

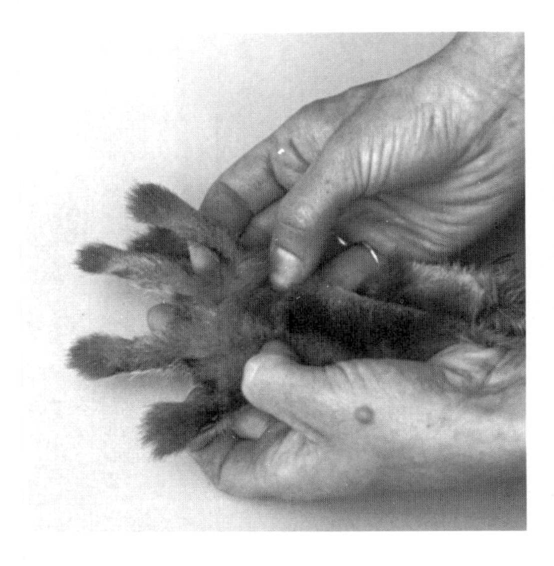

FIGURE 2.12 *(left)*
Because the snowshoe hare is able to spread its toes widely, its feet act like snowshoes, enabling it to elude predators in deep snow.

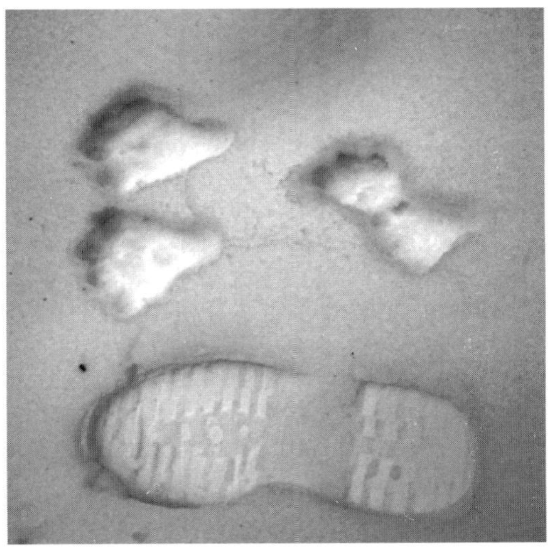

FIGURE 2.13 *(right)*
When the snowshoe hare spreads its toes, the hind tracks (top left) may widen to 4½″ across (its front tracks are to the right). The boot print (bottom) is 12″ long.

FIGURE 2.14
This is the trail pattern of a snowshoe hare in a dusting of snow. The toes of the two larger hind tracks (left) are not spread, creating a pattern like that of the cottontail.

FIGURE 2.15
Under certain conditions, snowshoe hare hind tracks can easily be mistaken for canine tracks. Here, the hind tracks dominate, and only faint impressions of the front tracks are visible (to the middle and lower middle of the hind tracks).

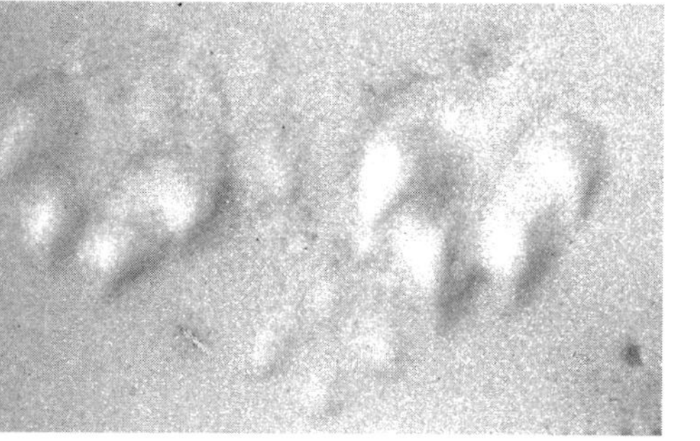

make forms, giving birth to their young on open ground.

White-tailed jackrabbits excavate their forms only to a depth of about two to four inches, hardly more than a depression in the ground, with the dirt piled in front of it. These forms are usually found near vegetation such as sagebrush, stunted pines or hemlocks, or low juniper bushes. In winter, jackrabbits may use tunnels in the snow as resting places.

SIGN: *Browse.* As mentioned earlier, lagomorphs have two sets of upper incisors, which help to distinguish them from members of the rodent family. Like rodents, however, they usually leave a perfect forty-five-degree-angle cut on twigs when they browse. If you know what the preferred winter food sources are for lagomorphs in your area, these plant species will be good indicators of the animals' presence. For instance, in some areas of Massachusetts, I have found that by inspecting rubus canes (such as those of the blackberry and raspberry); dewberry runners; and red maple, hawthorn, blueberry, and oak twigs, I can usually tell whether lagomorphs are present. Be careful with rubus canes, though, because porcupines go after them as well. Look for scat to determine species. In summer, most lagomorphs switch to herbaceous plants, including dandelion, plantain, clover, grasses, sedges, and garden crops.

Cottontail populations usually do not use conifers as food sources, but there may be some light use of hemlocks and white pines in some areas. Snowshoe hares may use white pines extensively. They also browse on white and red spruces and Douglas firs, which cottontails rarely use. In some areas, hares browse on tamaracks, but as far as I know, cottontails do not. Many people are surprised to find out that snowshoe hares also have been known to eat carrion.

A comparison of hare and deer browse (Figure 2.16) shows the difference between animals with sharp top and bottom incisors (lagomorphs) and those with no top incisors (deer). Rabbit and hare browse has forty-five-degree cuts that are so precise you'd think someone made them with a knife. The incisor marks themselves are sometimes evident on hardwoods such as oaks but usually not on softwoods. The largest diameter cuttings I have recorded are 7⁄16″ in diameter by cottontail on sumac, and nearly 3⁄8″ in diameter by snowshoe hare on balsam fir. Twigs that are browsed by deer are torn and pulled rather than snipped off. Notice the bulbous growth on the twigs in Figure 2.16. Rabbits will make plants grow in this way as they prune new growth year after year.

FIGURE 2.16
Snowshoe hares and rabbits have sharp top and bottom incisors, creating a clean, 45-degree-angle cut when they browse (bot. row). Deer have only bottom incisors, and as a result, when they browse, they pull and tear, leaving twigs looking tattered or squared off (top row).

FIGURE 2.17
In winter or early spring, rabbits and hares may debark apple trees (shown here) as well as other fruit trees. Incisor marks will not be well defined, and the gnawed area will have a ragged or rough appearance. Some studies indicate that debarking occurs most often when hare or rabbit densities are high. Voles also debark small saplings or branches close to the ground, but they do not bite into the wood as do rabbits and hares. With voles, you'll find the unused outer bark on the ground with their scat. Lagomorphs, however, eat the outer bark along with the inner (cambium) layer.

When rabbits or hares browse on fruit trees, such as apple trees (Figure 2.17), they will debark them and eat the small twigs and buds. This usually happens in winter or early spring. Porcupines also may dine on apple trees, especially on the fruit itself, as well as on its twigs and leaves. They are, however, less likely to do any debarking. Since rabbits are not known to climb, any work too high for a lagomorph to reach (taking high winter snow lines into consideration) is likely to be that of a porcupine. Look for droppings and the incisor work. Porcupine incisor work (Figure 1.71, page 71) tends to be more defined than that of a lagomorph, which has a more ragged appearance (Figure 2.17). Woodchucks may debark fruit trees as well, but this is uncommon.

SIGN: *Urine.* Snowshoe hares will urinate in the snow, and even on each other as part of their courtship ritual. One of the mating pair will leap into the air, and the other will run under the partner and be sprayed with urine. Scientists believe the females can sniff out pheromonal information from the urine that allows them to form an opinion about a particular male's reproductive efficiency, a very important consideration in lagomorphs. At any rate, the urine will appear as a faint yellow to reddish deposit on the snow. Urine on hard, crusted snow may spread to form a ring more than a foot in diameter and smell faintly of pine.

SIGN: *Scat.* The digestive system of lagomorphs is similar to that of deer, in the sense that they both digest their food twice. For lagomorphs, food goes into the stomach, is partially digested, and is defecated as a soft, jelly-like pellet. The animal eats this pellet, digests it, and defecates again. The second type of pellet is the harder pellet we see on the ground.

Cottontail scat and snowshoe hare scat may look different at times, but it also may look very much alike (Figure 2.18). Both droppings may appear wrinkled or shriv-

FIGURE 2.18
The scat of snowshoe hares (left) and cottontails (right) is not always this dissimilar. Notice that one of the cottontail pellets looks exactly like those of the snowshoe hare. You cannot rely on scat to differentiate between most of the rabbit family members.

eled, and both are often round, with some shape variations. Hare droppings have a tendency to be flatter, but this is not a reliable way to determine species. I've measured hundreds of snowshoe hare droppings and compared them to hundreds of cottontail droppings, but I've never been able to find a consistent way to determine which is which. (For the most part, jackrabbit scat cannot be differentiated from that of rabbits and other hares.) The best way to tell the two animals apart (besides tracks) is to check the habitat. Scat in open yet wooded country is probably that of a snowshoe hare (being faster, they take more chances by venturing out into the open), whereas that close to thickets and brush is probably from a cottontail. Don't forget the different food preferences discussed earlier.

Mounds of rabbit droppings can be confused with deer droppings, but lagomorph droppings are round or slightly flattened and sometimes wrinkled, whereas deer pellets are more elongated and usually smooth. Deer drop their pellets all at once, whereas rabbits drop their pellets one at a time. A pile of rabbit pellets indicates that the animal has been staying and eating in one spot, so look around to see whether there is rabbit browse near the pile. Look carefully, though, because rabbits will often eat the whole plant to just below the leaf litter or snow line.

CHAPTER 3: WEASEL FAMILY

Mustelidae

River otter tracks on the shores of Quabbin Reservoir, Massachusetts.

Long-tailed Weasel
Mustela frenata

Ermine
Mustela erminea

Least Weasel
Mustela nivalis

Although the long-tailed weasel is small, about twelve to seventeen inches long and weighing an average of only seven or eight ounces, it is a very efficient predator, killing animals up to five times its size, including muskrats, waterfowl, squirrels, and cottontails. It is an energetic, feisty fighter. In *Johnson's Natural History* (1865), there is an account of a kite catching a weasel in its talons and lifting it up into the air: "In a few moments the kite began to show signs of great uneasiness, rising rapidly in the air, or as quickly falling, and wheeling irregularly around. . . . After a sharp but short contest, the kite fell suddenly to the earth . . . with a hole eaten through the skin under the wing." Successful weasels often cache small rodents, especially mice, for future consumption.

The long-tailed weasel's range is located much farther south than that of either the least weasel or the ermine. It begins just north of the U.S. border, in Canada, and extends south throughout the continental United States and into Central and South America.

FIGURE 3.1
This ermine is in its winter pelage. The long-tailed and least weasels also turn white in winter, except in the southern parts of their ranges.

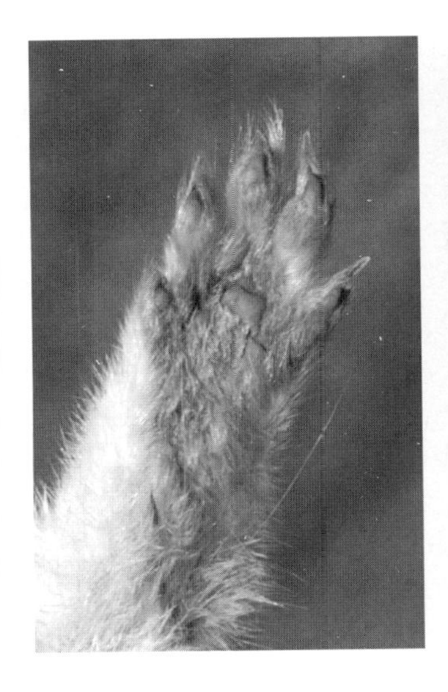

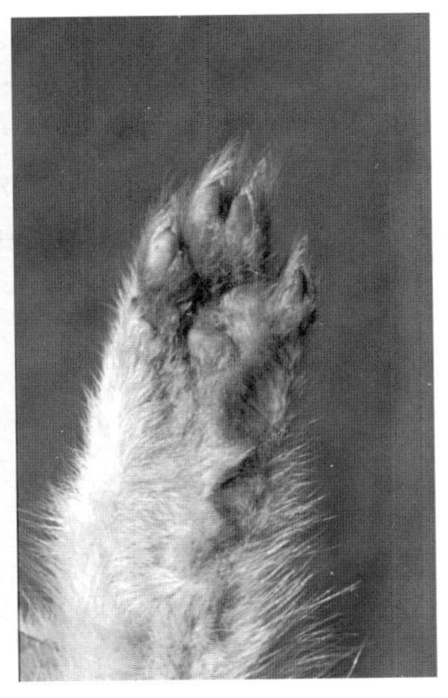

FIGURE 3.2 *(left)*
The long-tailed weasel's feet are small, less than ¾" wide. The front foot (left foot shown here) has five toes.

FIGURE 3.3 *(right)*
Weasel paws are heavily furred. As with the front foot, the hind foot has five toes. (The left foot is shown.)

The ermine is found throughout Alaska and Canada and south to Colorado in the West and Pennsylvania in the East. A study of the eating habits of 360 individual ermines in New York State indicated that their diet was 34.5% field mice (meadow voles); 13.1% rabbits; 11.3% deer mice; 11.2% shrews; 6.7% rats; 3.6% chipmunks; 3.2% birds, frogs, and snakes; and 16.4% undetermined mammals. It has a home range (the area in which it forages for food within any given year) of up to forty acres.

The least weasel is equally widespread in all but the most northern and eastern areas. It ranges throughout most of Canada, south to Iowa and northern Indiana in the Midwest, and east to West Virginia. It also lives in the southern Appalachian Mountains. As its name implies, it is smaller than all other weasel species. The male weighs one and one-fifth to two and one-fifth ounces and measures about six inches long (including its one-and-one-eighth-inch tail). It, too, is an important check on voles and mice. The weasel's generic name, *Mustela*, means "one who carries off mice." The word *weasel* comes from the Sanskrit *visra*, which means "to have a musty smell."

TRACKS AND TRAIL PATTERNS.

Weasel paws are small and hairy (Figures 3.2 and 3.3), with five toes, arranged in an asymmetrical pattern, on each foot. The inside fifth toe often does not register. You seldom get a track that illustrates the toes and pads clearly, especially in snow.

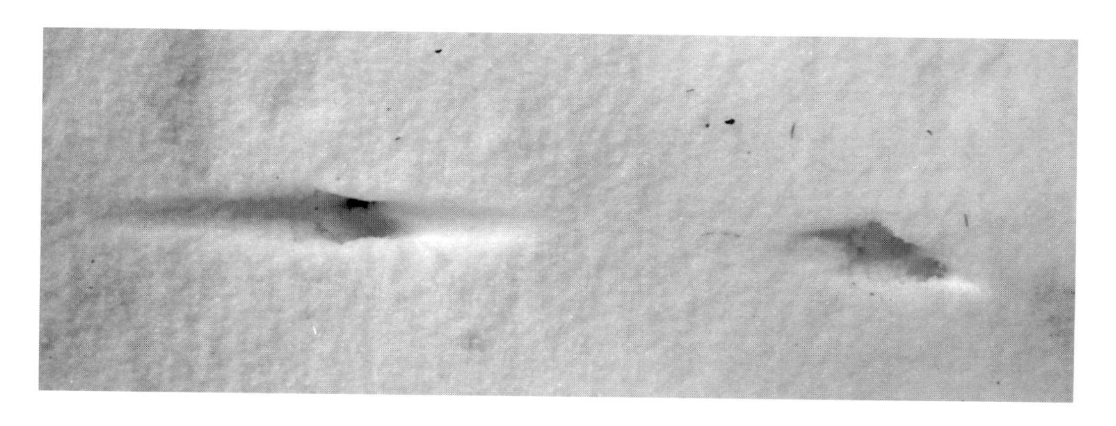

FIGURE 3.4
These long-tailed weasel tracks exhibit the drag marks that are often made as it bounds through the snow. The trail width is about 2″.

FIGURE 3.5
This is a typical long-tailed weasel trail in snow. The hind tracks registered directly on top of the front.

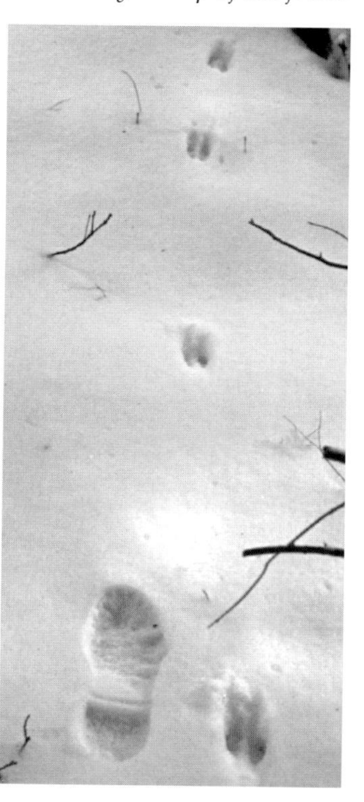

Weasels are land-based creatures, but they've been known to swim and climb trees. My wife, Paulette, and I were sitting by our kitchen window one day and saw a long-tailed weasel about twelve to fifteen feet up a white pine, coming straight down the tree. Research studies say that they spiral down the tree, but this one was heading straight down. It was quite agile and seemed very much at home. When it reached the ground, it scurried around in the snow for a while and then was off. Its tracks (Figure 3.4) were very small, less than 1″ by 1″, with a trail width of 1½″ to 2¼″.

The weasel's trail pattern (Figure 3.5) is a 2-2 pattern, with one track slightly ahead of the other as the animal moves along. Otters, mink, and fishers all leave this 2-2 pattern. The front feet come down, registering, and as the front feet leave the ground, the hind feet come in immediately thereafter, falling on the same set of tracks, or just behind, and causing a very slightly elongated 2-2 pattern. Note that the tracks are very small, with the trail width sometimes only 1½″. This is a bounding trail, where the stride is about 21″. Although the trail width is very close to that of a mouse, you can readily identify weasel tracks by the long bounds.

Measuring a weasel's trail width to determine species can be problematic, partly because of dimorphic overlap—the female of a larger species is about the same size as the male of a smaller species—and partly because of the tendency in most animals to narrow their trail as they increase their speed. For example, one female long-tailed weasel I followed varied its trail width from an average of 2⅛″, when its average stride was 11⅜″, to 1⅞″, when its average stride increased just 6″. Such variations, when

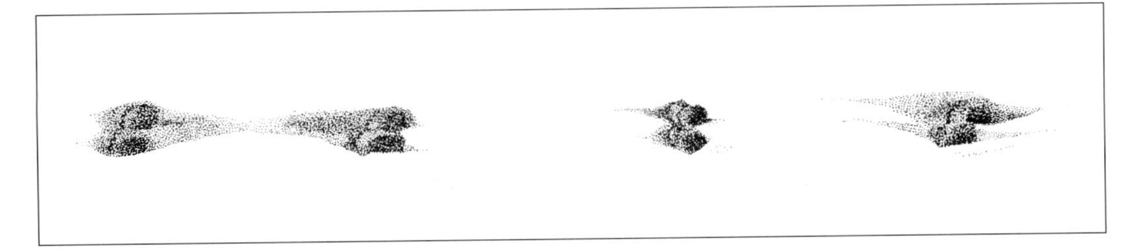

FIGURE 3.6
In snow, the ermine's shorter strides may be connected by drag marks (left) and often alternate with longer strides throughout the length of the trail.

added to those resulting from dimorphism and changes in substrate, can produce a complex jumble of figures that make identifying smaller weasels very difficult. Here are some averages to work with:

	Trail Width	*Stride*
Long-tailed weasel	1½″ to 3″	9½″ to 43″
Ermine	1″ to 2⅛″	9″ to 35″
Least weasel	⅞″ to 1⅝″	5″ to 20″

In each case, longer strides are possible.

To find out which species is frequenting your area, check the range information discussed earlier in this chapter. Least weasels, for example, do not live in Massachusetts. Also be aware of habitats. Least and long-tailed weasels may share some of the same wetland habitats, and both can be found in open woodlands or along field edges. Their difference in size, though, should prevent any confusion of the two. You are less likely to find ermines in wetlands, as they prefer more upland-type habitats, including meadows, woodlands, and mountains to thirteen thousand feet. Ermines also tend to avoid thick coniferous forests. If you have only long-tailed weasels and ermines, you may consider any trail widths consistently under 1¾″ to belong to ermines and those consistently over 2″ to belong to long-tails. Trail widths between 1¾″ and 2″ will require that you check other sign. In areas where all these weasels occur, trail widths consistently under 1⅛″ could be considered those of the least weasel.

The trail pattern provides more evidence in our effort to identify the weasel's trail. Often there will be drag marks connecting two prints in a short bound, leaving what some trackers call a dumbbell pattern: two dots, a dash, and two dots.This pattern is more typical of the ermine (Figure 3.6). Some of the long bounds can be up to 35″ long, and there will be no drag marks between them. This is caused by a very energetic, curious animal that can't make up its mind and is distracted by everything in its environment. To me, weasels are very exciting animals to track. Note the inconsistency of the bounds in the long-tailed weasel's trail shown in Figure 3.7: one long, three short, one long, two short, one long, three short, one long, three short, two long, three short, two long, about seven short. This is typical of long-tailed weasel trails: very erratic, all over the map, and inconsistent bounds.

Ermines tend to be more consistent in their bounding: one long, one short, one long, one short. This is not an infallible rule and will add only circumstantial evidence when trying to distinguish between these two animals. It also helps to distinguish the long-tail's trail from the mink's, which in its 2-2 pattern has a very consistent gait.

FIGURE 3.7
The long-tailed weasel's trail is very inconsistent, alternating between long and short strides and interspersed with short repetitions of either.

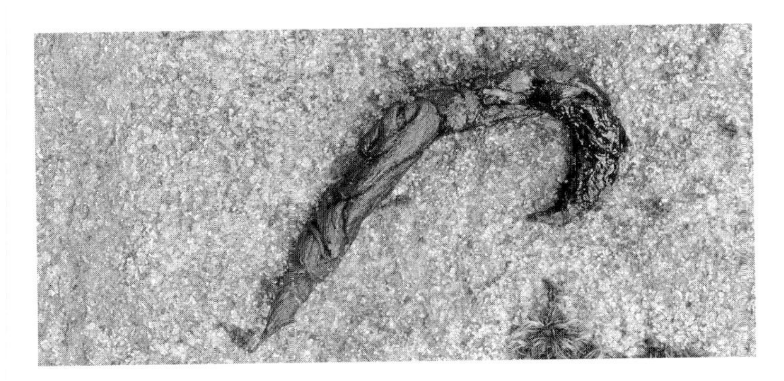

FIGURE 3.8
Weasel scat often looks twisted and is tightly wound and full of hair.

SIGN: ***Food Caches.*** Weasels may store large quantities of food. When confronted by many prey animals, their response may be to kill more critters than they can eat. (This is called surplus killing.) Their food stashes have turned up surprising numbers of mice and other small rodents. When food is plentiful, weasels often eat just the parts of the mouse they like best—the brains, inner organs, and muscles—so a pile of mouse carcasses with only the heads missing or the remains of just legs and tails is a reliable sign of weasel predation. Occasionally, they may invade a chicken coop, for example, and in response to the actions of the prey, go on a rampage of killing, leaving whole chicken carcasses behind with only the heads eaten. A neighbor of mine has had this unfortunate experience. I also have heard that weasels sometimes enter a henhouse and take out only the rodents, leaving the chickens undisturbed.

SIGN: ***Scat.*** Weasel scat (Figure 3.8) is very small, rarely exceeding ⅛″ in diameter and 1¼″ in length, and is often found on stone walls or on objects in trails. The scat is usually very tightly wound, looking as though it has been twisted, and contains fur. Weasels are not known to eat berries, so if you find weasel-like scat containing seeds, it's probably that of a marten.

Mink

Mustela vison

FAVORING ALL KINDS of wetland areas, this small (two to three and a half pounds for the male, one and a half to two and a half pounds for the female), dark-furred member of the weasel family is often found in habitats similar to those of the muskrat, on which it occasionally dines. Both its names mean "weasel"—*Mustela* in Latin, *vison* in Swedish—suggesting that the mink is the quintessential family representative. A confirmed and opportunistic carnivore, it will eat any meaty creature that comes its way: fish, frogs and crayfish, small rodents, waterfowl and their eggs, worms, and insects. One study found that common prey species included meadow and deer mice and even cottontails.

The mink is equally at home on land or in water and has survived even in areas where it shares its habitat with man. Its range includes most of Canada south of the tree line and all of the United States, except Arizona and other dry areas in the Southwest. Because of the fish they eat, mink are susceptible to waterborne environmental contaminants such as mercury and PCBs, which severely affect their reproductive systems. A healthy breeding mink population *might* be a good indication of a habitat free of these pollutants.

Years ago, when I lived near the Middle Branch of the Swift River in Massachusetts, I ventured into the woods for a walk just after a heavy snowstorm. This is one of my favorite times, when pristine snow covers the forest and its quiet beauty makes me hesitate to disturb it. When I find a trail in that type of snow, I know it's fresh. If I'm not backtracking, I know I'm just behind the animal and will follow its trail for miles. This time I was just walking—there were no tracks—when I came upon a black object. Although I knew there were mink in the area, I was not prepared to find one lying motionless on top of fourteen inches of new snow. No tracks led up to it. A light dusting of snow covered its body.

I stood without moving for some time, studying the situation. It couldn't have been there long. I approached cautiously, looking for signs of life, and did see some movement. The animal was still alive but barely. I should have known better and left it there, but I thought it had become exhausted during the storm. I figured I'd bring it in the house just long enough to let it recover and then release it.

FIGURE 3.9
Aquatic animals, mink are usually found investigating the edges of various wetland habitats.

It moved around a bit as it warmed up, but it died soon after. It may well have died anyway, but sometimes when you think you're helping animals, you aren't. Since that incident, I've learned that wild mink cannot live in captivity, so it's not a good idea to bring one indoors, regardless of its condition.

Like all weasels, the mink has an anal musk gland that discharges a liquid as malodorous as that of a skunk, especially during the mating season. It cannot spray like a skunk, however, and its scent does not carry as far. Its greenish yellow urine also has a distinct, musky odor. The word *mink* comes from the Swedish *menk,* which means "the stinking animal from Finland."

The mink sometimes has a very casual, almost reckless nature. I've often stood on the banks of a stream or beside a swamp, being careful not to move, and had mink run right up to me or pass within five feet of me and seemingly not notice that I was there. Mink are very powerful for their size, but they may occasionally be preyed upon by gray and red foxes, coyotes, wolves, fishers, bobcats, lynx, and great horned owls.

TRACKS. You never know where a mink's tracks will turn up, but you can be fairly sure you'll see them next

FIGURE 3.10 *(left)*
The mink's front foot has five toes and short nails, five loosely connected palm pads, and a single heel pad.

FIGURE 3.11 *(right)*
This front track shows all the toes, most of the palm pads, and the heel.

FIGURE 3.12
The hind foot has five toes, short nails, and four palm pads. The heel area is completely covered with hair.

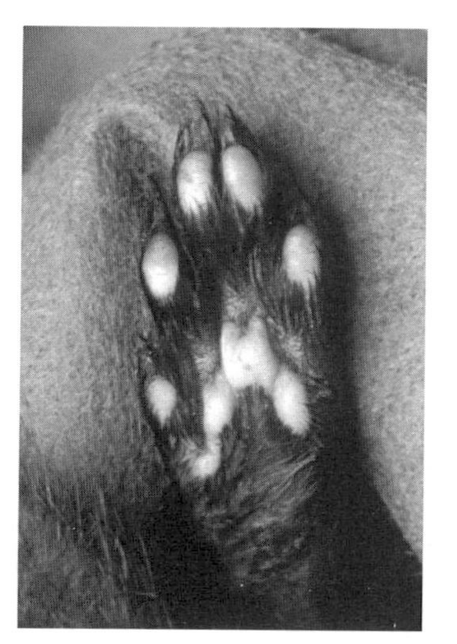

to a body of water. The mink's trail will take you on a wondrous adventure through the thickest underbrush, over dreadfully thin ice, into ice-cold streams, and across marshes and swamps. Tracking mink requires all the outdoor skills you can muster, but doing so is almost always worth the effort.

Like all weasels, the mink has five toes on each front foot and five on each hind foot. The front foot (Figure 3.10) has five loosely connected palm pads, each related to a toe, and one heel pad farther back that doesn't usually show in the track, although it may (Figure 3.11). The hind foot (Figure 3.12) has four palm pads, also related to the toes. The heel is all hair and rarely shows up in the track.

The front track of the mink measures 1¼″ to 2″ long by 1¼″ to 1¾″ wide. It can be 2⅝″ long if the heel pad registers. The hind track measures 1¼″ to 1⅞″ long by 1⅛″ to 1¾″ wide. The mink's hind heel (all hair) does not usually register, but when it does (Figure 3.13), it can add ¾″ to 1″ to the track's length. Toes, especially those of the front foot, often spread out; the small inside toe of the mink, as of other mustelids, usually does not show in the track, though it sometimes does (Figures 3.11, 3.13, and 3.14).

TRAIL PATTERNS. The trail of a mink (Figure 3.15) shows a definite 2-2 pattern, which all weasels leave. But unlike the ermine and long-tailed weasel, the mink leaves more consistent strides. There's no short, long, three short, one long, and so on; rather, the mink's gait tends to be more evenly spaced throughout the trail. Strides of the

2-2 pattern are 11″ to 38″, measuring nail to nail, and the trail width is usually 2″ to 3⅜″. Trail width alone, however, is not a reliable indication of a mink trail because the male long-tailed weasel also can have up to a 3″ trail width, so you must check the gait.

Other mink trail patterns include tracks in groups of three and four (Figure 3.16). Such variability in pattern does not seem to have much to do with the animal's speed, as the pattern often changes without a corresponding change in stride. The best way to determine speed is to

Figure 3.13
The sequence of these tracks is (from right to left) front-hind-front-hind. The bottom tracks are of a raccoon.

FIGURE 3.14
From left to right, these are front-hind-hind; the middle track is a hind track superimposed over a front.

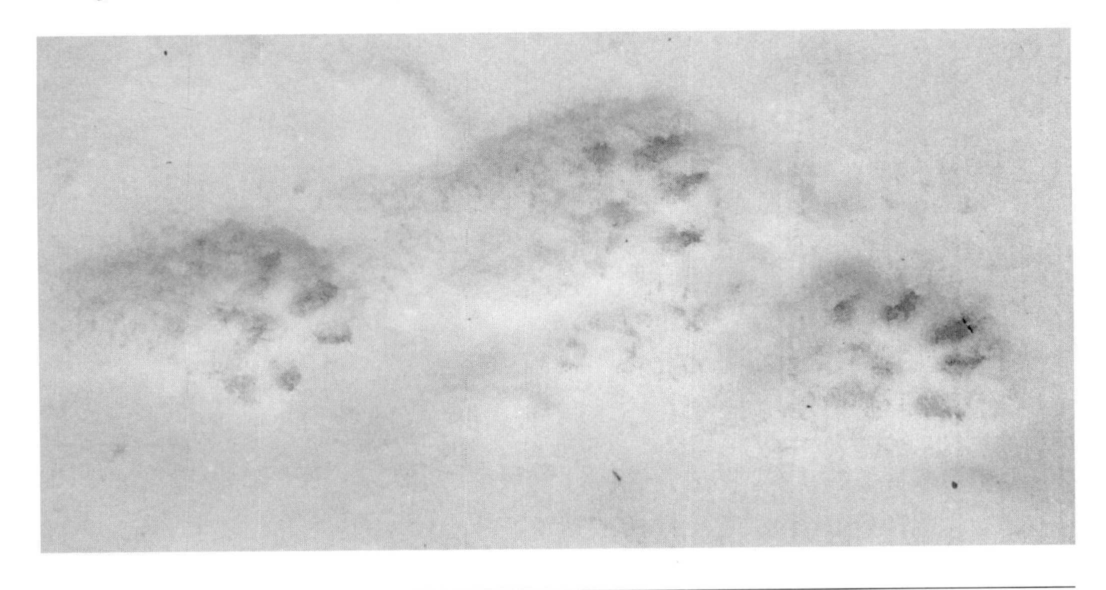

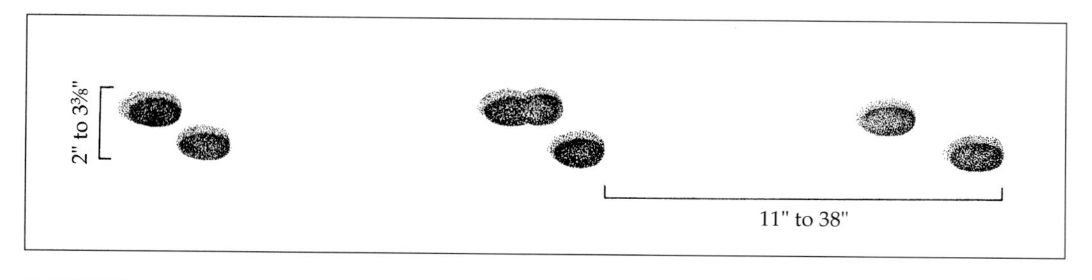

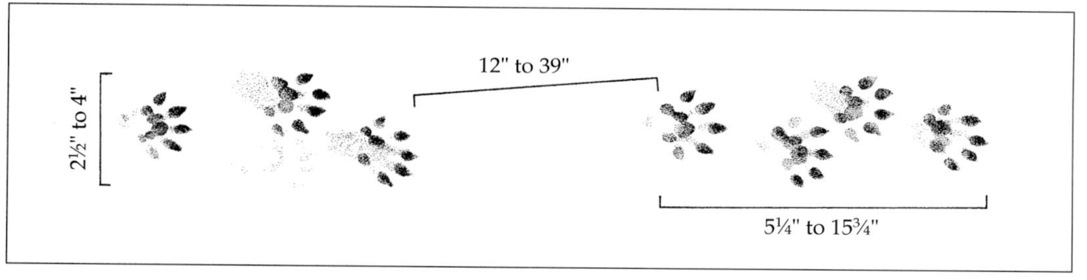

FIGURE 3.15 *(top)* *The mink often uses the 2-2 pattern. Its tracks are usually more evenly spaced across the trail than those of the smaller weasels.*

FIGURE 3.16 *(bot.)* *The trail pattern of the mink varies to include groups of three and four tracks as well as groups of two. They do not appear in any particular order.*

measure the stride—the distance between the groups of three or four. The group measurement in these patterns will be 5¼″ to 15¾″; those of the otter will be somewhat larger, 10″ to 17″. The stride of the groups of three or four usually ranges from 12″ to 39″ but may occasionally reach 45″. Trail width is 2½″ to 4″. In soft snow that is three or more inches deep, the mink will sometimes plow through, leaving a trough rather than a clearly defined trail pattern.

SIGN: *Dens.* A mink is a highly efficient predator. It will follow a muskrat to its lodge or bank burrow, kill and eat its young (and possibly the adults as well), and then move into its home. It also will inhabit abandoned beaver lodges. In addition, it may dig its own den in soft soil, going down one to three feet deep and making a tunnel eight to twelve feet long. Dens are usually along waterways, with two to five openings along the water, often under tree roots. Each opening is about four to six inches across. Away from water, a mink will move into abandoned woodchuck holes, crevices in rocks, or tree cavities. In most cases, it will use one den for several days and then move on to another.

SIGN: *Slides and Dive Holes.* Mink are almost as playful as otters at times, and when walking in soft snow, they will suddenly dive under it, as if they were frolicking in water. The tunnels they make are usually not very long, about 4′ to 7′. Mink also will slide for a short distance—sometimes into water, sometimes just on hills. The

FIGURE 3.17
Mink scat is usually black and twisted and is often composed of small rodent hairs. It also sometimes looks folded over.

slides are usually around 3″ wide, occasionally to 5″ wide, but still narrower than otter slides.

SIGN: ***Scat.*** Mink scat (Figure 3.17) is generally larger than that of weasels and is usually black and twisted. Sometimes it has an overlapping appearance, looking as though it has been folded. It often consists of very short rodent hair, but because mink also eat birds and fish, you may see remnants of these as well. If the scat contains feathers, it will be lighter in color; if it contains fish, it will be black and shiny. On the whole, it's almost always twisted, folded, tapered on the ends, and black in color. Look for it on prominent objects, such as rocks, stumps, and logs. Mink also will make latrines near denning sites. These are more common than you might think. If there is a dense population, you should be able to find a latrine by following a mink's trail or searching the edges of wetlands.

It is easy to confuse mink and fisher scat, as both may be the same size. Mink scat is usually very tightly knit and composed of the short, fine hair of small rodents. Fisher scat usually contains the coarser and longer hair of larger mammals.

Marten
Martes americana

THE MARTEN (Figure 3.18) is a much smaller version of the fisher. It has a more pointed, weasel-like snout and more prominent, rounded ears. It also is much lighter, weighing on average one and a half to two and three-quarters pounds, compared to the fisher's average of six to twelve pounds. The marten's color varies considerably from animal to animal and from season to season. The most common color is a yellowish or golden brown, with an orange-yellow patch covering the throat and chest and a darker brown on the legs. The head is lighter, and the ears are edged with white. The marten's long and lustrous pelage made it one of the prizes of fur trappers. Peter Matthiessen, in his *Wildlife in America,* reported that like the fisher, the marten was all but exterminated from its northern woodland habitat by a combination of "lumbering, forest fires, and the elimination of pregnant females in winter trapping." In 1743, Matthiessen notes, the French port of Rochelle received shipment of 127,080 beaver pelts, 30,325 martens, 1,267 wolves, 12,428 otters and fishers, 110,000 raccoons, and 16,512 bears. Not even a territory as

FIGURE 3.18
Martens seem to prefer undisturbed forests, usually conifers, but frequent other habitats as well. Arboreal hunters, they may often be found in trees.

vast as North America could sustain such depletion for long. Today the marten is staging a comeback in its southern range, especially in Maine, thanks to regulatory agencies setting needed restrictions on trapping. It is important to note, however, that the logging of old-growth forests could have detrimental effects on marten populations, since they require more specialized forest conditions than do fishers. Martens now range throughout most of Canada and Alaska to tree line and south to northern California and through the Rocky Mountains. In the East, they are found in the northernmost parts of New England and New York.

FIGURE 3.19
The marten's front foot (right shown) has five toes and is heavily covered with hair. Its palm and heel pad development can vary considerably.

Being so light, martens seldom leave a good track other than in snow, but they do move around a lot. Dens are usually found high in hollow trees, on the ground under the snow, in rock crevices, or in some other convenient shelter. They may be lined with soft vegetation. Away from their dens, martens are constantly scrutinizing rock and brush piles and any other nook or cranny within their home range—reported to be less than one-tenth of a square mile for a female and up to about three square miles for a male. This investigative activity often obliges the marten to cross its own trail, a characteristic that can help the tracker identify it. On these excursions, the marten will hunt down squirrels, chipmunks, and other small rodents, especially red-backed and meadow voles. They consume birds and fish to a much lesser degree. Although martens may eat fruit, especially blueberries, in large quantities when available, they are primarily carnivores. Even in high summer, their diet may be 80% meat.

Martens have been known to jump out of trees or off stumps sticking up out of the snow. They also tunnel under the snow on their hunting forays. Look for a tunnel entrance near a stump or a log for this sign of marten activity.

TRACKS. The bottom of a marten's foot can be very hairy, to the point, in some cases, of obscuring all the pads. Figure 3.19 is fairly demonstrative of a marten's right front foot: five toes with nonretractable nails and very hairy. The palm pads are not well developed; any development is more pronounced to one side of the foot. This is well demonstrated in Figure 3.20 (two front tracks, bottom right). In many martens, there will be hardly any development of

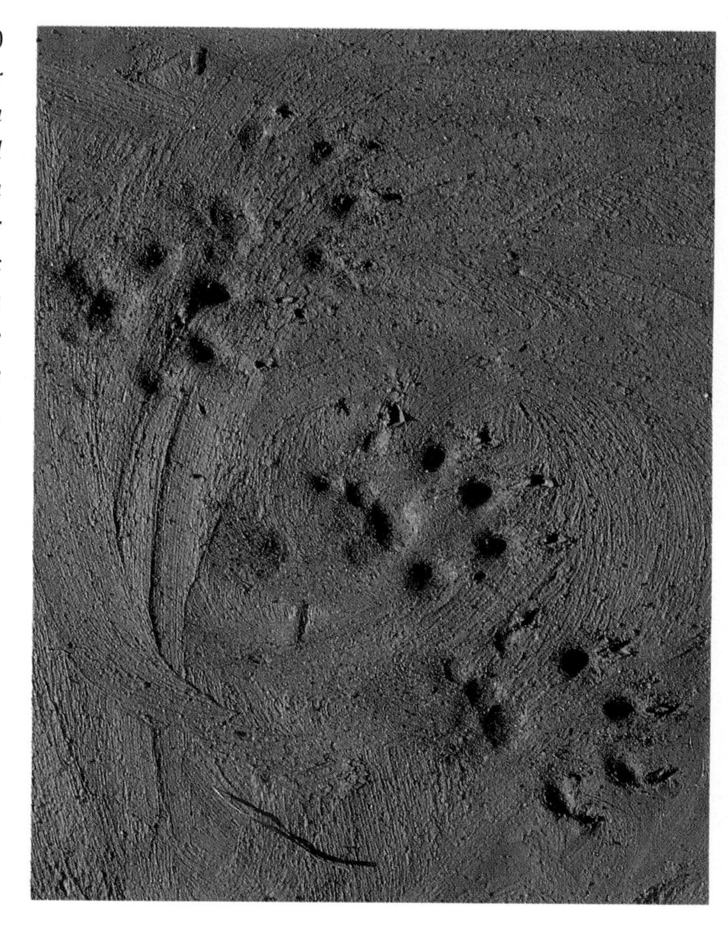

FIGURE 3.20
The two front tracks (lower right) show five toes, palm pads, and even the heel pad on the upper front track. In the two hind tracks (upper left), not all the toes register; one is missing completely on the upper hind track, and the inside toe at the bottom of the lower track is barely visible.

FIGURE 3.21
Heavily furred, the marten's hind foot (right shown) has five toes. The palm pads, obscured by hair, have limited development. The heel is completely covered with hair.

palm pads and sometimes no heel pad, just hair. Individual martens vary in the extent of pad development. The hind foot (Figure 3.21) shows five toes with nails but no heel pad, just hair. It has even less development of palm pads than the front foot. Again, any development is usually to one side of the foot. Like most of the weasel family, the marten's fifth inside toe often does not register.

The lack of pad development and a hairy foot can help in identifying this species. Note that in Figure 3.22 (right front track), there is very little pad development. In the hind track (Figure 3.23), just hair shows in the heel area.

Because of dimorphism inherent in this species, there is quite a variation in track size. Measurements range from 1⅝″ to 2⅝″ long by 1½″ to 2⅝″ wide. When measurements fall to the low side of the parameters, it usually indicates a female; when they are on the high side, it is probably a male. There is no significant difference between front and rear track sizes.

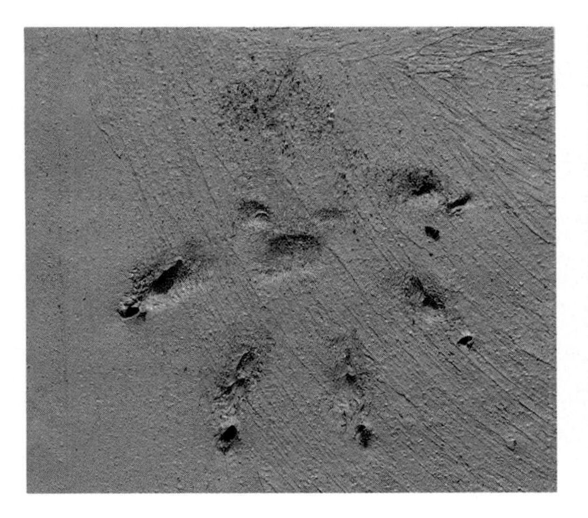

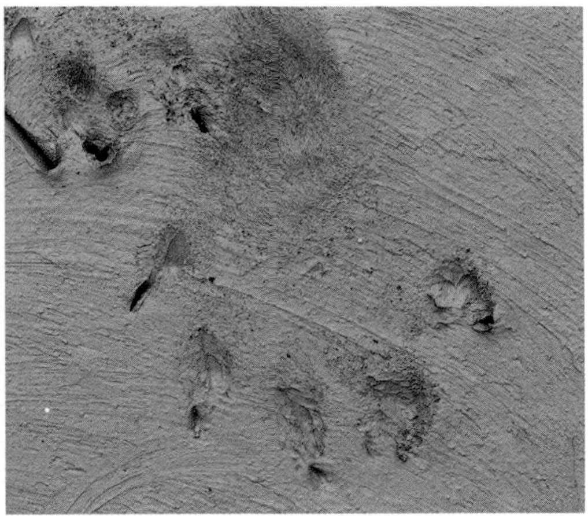

FIGURE 3.22 *(left)*
This right front track shows five toes, some palm pad registration, and a slight indication of a heel pad.

FIGURE 3.23 *(right)*
All five toes registered in this left hind track, but the heel area shows hair marks only.

In track size, the marten is between the fisher and the mink. This causes overlap in both directions. Remember that the marten has very hairy feet and is much lighter in weight than the fisher or even a large male mink. Martens also spend a lot of time in trees, whereas mink are not known to climb and prefer to be in or near bodies of water.

TRAIL PATTERNS. The marten's trail patterns are similar to those of the mink and fisher and fall between the two in size. One of its more common patterns is the 2-2 (Figure 3.24). Its stride in this pattern is usually 15″ to 33″ (occasionally to 40″), with a trail width of 2⅞″ to 4¼″. Do not use the tracks in this pattern for measurement, since each track is a combination of two, one on top of the other, which can exaggerate the size of the track. More accurate track sizes can be taken from the 3-4 pattern. Look for individual tracks.

Like the fisher and the mink, the marten may occasionally use an alternating walking pattern (Figure 3.25). Each track is a combination of a front and a hind foot registering in the same spot. Strides are 4½″ to 9″, while trail widths are 3″ to 4″.

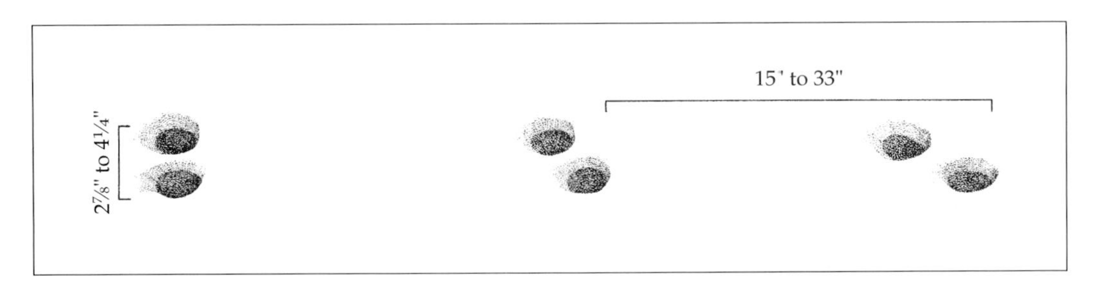

FIGURE 3.24
One of the marten's more common gaits is the 2-2 pattern, usually indicative of a moderate speed. Each dot represents two tracks, a hind superimposed over a front.

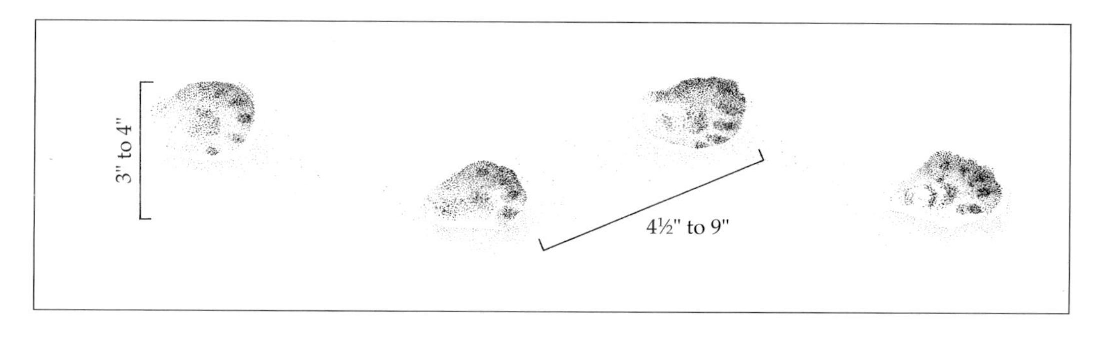

FIGURE 3.25
The alternating walking pattern is a slower gait than the 2-2 pattern (Figure 3.24). Each of these footprints represents a hind track over a front track.

The marten may leave a 3-4 pattern, although groups of three may be less common than groups of four. Strides in this pattern range from 9″ to 28″. Group lengths fall between those of the fisher and the mink.

SIGN: *Scat.* The scat of a marten (Figure 3.26) is similar in size and composition to that of the smaller weasels. You'll often find marten scat on trails, even hiking trails, on prominent rocks, stumps, or other objects that are raised off the ground. Martens sometimes eat fruit, so there may be some seeds in their scat. This distinguishes it from the scat of smaller weasels and mink, but not from that of the fisher. Marten scat is long and thin, sometimes twisted, and composed primarily of small rodent material (vole, mouse, or squirrel hair, plus small bones).

FIGURE 3.26
This scat is ¼″ in diameter and 1½″ long (lower specimen). Marten scat may contain the hair of small rodents and sometimes the seeds of various fruits.

Fisher

Martes pennanti

THE FISHER, sometimes called the fisher-cat, black cat, wejack, and pekan, is such a capable predator that it has been referred to as a furred snake. It's a mustelid, or weasel, whose size and temperament fall between those of a marten and those of a wolverine. Males can weigh up to twelve pounds, though occasionally as much as twenty. They are long, sleek, and built very low to the ground; they move with determination and efficiency. The fisher is the only animal except the puma that actively preys on porcupines, which it kills by continuously circling the animal and lashing out at its head until the porcupine slows down or passes out from loss of blood. If the ponderous porcupine seeks refuge in a tree, the fisher—an arboreal weasel—will dash up the tree and attempt to force the porcupine to the end of a branch, where it may fall to the ground. Then the fisher will resume its circling attack. The fisher has a swivel joint in its hind ankle that allows it to descend trees headfirst. When prey is abundant, it may kill more than it can eat. Although it doesn't make large caches, as the smaller weasels do, it will cover its leftovers with leaves and forest litter and scent them, returning when food sources are low.

After tracking a fisher in the forest for several days, you'll get to know all the old, rotting snags. Fishers climb these and literally tear them apart looking for flying

FIGURE 3.27
This fisher kit, like most young animals, is curious about its environment. Once thought to be strictly a deep-woods animal, the fisher's range has expanded to include a variety of habitats, including wooded areas in some suburban towns.

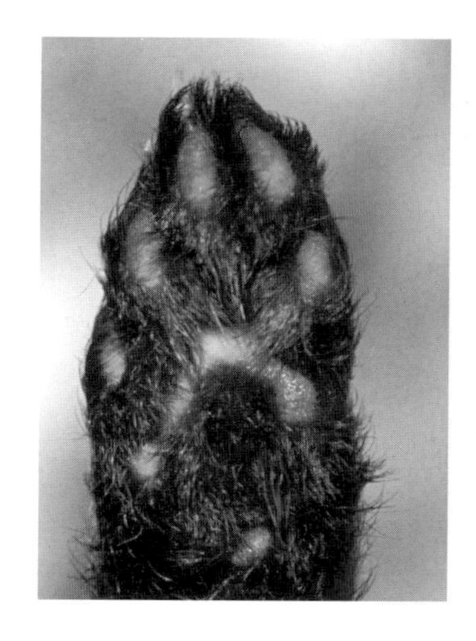

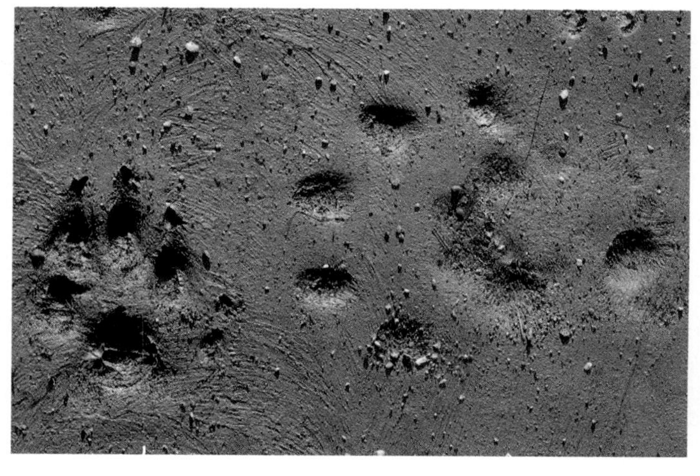

FIGURE 3.28 *(top left)* *The fisher's front foot is hairy and has five toes, a C-shaped palm pad, and a small heel pad.*

FIGURE 3.29 *(top right)* *The splayed front track (right) is of a large male fisher; on the left is a female's hind track.*

FIGURE 3.30 *(bot. left)* *The fisher's hind foot has five toes and a palm pad; the heel area is covered with hair.*

FIGURE 3.31 *(bot. right)* *This hind track in snow is superimposed over the front track.*

squirrels, red squirrels, or mice. Fishers also seem to relish snowshoe hares, beaver kits, chipmunks, martens, mountain beavers, birds, frogs, insects, nuts, and berries. Despite its name, it does not fish, but it will eat spawned-out salmon and trout.

French Canadian trappers called the fisher, or more accurately its fur, *pékan*, from the Abnaki word for the animal, *pekane*. The fisher was all but eradicated from most of North America for its pelt, but there are welcome signs that it has made a comeback in much of its original range. Fishers are found in the southern tier of the Canadian provinces and south to northern California and the northern Rocky Mountains in the West and to New England and northern New York in the East.

TRACKS. Each front foot of the fisher (Figure 3.28) has five digits, and its palm pad is C-shaped, curving away

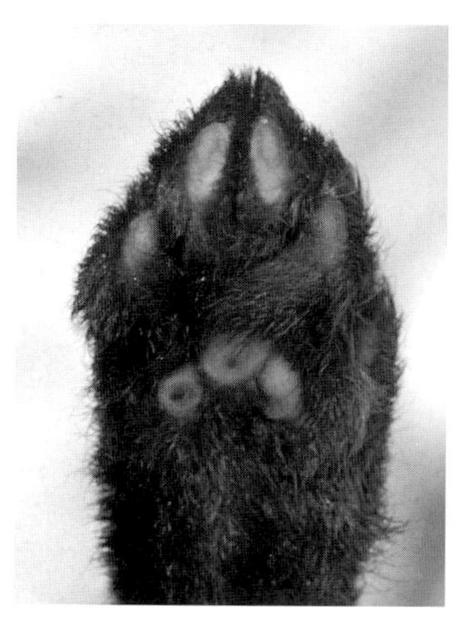

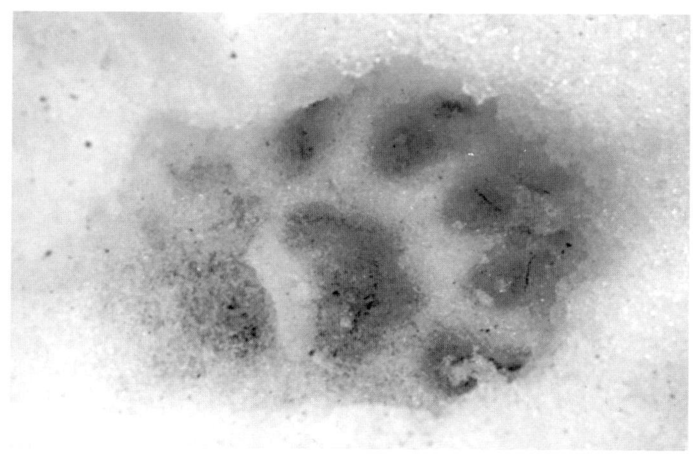

from the toes. It also has a small heel pad that sometimes shows up in the front track (Figure 3.29). Each rear foot (Figure 3.30) has five toes and a palm pad but no heel pad. Both the front and rear feet demonstrate some asymmetry, with the smaller toe set off to the side and often not showing in the track. If the fifth toe of the fisher fails to register, its track (Figure 3.31) could be confused with that of a bobcat. It's a truism of tracking that sooner or later, any animal will leave the perfect track of a totally different animal. In this case, look for nail marks. Bobcats have retractable nails that usually don't show in its track. Also look for the fisher's 2-2 pattern, which a bobcat does not leave.

Fishers are sexually dimorphic—the male is almost twice as large as the female—and this clearly shows in the tracks. Finding male and female tracks together is very unusual because males are solitary except during mating, which takes place during only two or three days in April. Like other members of the weasel family, fishers have a long gestation period (338 to 358 days) due to a process called delayed implantation. This means that the eggs are fertilized after mating but do not implant in the uterus until 9 to 10 months later. After implantation occurs, development of the embryos recommences, and 30 to 60 days later, the young are born, nearly a year after successful mating. When you do find tracks of a pair of fishers (Figure 3.29), they might lead you to believe you're seeing a fisher moving with a marten. In this photo, the male fisher's track is to the right and the female fisher's to the left. Note the five toes in each track. The fifth toe of the female is harder to see; it's to the right of the track itself, at the bot-

FIGURE 3.32
Depending on substrate conditions, the fisher's tracks can look quite different. The sequence of this set of tracks in hard snow is front-hind-front-hind, from left to right.

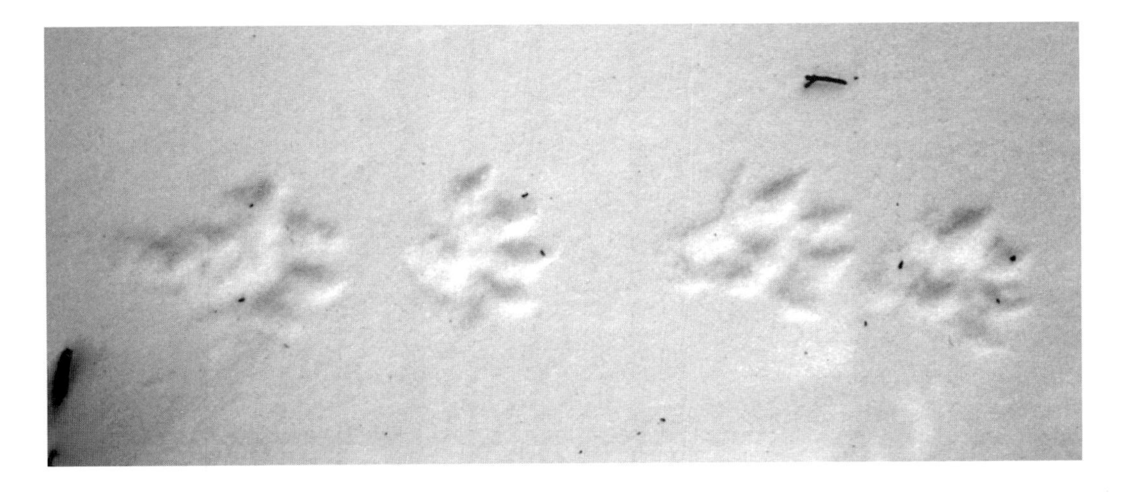

FIGURE 3.33
This set of tracks in a dusting of snow is front-hind-front-hind, from right to left.

tom. You can see that the splayed front track of the male shows a heel pad, and the closed hind track of the female has a rather oddly shaped palm pad, which, though not typical, is sometimes visible in good, clay substrate. Because we are comparing a splayed front track to a closed hind track, the size difference is somewhat exaggerated.

When measuring tracks from the different members of the weasel family, it is important to keep in mind the dimorphism between males and females. For fishers, the front track usually measures 2⅛″ to 3⅞″ long by 2⅛″ to 3¼″ wide, but can splay up to 4″. The front track can be much longer if the heel pad registers, adding about an inch to the length. The hind track is usually 2⅛″ to 3″ long by 2″ to 3″ wide. Tracks under 2½″ wide are probably those of a female fisher or a male marten, while tracks 2½″ to 2⅝″ wide could be those of a female or male fisher or a male marten.

FIGURE 3.34
A fisher will occasionally jump out of a tree, leaving an impression of its body (including the tail) in the snow.

If you see tracks over 2⅝″ wide, you can be fairly certain that you have a male fisher. Don't forget weight differences, as heavier animals will make a deeper impression in the substrate. There is a weight difference not only between the male and female fisher but also between the marten and the fisher. Martens weigh only one and a half to two and three-quarters pounds; fishers usually weigh six to twelve pounds, sometimes more for large males.

Fisher tracks look somewhat different in snow than they do in mud or clay (Figure 3.29). See variations in Figure 3.32—front-hind, front-hind, from left to right; Figure 3.33—front-hind, front-hind, from right to left; and Figure 3.31—rear track superimposed on front track. I measured the tracks of a male fisher in one-quarter inch of snow and found they varied from 2¹¹⁄₁₆″ to 3¾″ wide. These were individual tracks taken from a 4-4 pattern. The 2-2 pattern will exaggerate the size of the tracks because there are actually four prints there: two hind tracks superimposed over two front tracks. If you are careful to measure just the tracks of individual feet, your measurements will be much more helpful in determining sex and/or species.

The most interesting track I've seen a fisher make is the impression of its entire body (Figure 3.34). When a fisher leaves a tree, it does so in one of two ways: It can climb down headfirst, or it can jump. When it jumps, it lands belly down in the snow, leaving the impression of

FIGURE 3.35 *(top)*
The alternating walking pattern of the fisher is one of its slower gaits. Each track here represents a hind track on top of a front track.

FIGURE 3.36 *(bot.)*
This illustrates the fisher's 2-2 bounding pattern, common to most of the weasel family.

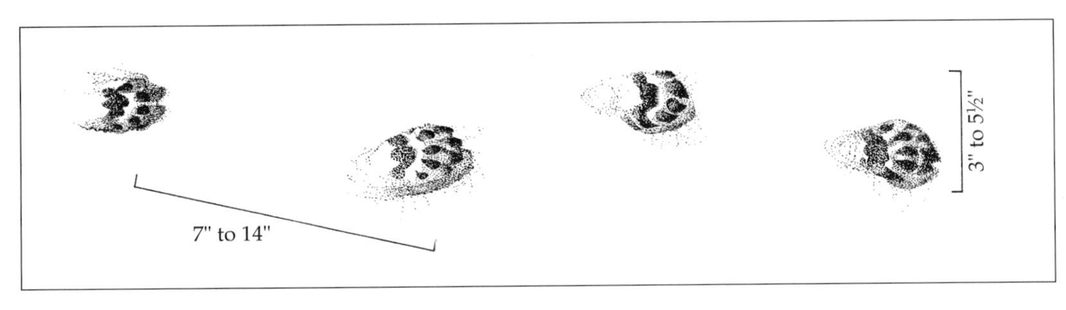

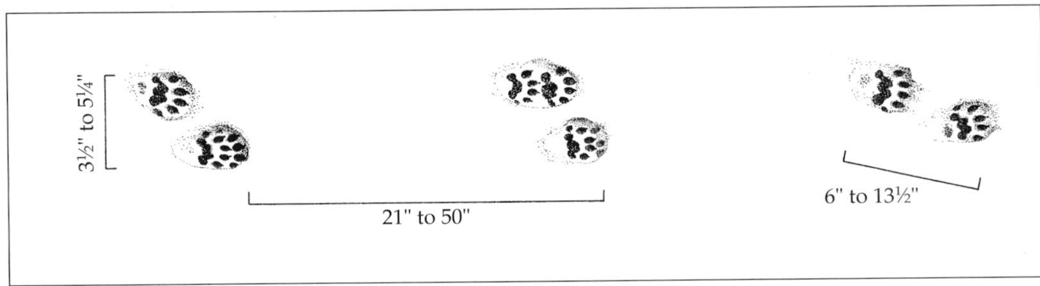

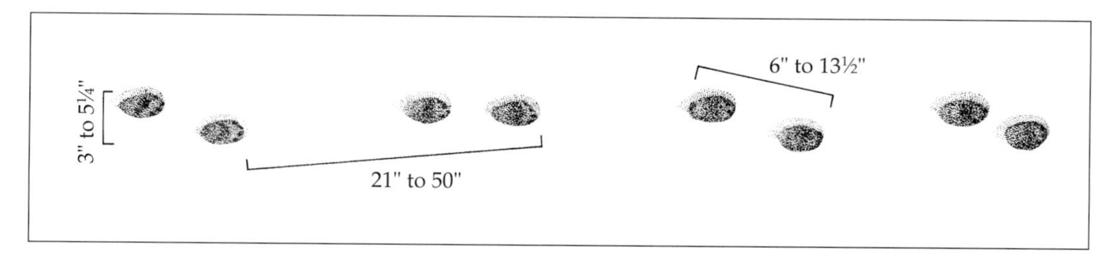

FIGURE 3.37
In this variation of the common 2-2 pattern, note the elongation of the sets of two tracks; also, one track may be almost or directly in front of the other.

FIGURE 3.38
A fisher's tracks and trail patterns can be confused with those of a raccoon. A and B represent patterns common only to raccoons; C, the fisher's 2-2 pattern. The raccoon has a large and a small track, which change sides in each set of two tracks; usually the angles of its trails also differ from the fisher's. *See figure 5.8.*

its body and tail. A male fisher can be up to three and a half feet long, including the seventeen-inch tail, which is about the size of a fox. Body track measurements, however, do not always represent the actual size of the animal, since it may be hunched over when it hits the ground.

TRAIL PATTERNS. The fisher's trail meanders and zigzags through the forest. The trail will lead you to the oldest trees, which often are dead or hollow and decaying. Fishers gravitate toward these, sometimes climbing and tearing at them along the trunk and at the base. The trail also will lead you in large circles covering several acres. Fishers have corridors that they travel repeatedly, and once you learn these paths, you should have no trouble locating the animal's trail.

If you spend enough time following fisher trails, you will soon discover that the fisher has three basic trail patterns. The first is an alternating walking pattern (Figure 3.35). This pattern usually has a stride of 7" to 14" and a

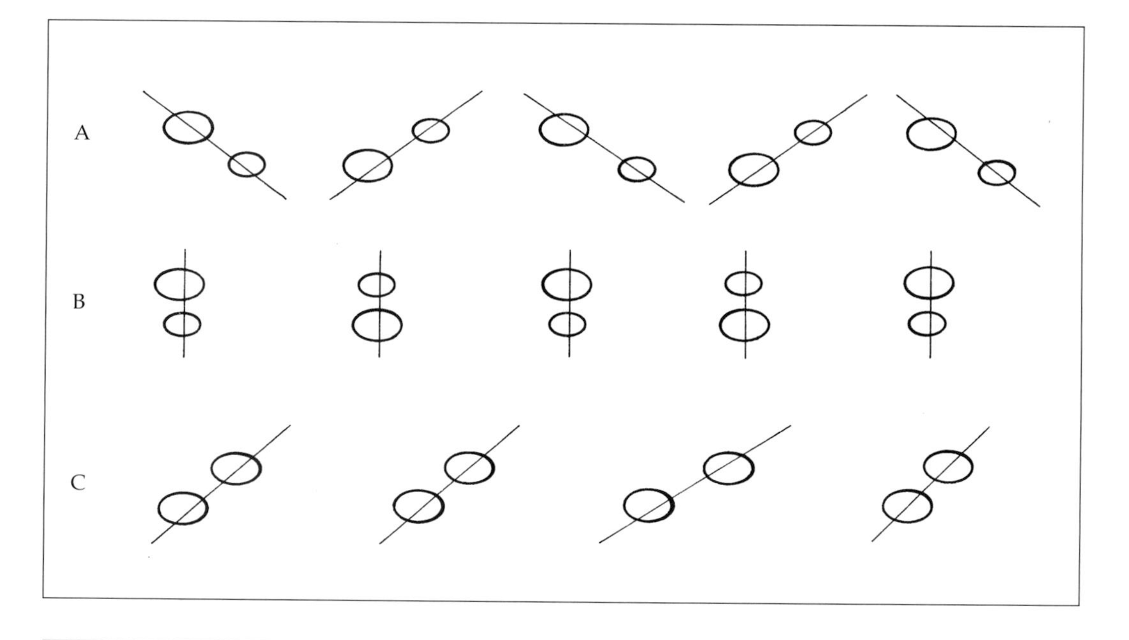

trail width of 3″ to 5½″. It is a direct- and sometimes double-registering pattern. It may be confused with the trails of the raccoon and the marten. Any stride over 10″ is probably a fisher; under 7″, a marten; in between, either. If most of the strides fall between 7″ and 11″, it is probably a female fisher or a male marten. If most strides fall between 11″ and 14″, it is probably a male fisher. Remember to take at least seven measurements before you compare them to those in this book.

The second pattern is the 2-2 pattern (Figure 3.36). This bounding pattern is common for the mustelids: two sets of tracks, one slightly ahead of the other, followed by another set of two, etc. The two sets of tracks, however, represent four prints. This is because the hind feet directly register on the tracks left by the front feet. The 2-2 pattern has a common stride of 21″ to 50″, with longer strides possible. Trail width is usually 3½″ to 5¼″. Although this is a common pattern for the weasel family, there is something peculiar about the fisher. For the fisher, the two prints of a group are sometimes spread farther apart and one in front of the other (Figure 3.37). Although for the fisher this is the exception rather than the rule, this variation seems to occur more frequently in the fisher's trail than in the trails of the rest of the weasel family. In the 2-2 pattern, the groups of two tracks can measure 6″ to 13½″—quite long. Some people confuse this pattern with the raccoon's 2-2 pattern, but the raccoon's pattern will usually alternate (Figure 3.38A) or stay parallel (Figure 3.38B), while the fisher's will remain slanted (Figure 3.38C).

The third pattern is the 3-4 pattern, another bounding pattern that is also common among the other members of the weasel family. At times the fisher may use it consistently over fairly long distances. A group of three tracks may show up, then a group of four; this may change back and forth in no particular order (Figure 3.39). Common strides usually are 6″ to 30″, trail widths 3″ to 4⅞″, and groups 12″ to 25″, occasionally to 39″.

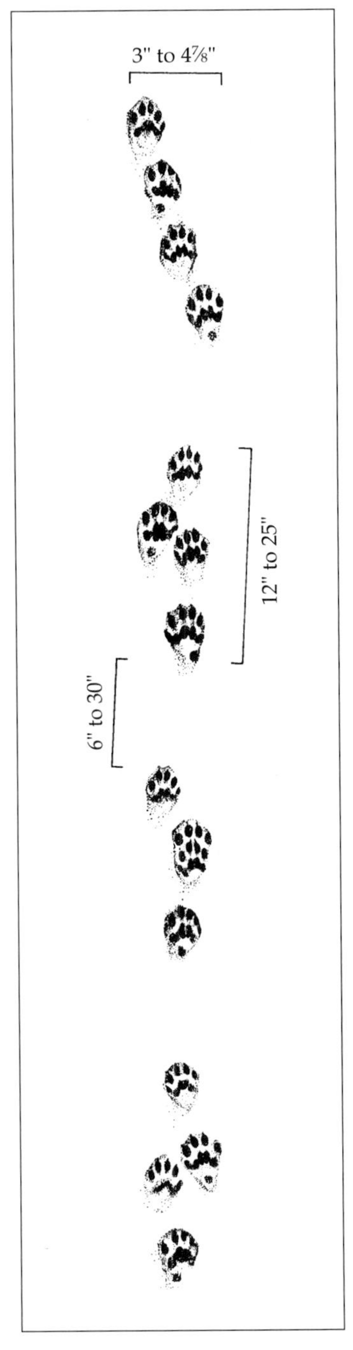

FIGURE 3.39
This illustrates the 3-4 pattern common to the fisher, where groups of three and four tracks are in different configurations, with no apparent order.

SIGN: ***Kill Sites.*** The fisher is an exciting animal to track because it's such a capable predator and leaves so much evidence of its passing. In one of our deep-woods tracking expeditions, we came upon an oak that had a lot of tracks around it—fisher tracks and those of a squirrel

NOTE: *Fishers normally cover their kills with materials that are immediately available—leaves, pine needles, snow, and so forth. Trackers have also discovered the remains of rabbits, a raccoon, and a red fox stashed up in trees by fishers.*

that had been moving back and forth from tree to tree. The fisher tracks came in from the left, and there was a small matted spot near the tree that, when we took a closer look, showed some traces of blood and the tip of the squirrel's tail. Obviously, the fisher had caught itself a meal, but the story wasn't over. The fisher went on to visit a few snowshoe hare trails, and when we came to a place where snowshoe hare tracks were heavy, the fisher's tracks also seemed to become more intense, exploring every crevice. We found no evidence of a snowshoe hare kill, but as we followed the fisher, we came to another area that was covered with tracks and activity and saw a place where some feathers were sticking out of the snow. The fisher had buried something and scented it. We dug it up and found the remains of a wild turkey. We didn't know whether the fisher had killed the turkey or just come across the carcass.

We continued to track the fisher, following it up a hill to another giant oak, which was hollow at the bottom. At the base of the tree, the fisher tracks were so numerous I could hardly read what had happened. I could see, however, that a fox had been there that morning. It had walked up to a rock, climbed the rock, and sat down as if watching something. It had sat there so long that I could still smell its scent in the snow. When I walked around the tree to examine the hollow, I saw an animal's leg sticking out of it and realized that I was looking at a raccoon's leg. The fisher had tried to pull the animal out of the tree, but the raccoon's upper leg was folded inside, and it was too big to come out of the small hole. It seemed as though the fisher had tangled with the raccoon in the tree and had killed it but could not pull it out. So the fisher buried it in the hollow.

I kept tabs on the partially buried kill, checking it for a week or so. The fox never visited the area again. Although I'm sure it wanted food, it wanted even less to confront the fisher. Soon the snow melted, and I could no longer tell what animals were visiting the tree. After a while, it looked as though some small rodent had been gnawing at the raccoon's foot. A few days after that, the raccoon was completely removed from the tree. There was no evidence to tell me what animal had removed it or how, but I have a hunch it was the fisher, a little disturbed at the fact that his cache was being eaten by someone else.

FIGURE 3.40
Fisher scat, shown here in four variations, is usually dark in color, twisted, and overlapping. Fishers deposit very tiny scats on scent posts and also make latrines next to old trees, as do raccoons.

SIGN: ***Scent Posts.*** Fishers have certain corridors along which they travel to and from their hunting grounds. One corridor may lead to a snowshoe hare habitat, another may lead from there to a good squirrel area, and so on. They cover a lot of ground, zigzagging from hunting area to hunting area. Males may cover a home range of eight to fifteen miles in diameter, females slightly less. Hunting circuits may be sixteen miles long. If you find one of these corridors, you can expect to find a fresh fisher track every two or three days.

On these corridors, especially where two intersect, you will often find a scent post, a place where the fisher defecates and urinates. This may be an old stump on the forest floor, a raised hump, or anything that stands out and can be used as a scent carrier. The fisher often rubs or rolls against these objects as an additional means of scenting.

SIGN: ***Scat.*** Fisher scat (Figure 3.40) is very similar to mink scat, and the two can easily be confused. Although fisher scat is usually larger, there can be a lot of overlap in size. Look at the composition of the scat to differentiate between the two. Mink scat is tight and compact, with very fine hair (its diet consists mainly of small rodents). The fisher usually kills much bigger animals with coarser hair, so its scat is not as tightly packed.

River Otter
Lutra canadensis

NEXT TO THE SEA otter, the river otter (Figure 3.41) is the most aquatic of the mustelids. Weighing up to thirty pounds and measuring three to four and a half feet in length, counting its sleek tail, the river otter provides rare but completely entrancing wilderness experiences. Its range extends across North America, and it is found wherever there is enough water to support it, although it is rare in the arid western states and much of the Midwest east to Pennsylvania. A male otter needs fifteen to thirty miles of shoreline for its home range; females with young have much smaller ranges, sometimes as little as one mile. All otters spend most of their time in the water, but they do go ashore occasionally and can travel great distances overland.

FIGURE 3.41
Having spent many hours tracking, observing, and photographing otters, I can certainly attest to their playful and fun-loving nature. I also am impressed by their seeming intelligence.

The otter's diet consists largely of fish, though not exclusively trout and salmon, as some frustrated fishermen have claimed. Otters also eat a wide variety of slower-swimming, nongame fish, including minnows, sculpins, perch, and catfish as well as crayfish, turtles, frogs, and even lamprey eels, which are the scourge of the Great Lakes fishery. Otters are extremely capable fishers, sometimes even teaming up to increase their efficiency. They've been known to herd schools of fish into coves and catch them as the fish try to swim out.

Otters' efficiency at catching fish leaves them a lot of time to play. They often seem to be carefree and full of fun, sliding down beaver lodges and riverbanks and rolling in the water. They'll pick up a small pebble from the bottom of a pond, bring it to the surface, and drop it, then swim down under it and catch it on their forehead. They'll flip and turn with the pebble still on their forehead as they bring it up to the surface, then drop it again and start all over. They're very comical to watch. They're curious as well, like most weasels, and can be very secretive. Recently they've managed to increase their number in the northeastern United States, where human populations had nearly decimated them in some areas.

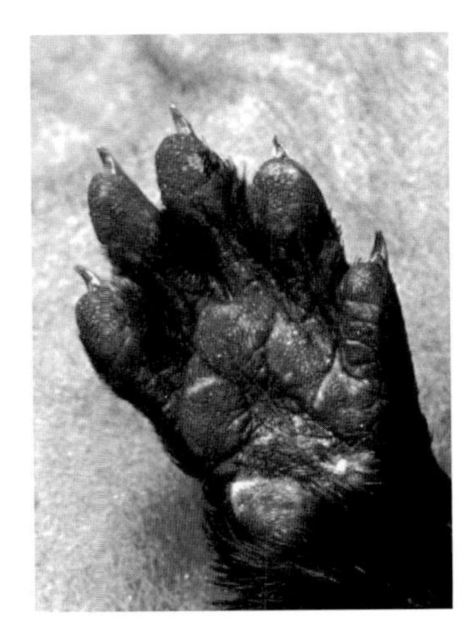

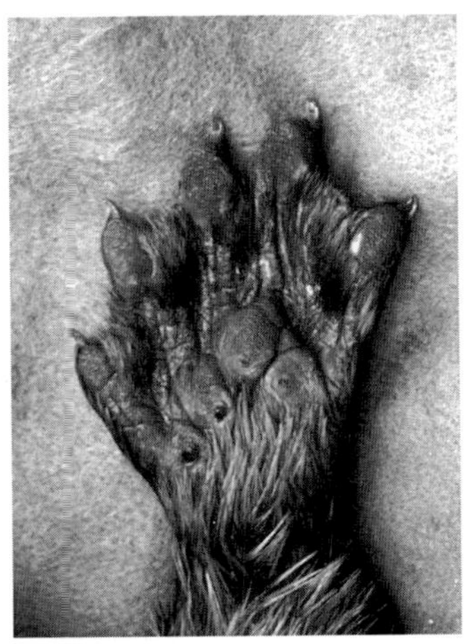

FIGURES 3.42 *(top)* and 3.43 *(bot.)*
Otters have five toes on their front (top) and hind (bot.) feet, all of which are partially webbed, and have some hair between the toes. Palm and heel pads are not well developed.

TRACKS. Otters have five toes on both the front (Figure 3.42) and hind (Figure 3.43) feet. The feet have some hair between the toes and are webbed, though the webs are much more apparent in the hind tracks. Like most mustelids, the placement of the toes is asymmetrical. But unlike other mustelids, all five toes on both feet are well developed. A comparison of otter tracks (Figure 3.44) to those of the fisher (Figure 3.45) shows that the toes in the fisher tracks tend to look bulbous, while those in the otter tracks appear to be elongated. There is also more of a structure connecting toe pad to palm pad in the otter. One of the best ways to distinguish between the two is to look for the otter's heavy tail drag (Figure 3.46). The fisher rarely leaves a tail drag, and when it does, the drag mark is very light, registering only in very sensitive snow.

When otters travel overland, they usually move straight from one body of water to another. In contrast, fishers take a lot of sharp turns, circling constantly, going from tree to tree. Remember, too, that there is a substantial

FIGURES 3.44 *(top)* *and* 3.45 *(bot.)*
The tracks of the otter (top) and fisher (bot.) are similar in size and shape; however, the otter has more well-developed, elongated toes, which at times show up in the track, and its hind track is usually longer. Look for webbing in otter tracks to distinguish them from the fisher.

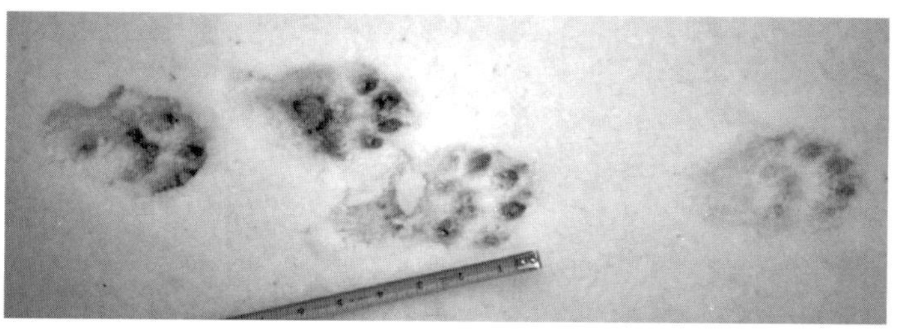

FIGURE 3.46
The otter's tail drag in this bounding pattern is a dead giveaway to its identity. The tail drag, however, does not always register.

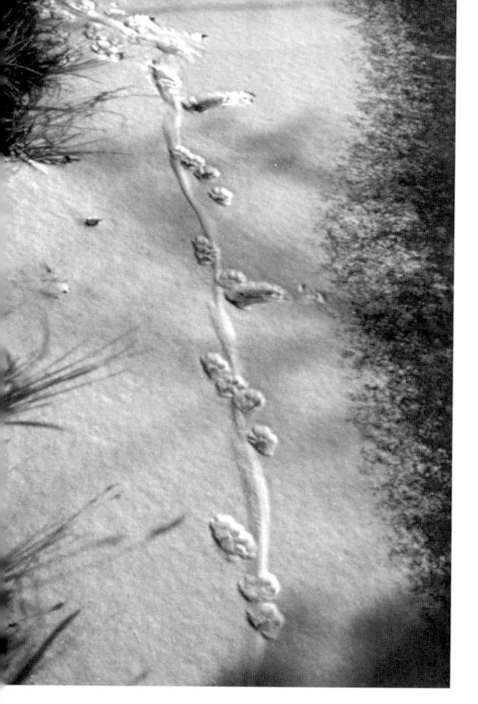

weight difference between the two animals. The fisher weighs six to twelve pounds and the otter fifteen to thirty. Consequently, the otter makes deeper tracks.

The otter's front track measures 2⅞″ to 3¼″ long by 1⅞″ to 3″ wide. The rear track is 3″ to 4″ long by 2¼″ to 3¼″ wide, although the length of the hind track may vary depending upon whether the entire foot registers. In certain snow conditions, the toes may splay, widening the track to 4″. Figures 3.47, 3.48, and 3.49 show different track variations.

TRAIL PATTERNS. When otters run, they bound, hunching their back as though their hind legs are trying to catch up to their forelegs. One of their most common trail patterns is the 3-4 pattern (Figure 3.50). This may show a lot of variation, with elongated groups of three or four tracks registering various combinations. Strides are usually 6″ to 23″, groups 10″ to 17″, and trail widths 4″ to 6″. Otters also may leave the consistent elongated 4-4 pattern demonstrated in Figure 3.51 or an elongated, slanted 4-4 pattern (Figure 3.52), the latter of which can have strides up to 14″, groups from 13″ to 17½″, and trail widths from

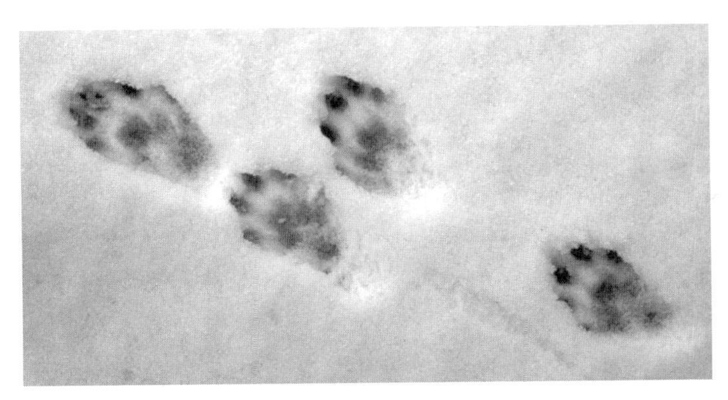

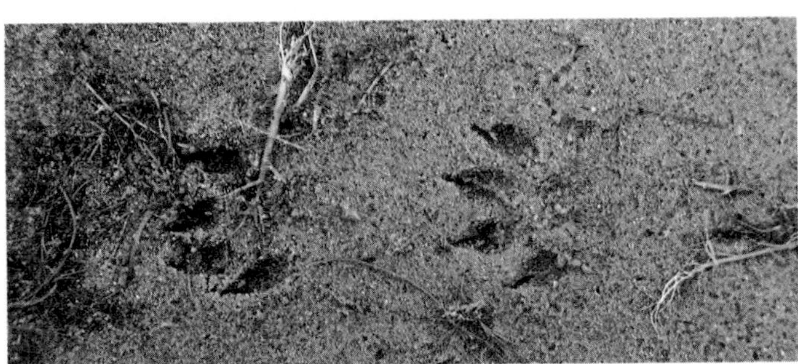

FIGURES 3.47 *(top)*, 3.48 *(mid.)*, and 3.49 *(bot.)* *Tracks vary according to the substrate in which they are placed. The top tracks are in wet snow, the middle and bottom in firm sand.*

FIGURE 3.50 *Otters often use this 3-4 bounding pattern.*

4″ to 8″. Another pattern, in short groups of four (Figure 3.53), usually indicates a faster speed, with strides of 14″ to 23″ or longer and trail widths from 5″ to 6¾″.

The otter also uses the 2-2 pattern so popular with the rest of the weasel family (Figure 3.54). Strides are 15″ to 29″ or longer, trail widths 4″ to 6¾″, and groups 6½″ to 12½″. Its alternating walking pattern (Figure 3.55) has strides of 6″ to 14″ and trail widths of 4½″ to 6½″. In both the walking and 2-2 patterns, the otter's trail may be wider than the fisher's, while the fisher's 2-2 stride may be longer than the otter's. Also, unlike the otter, the fisher may occasionally

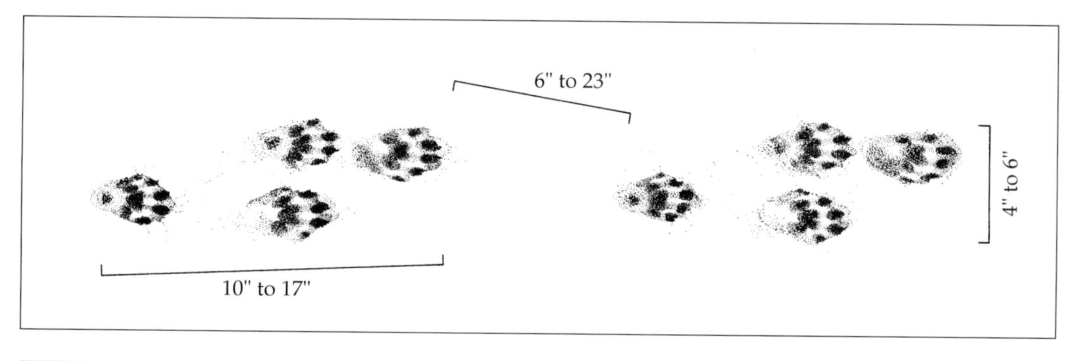

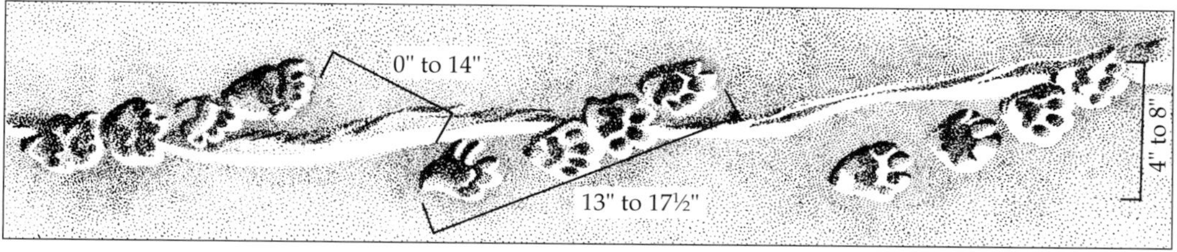

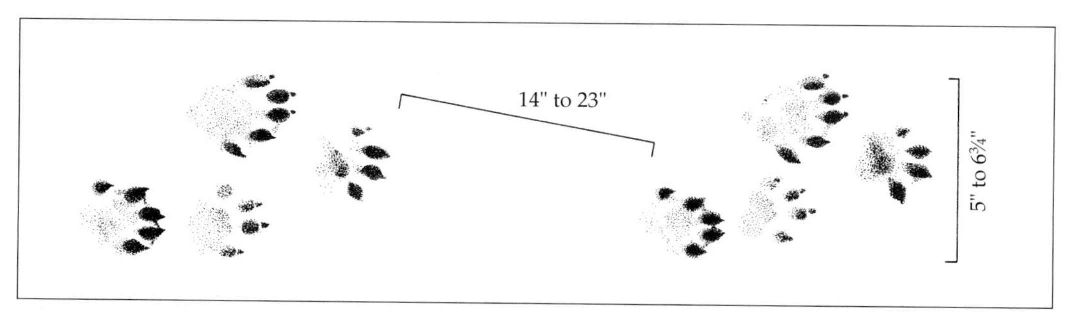

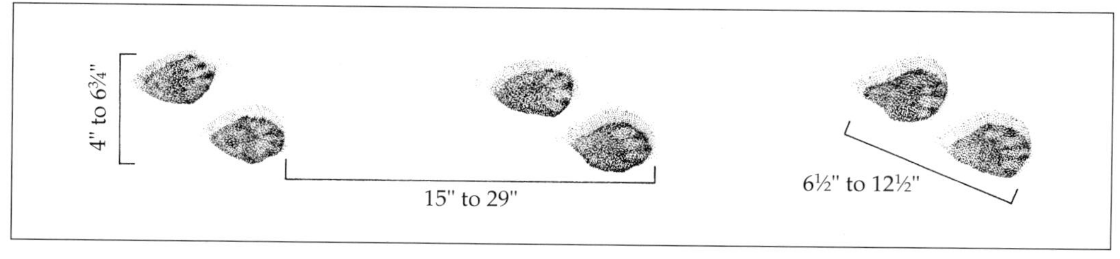

FIGURES 3.51 *(top)*, 3.52 *(top mid.)*, 3.53 *(bot. mid.)*, and 3.54 *(bot.)* *These are the otter's other bounding gaits: At top is an elongated 4-4 pattern, below that is a slanted 4-4 pattern with tail drag, and below that is a short 4-4 pattern. At bottom is the common 2-2 pattern.*

put one foot in front of the other while in this pattern. See Figure 3.37 for the fisher's 2-2 pattern variation.

SIGN: *Slides.* When otters move on snow, they tend to bound a few steps, then get down on their belly and slide, pushing themselves along with their short legs as though they were swimming. This leaves a distinctive pattern in the snow (Figure 3.56). The slide is usually 6″ to 10″ wide. Mink sometimes slide for short distances, but their slides are usually 3″ wide. Although fisher tracks are occasionally confused with otter tracks, I've never seen a

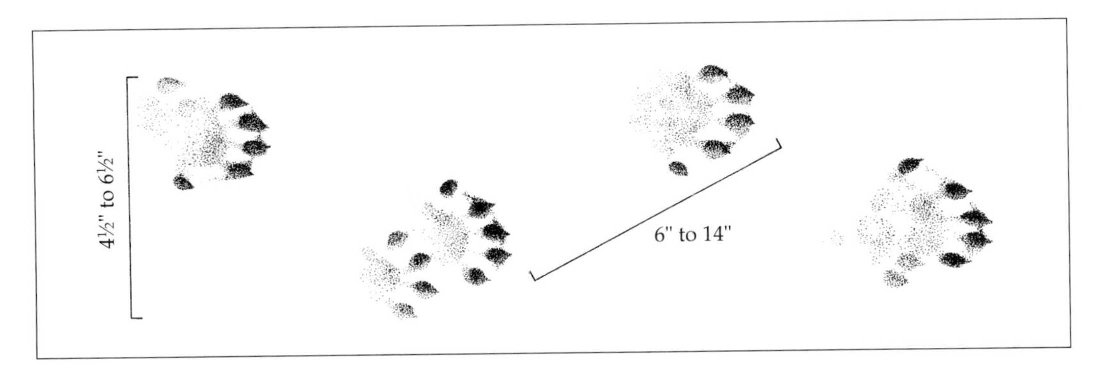

FIGURE 3.55
This illustrates the otter's alternating walking pattern.

fisher slide. I have seen them push themselves down in the snow for 3′ to 4′, but this is unusual and lacks the neat appearance of otter slides. Often an otter slide will lead down a hill and straight into water. The length of the slides seems to be restricted only by the length of the hill. They can be very long, with tracks in them where the animal has given itself an occasional push.

SIGN: *Otter Rolls.* Otter rolls, also called haul outs or scenting areas, have always fascinated me. They can turn up in the most unexpected places—in fairly densely populated areas or even near small ponds in people's backyards.

Otters maintain these roll areas where they come out of the water. Rolls are usually located within twenty feet of the water's edge, though they have been recorded as much as one hundred feet away. The otters roll on the ground, matting down the vegetation, scraping up the earth, and pushing vegetation aside to make a bowl-shaped depression. Sometimes they will defecate at the periphery of this bowl, on top of the pushed-up vegetation; other times they may simply scrape the area and push the vegetation into a mound. There may be scat on the mound, but not necessarily. Look for any combination of these activities at a given roll site.

I have found that otters maintain the same roll area year after year. An otter may not use exactly the same spot, but the new roll often will be adjacent to or only several feet away from the old one. I also have discovered that otters do not always perform the same ritual in the same place. Over a number of years, an area may be used at one time as a scrape and at another time as a roll.

Although the details are not well understood, olfaction plays an important role in the social lives of otters.

FIGURE 3.56
The two otters that made these trails just had to take a few slides, even though they were on a flat surface. Such slides are easily interpreted as otter trails.

FIGURE 3.57
Otter scat most commonly consists of fish scales and bones, and it often appears scattered about. It may be deposited in lumps and, very occasionally, in a tubular shape.

Their scent glands are located in the anus, so scat deposited in roll areas probably serves as a message to other otters. Scent from an otter's anal glands is especially noticeable during moments of rage or fear.

When looking for these scent areas, check along the banks of streams or ponds, especially in places where two bodies of water meet—for instance, where a stream enters a river or where a brook or river enters a pond. Roll or scent areas may be found on peninsulas as well, or where the animals have to cross a short piece of land to get from one body of water to another. I have seen otters using beaver lodges, especially abandoned lodges with holes at the top, as scent areas. A hole in the bank of a river or other body of water is a big attraction for otters. They move in and out of the water through these holes, and the areas around them are good candidates for rolls. An easy way to determine whether an otter is using one of these areas is to check for scat.

SIGN: *Scat.* In its common form, otter scat (Figure 3.57) is simply a pile of fish scales; if fresh, it will smell like

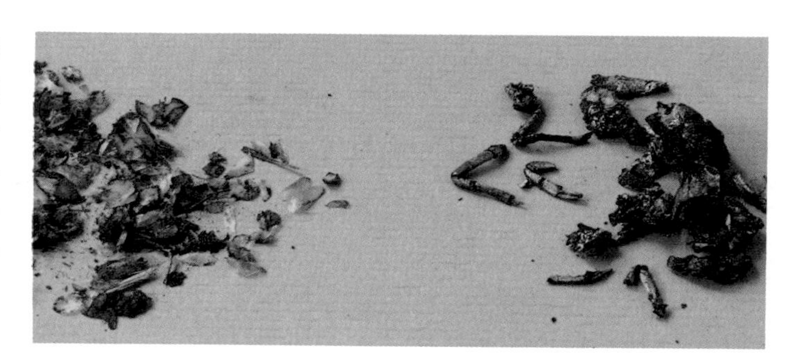

FIGURE 3.58
Sometimes otter scat is dull red in color. This is the result of a diet of crayfish, the remains of which are in the scat on the right. The scat on the left is from a more typical fish diet.

FIGURE 3.59
This black mucuslike substance is another variation of otter scat.

fish. Seldom tubular, it appears scattered. At times it may contain crayfish parts and have a dull red appearance. Figure 3.58 shows otter scat with fish parts on the left and crayfish on the right.

The scat also may be a black, mucuslike material (Figure 3.59), which might be caused by the otter's occasional diet of amphibians. This form also may have something to do with a white, mucuslike substance that otters secrete (Figure 3.60). Some biologists believe that the anal gland secretes this substance, others that it comes from the intestinal tract.

Look for otter scat at rolls and scent areas. The scat itself is easily missed, but more noticeable are areas called brownouts, vegetation that has turned black or brown due to the acid in otters' excretions. Check brownouts near bodies of water carefully. Also check disturbed areas, scrapes, mounds, and flattened vegetation for scat.

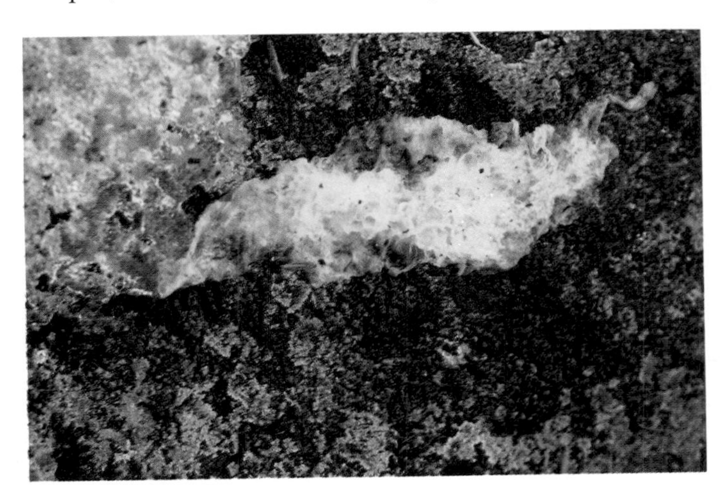

FIGURE 3.60
Otters also secrete a white mucuslike substance. This sample is almost 1″ across, larger than that more commonly seen.

Wolverine
Gulo gulo

ALTHOUGH THE wolverine looks like a small bear and its name implies a connection with the canids, it is the largest of North America's mustelids, or weasels. Its range extends from Alaska across the northernmost parts of Canada, then dips south into Washington, with limited populations in Oregon, Montana, Idaho, California, and Colorado. The wolverine's reputation for ferocity is widespread. The Montaignais of eastern Canada called it *qua-qua-sut*, or "devil of the woods," and some of its later epithets include beaver eater, glutton (its Latin name, *Gulo,* comes from *gulosus,* meaning "gluttonous"), Indian devil, and mountain devil. It also was called skunk bear because it has stripes like a skunk, looks like a bear, and scents its kill with a musky secretion that some say renders it unpalatable to other carnivores. This is probably done to mark possession. Although the wolverine is strong enough to bring down an adult caribou in deep snow, it would rather eat carrion than go to the trouble of procuring fresh meat. Throughout its range, it is more often cursed as a cache robber than feared as a predator.

Wolverines weigh twenty-one to forty-seven pounds, males being larger than females, and are about three to four feet long, counting their seven- to ten-inch tail. (Their long,

FIGURE 3.61
The wolverine, the largest and most powerful member of the weasel family, is the object of many campfire stories and is truly a legend of the north woods.

FIGURE 3.62
Wolverines have five toes on both the front and hind feet, although sometimes only four register. The toes tend to spread out in the track, occasionally making it as wide as it is long. The palm pads are curved and the heel pad often does not register.

black-tipped, bushy tail is one way to tell a wolverine from a bear cub.) They are two-tone, with black legs, face, and rump and lighter brown across the shoulders, along the sides, and halfway down the tail. They have a light or grizzled mask above the eyes and at the ear tips. Their feet are extremely furry, and hair marks will show up in clear tracks.

TRACKS (Figure 3.62). The wolverine has five toes on its front and hind feet, although the fifth toe on the front feet does not always register. Its palm pads are curved across the front, and the toes tend to splay out. The front heel pad usually registers as an oblong dot behind the palm pad, and the rear heel pad usually does not register at all. Front tracks with the heel pad showing are 4½″ to

7½″ long; rear tracks without the heel pad are about 3½″ long. Individual tracks tend to be almost as wide as they are long and usually show nails and drag marks. Larger wolverine tracks tend to look a bit like bear tracks. More often, though, they are confused with those of the wolf. The fifth toe (when it shows), the hairy feet, the difference in the shape of the heel (palm) pads, and the fisherlike track pattern ought to make it easy to differentiate between wolverine and wolf tracks.

TRAIL PATTERNS. Wolverines seem to exhibit trail patterns similar to those of other weasels, especially the fisher: the alternating walking pattern, the common 2-2 pattern, and the 3-4 pattern so often used by the fisher. Look for belly drag in deep snow to identify the wolverine. The 2-2 pattern strides are 10″ to 40″, and the trail width is 7″ to 9″. The 3-4 pattern strides are shorter—7″ to 20″—with the same trail width, 7″ to 9″. Groups are 25″ to 47″.

SIGN: *Marking Trees.* Wolverines mark trees in a fashion similar to that of bears—biting and clawing the bark and scenting the trunk. Most trees chosen for this are small. Boulders and other objects also may be scented.

SIGN: *Scat.* Wolverine scat is similar to fisher scat except that it is larger—sometimes measuring more than 5″ long. It is the largest of all weasel scat.

Striped Skunk
Mephitis mephitis

Eastern Spotted Skunk
Spilogale putorius

Western Spotted Skunk
Spilogale gracilis

THE STRIPED SKUNK is the most common of the three skunks encountered in the northern forest; it has a wider and more northerly range than do the spotted skunks. It is found across the southern Canadian provinces, from Nova Scotia to British Columbia, and throughout the continental United States, except for the desert regions of the Southwest. The eastern spotted skunk ranges throughout the Midwest, from Minnesota through the Great Plains, and south to the Mexican border, as well as in most of the southeastern United States, though not along the Atlantic Coast, except for Florida. The range of the western spotted skunk extends from southwestern British Columbia to Mexico, but only in a small portion of southeastern Montana and western South Dakota.

All three skunks are black with a series of white stripes along the back. In the striped skunk, the white stripe begins at the nape of the neck and usually forms a V as it runs down the back. The eastern spotted skunk has a series of six white stripes along the back and sides. The western spotted skunk resembles the eastern, except that its white markings are more extensive. Striped and spotted skunks also differ in their size and facial markings. The striped skunk is much bigger than the spotted, averaging about six pounds but sometimes reaching twice that, and has a long white line running from the tip of its snout and up between its eyes to its forehead. The eastern spotted skunk, which rarely reaches more than two pounds, has a white spot on the forehead and under or next to each ear; white markings are more extensive on the western spotted skunk. The spotted skunk (sometimes called the civet cat, though it is

FIGURE 3.63
Both striped and spotted skunks are black with a series of white stripes along their backs The striped skunk, pictured here, is the type most commonly encountered in the northern forest.

unrelated to the European civet cat) is more agile than its striped relative and is able to climb trees.

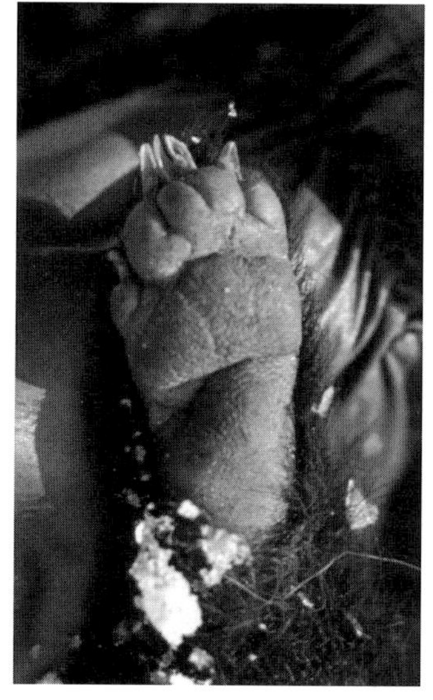

FIGURE 3.64 *(top)* *Used for digging, the long, curved nails on the skunk's front feet are its identifying feature.*

FIGURE 3.65 *(bot.)* *The hind foot has short nails, but smooth and almost continuous palm pads, like the front foot.*

The skunk's defense system is so well established that even people with little knowledge of wildlife know enough to keep their distance from these animals. It is commonly thought that the skunk's musk comes from its urine, but in fact it is emitted from a nozzlelike duct that protrudes from the animal's anus and can direct a jet of its sulfur-alcohol solution, known to chemists as butylmercaptan, with pinpoint accuracy up to nine feet and a fair degree of accuracy up to sixteen feet. Striped skunks differ from spotted skunks in their method of delivery. Striped skunks approach their attacker head-on, tail raised in warning, then twist their hind ends around so that sphincter and snout are both aimed at the predator. Spotted skunks charge their victim running on their forelegs, hind end raised in the air and back arched so that the spray is directed forward, a most impressive acrobatic display that would be fascinating to watch were it not the prelude to such a noxious performance.

The spray of a skunk is so powerful that it is noticeable up to half a mile away. A direct hit can produce temporary blindness and involuntary vomiting. Native Americans used to eat skunks, but they were careful to cut out the small musk gland in the anus first. When eaten, the skunk's poison attacks the nervous system and can cause death. John James Audubon told of a clergyman he once met who kept the scent glands of a skunk in a small glass container, and whenever he suffered an attack of asthma, he would remove the stopper and take a deep sniff, after which he felt greatly relieved.

TRACKS. Like all members of the mustelid family, skunks have five toes on their front and hind feet (Figures 3.64 and 3.65), very smooth, continuous palm pads, and relatively small heel pads. The pads are not segmented as they are on squirrels. Skunks also have very long front nails, which they use for digging. Their feet are very small, and the toe pads are tiny. Most people misidentify skunk tracks because they don't expect a large-looking animal to leave such delicate tracks. The track of the forefoot usually measures only 1⅞″ to 2³⁄₁₆″ long (including the nails and heel) by 1″ to1⅛″ wide; that of the hind foot measures 1¾″ to 2″

long (including the nails, which are much shorter on the hind foot than on the forefoot, and the heel) by 1″ to 1⅛″ wide. The spotted skunk's feet are even smaller. The track of the front foot is 1⅜″ long (including nails) by 1″ wide; that of the hind foot is 1¼″ long (including nails) by 1″ wide. The long nails of the forefeet are the skunk's identifying feature. Look for small dots well ahead of the toe pad impressions, sometimes so far ahead you may not realize they are part of the track.

FIGURE 3.66
In this set of four tracks, the striped skunk's two front tracks are at the bottom and two hind tracks are at the top. Notice the long nail marks left by the front feet.

Figure 3.66 shows skunk tracks in sand. The two tracks at the bottom left are from the front feet. Just the palm pads show. Even though skunks are technically plantigrade, which means they walk on the soles of their feet, the heel pads often don't show in the tracks of the forefeet. The right rear track shows both heel and palm pads, while the left rear track shows only the palm pad. Both hind tracks show very short nails; the nails of the front tracks are much longer.

TRAIL PATTERNS. One of the striped skunk's walking patterns (Figure 3.67) is very irregular. The individual tracks are unevenly spaced and wavering all over the trail—so much so that I've come to consider no pattern at all as an identifying characteristic of this particular walking pattern. The trail width is 2¾″ to 4¼″ for the striped skunk and under 3″ for the spotted skunks in their similar pattern. The stride is extremely variable, measuring 4″ to 8″. The spotted skunks' stride is usually not more than

FIGURE 3.67
The slow walking pattern of the striped skunk is very irregular and typical for skunks. The direction of travel is from left to right.

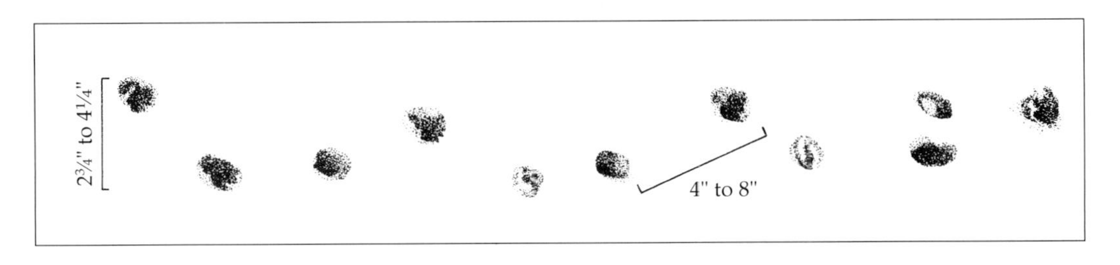

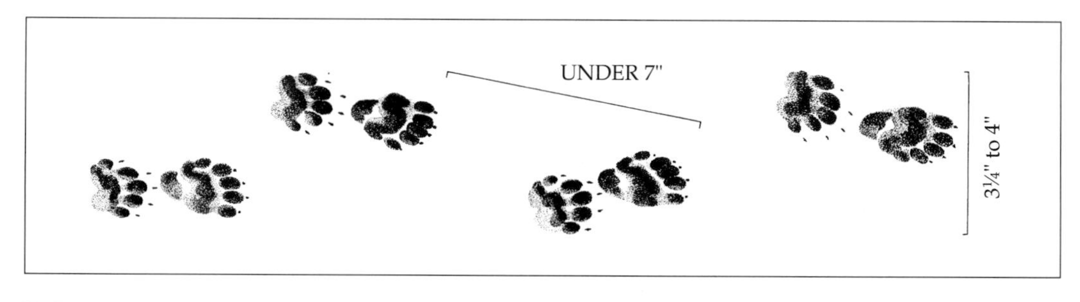

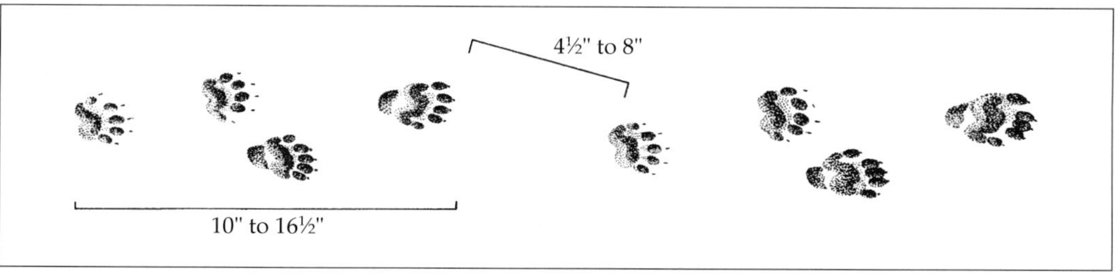

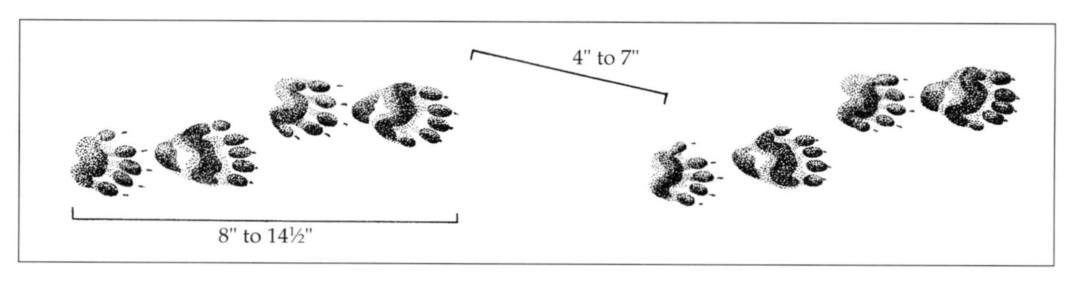

FIGURE 3.68 *(top)* *The striped skunk's common 2-2 pattern, with the hind feet overstepping the front, is another walking pattern.*

FIGURE 3.69 *(mid.)* *This 4-4 running pattern of the striped skunk is similar to the galloping pattern of canines, with a track sequence of front-front-hind-hind.*

FIGURE 3.70 *(bot.)* *The slanted 4-4 gait is similar to canines' loping gaits, with a front-hind-front-hind track sequence.*

5″. The 2-2 trail pattern in Figure 3.68 also is fairly common and looks more consistent than the walking pattern mentioned above. In it, the hind and front tracks are close together, producing a double pattern with two tracks, one slightly ahead of the other. In snow, the two tracks may join to make one elongated track.

The striped skunk has several running patterns, two of which are most commonly seen. One (Figure 3.69) I call the 4-4 pattern, showing groups of four tracks in a front-front-hind-hind or front-hind-front-hind configuration and measuring approximately 10″ to 16½″ overall; the distance between each set of four tracks is 4½″ to 8″. Figure 3.70 shows the second typical running pattern. This pattern is a slanted 4-4 (front-hind-front-hind), very similar to the second walking pattern above, except this time the groups of two have formed into groups of four. The stride and group length are about the same as for the 4-4 pattern.

SIGN: *Dens.* Both striped and spotted skunks seem to prefer to take over the abandoned dens of other animals, whether in rocks or caves, the crevices of trees, or under buildings. Striped skunks also dig their own burrows, some of which may be up to fifty feet long, although the average is probably closer to six to eighteen feet. These burrows have two or three chambers, which are sometimes lined with grass or leaves, and there may be up to five entrances. The striped skunk also dens communally in winter. In summer, it usually leaves its den and beds in open, aboveground sites.

FIGURE 3.71
Skunk digs are small, round pits. The opening here is about 1½" in diameter, but some holes may be as large as 8′ in diameter and narrow as they get deeper.

SIGN: *Digs.* Many animals make digs, and identifying which animal made a particular dig is a tricky business. Skunks are omnivores; they feed on small mammals, amphibians, reptiles, birds and their eggs, insects, or fruit and other plant material depending on what is available at a certain time of year. The striped skunk is larger and slower than the spotted skunks, and it tends to wait in ambush for its prey or to seek food items more accessible to slow-moving animals. In spring and summer, its diet consists of 43% insects and may at times be more than 90% insects. It likes to dig up ground bees, leaving discarded hive material scattered about the dig site. These digs are similar to those of raccoons (see Figure 5.12 on page 169). It appears that both striped and spotted skunks are able to tolerate bee stings, perhaps the way the opossum can tolerate certain snake venom. The most common skunk digs are illustrated in Figure 3.71. They are a result of skunks digging for insects and/or their larvae. Many lawn owners, to their frustration, find these types of digs in their yards. Look for scat nearby to confirm species.

The spotted skunks, being much more athletic than the striped, eat prey that are more difficult to catch. In winter, their diet may be as much as 90% small mammals, including cottontails in areas where the rabbit population is particularly dense. In summer, however, insects may make up 90% of the spotted skunks' diet. In general, studies show that the striped skunk will eat twice as much insect material as the spotted skunks and the spotted skunks will eat four times as much mammal material as the striped skunk. Larger skunks tend to eat more insects than smaller skunks.

SIGN: *Bird's Nest Predation.* Both striped and spotted skunks steal eggs from birds' nests, using various

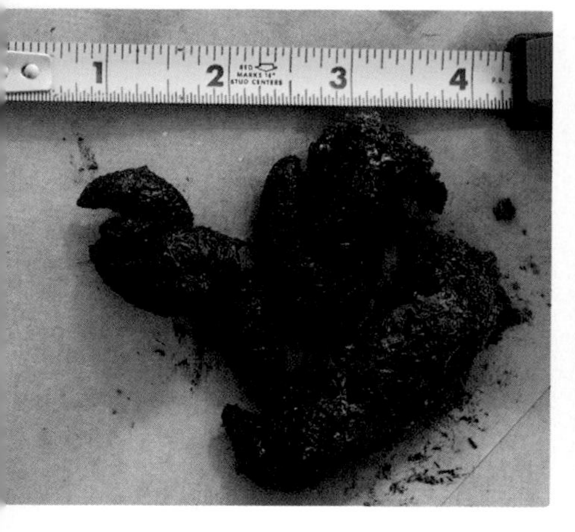

FIGURE 3.72 *(left)* *Striped skunk scat is usually black and composed of insect parts.*

FIGURE 3.73 *(right)* *This scat has been broken apart to reveal its contents.*

methods of opening the eggs depending on their size. Some researchers have reported finding crushed eggs and nests torn to shambles; others have seen nests and eggs left pretty much intact (more common with the spotted skunks than with the striped). Skunks usually don't carry eggs off to be enjoyed elsewhere, so whether they are crushed or not, the depredated eggs should be near the nest site.

A note on egg depredation in general: When you find a pile of eggshells in a nest, it is often difficult to determine whether the eggs were depredated or whether they hatched normally. With successfully hatched eggs, the eggshell, even after being weathered, can easily be separated from the lining. With eggs that have been broken before hatching, it will be hard to detach the eggshell from the lining.

SIGN: *Scat.* Skunk scat (Figure 3.72) is usually composed entirely of insects and is much bigger than you might think. Most spotted skunk droppings measure ⅜″ to ⅞″ in diameter, and striped skunk droppings may reach ¾″ or more in diameter. When you separate skunk scat with a stick (Figure 3.73), you can see that all the components are insect bodies and wings. Skunks also eat small animals and fruit, in which case the scat will contain remnants of hair and seeds, respectively.

Skunks defecate randomly, but scat often may be found in feeding areas. If spotted skunks are inhabiting a barn or outbuilding, you may find their scat under the rafters. In areas where they have been living for a long time, their scat may accumulate to a depth of two inches.

CHAPTER 4: VIRGINIA OPOSSUM

Didelphis virginiana

Two Virginia opossum trails meandering along a frozen brook.

Virginia Opossum

Didelphis virginiana

ALTHOUGH THE Virginia opossum was once believed to be among the oldest mammals on earth (Will Barker, in his *Familiar Animals of America*, says that "it has changed but little since it roamed the continent with dinosaurs 70 to 100 million years ago"), recent genetic evidence suggests that opossums have evolved only since the last Ice Age, which ended ten thousand years ago. They are marsupials, native to Central and South America, where today there are about seventy-five species of opossum. The Virginia opossum gradually moved north into the United States at about the same time European settlers did, reaching Canada in the twentieth century. Its current range extends throughout most of the eastern United States (except for Maine and parts of northern Michigan and Minnesota) and west to Colorado and Texas. Populations of opossums also exist along the West Coast, from southern British Columbia to southern California, and in a few scattered locations in the West; these animals are a result of transplants from the eastern United States.

The opossum is one of the most adaptable animals in existence. It can tolerate all but extremely cold climates and will eat just about anything, from carrion to fresh fruit, crickets, and skinks. Like the raccoon, the opossum has adapted well to human habitations. Over the past three hundred years or so, it has become one of North America's most prolific arboreal creatures.

The word *opossum* comes from the Algonquian word for the animal—*apasum*, meaning "the white animal." With their white fur and black markings; pointed, piggish snout; ratlike tail and eyes; and long, delicate fingers (complete with opposable thumbs on the hind feet), opossums are one of the oddest-looking animals in the forest. Being marsupials, they rear their young in little pouches, like kangaroos, and seem to have borrowed body parts from many different animals. "Here was a strange animal," wrote John James Audubon in the mid-nineteenth century, "with the head and ears of a pig, sometimes hanging on a limb, and occasionally swinging like a monkey by the tail! Around that prehensile appendage a dozen sharp-nosed, sleek-headed young had entwined their own tails and were sitting on the mother's back. The astonished traveller approaches this extraordinary compound of an animal and touches it cautiously with a stick. Its eyes close, it falls to

FIGURE 4.1
The opossum seems to thrive near human habitation. Its pointed snout; white, grizzled coat; and long, ratlike tail make it one of the easiest animals to identify.

the ground, it ceases to move, and appears to be dead! He turns it on its back, and perceives in its stomach a strange apparently artificial opening, a pocket with another brood of a dozen or more young, each scarcely larger than a pea."

Audubon was referring to the opossum's supposed habit of "playing possum" when under attack. I once witnessed this peculiar defensive strategy. While driving up a road at night, I turned a corner and saw two dogs attacking an opossum on the road. The opossum literally fell over, as if dropping dead of a heart attack. Saliva drooled from its mouth, and a very unpleasant smelling liquid began oozing from its anal glands. The dogs seemed entirely befuddled by this behavior. They didn't know what to do. They barked at it, then nipped at it a few times, but the opossum didn't move a hair. In the end, I chased the dogs off. After several minutes, the opossum got up and walked into the woods.

An opossum also is capable of responding in quite the opposite way. Sometimes it will turn on its tormentor, baring its fangs and growling or hissing aggressively. Because opossums have a lot of teeth—more than most animals—their bite can be worse than their bark. A wound inflicted by an opossum can become quickly and seriously infected, so beware.

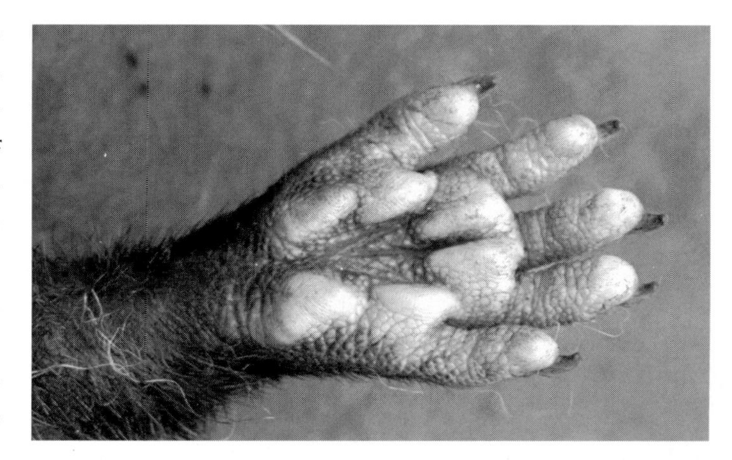

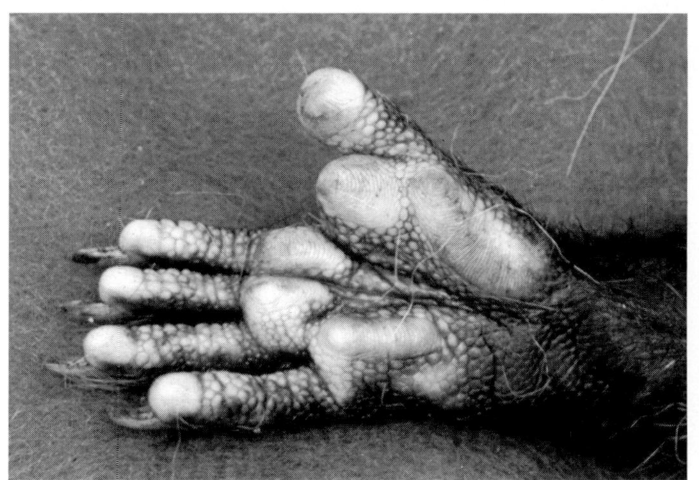

FIGURE 4.2 *(top)*
The opossum has five elongated toes on both its front and hind feet, the bottoms of which are naked. The front foot has four palm pads and two heel pads.

FIGURE 4.3 *(bot.)*
The opossum's hind foot has four palm pads and a single heel pad. The inside toe, or "thumb," of the hind foot is opposable and has no nail.

In a curious way, the opossum's tail has enabled the animal to increase its northern range by acting as a kind of fat storage area. In the fall, opossums increase their fat content by about 35%, much of which is stored in the tail. Still, that unprotected tail can be a handicap, for I've seen opossums with part of their tail missing because of frostbite. Since the tail acts as a kind of fifth leg when the animal climbs, losing it can inhibit the animal's movement. Opossums use their tail in other ways as well. I have a friend who watched an opossum gather a bunch of leaves with its forefeet, push them under its body with its snout toward its rear, and wrap its tail around them. Then the animal used its tail to carry the pile under my friend's porch, where it was lining its den.

Female opossums, like other marsupials, bear their young—one litter a year in the North, two in the South—after only thirteen days' gestation, when the newborns are

FIGURE 4.4
This close-up of the opossum's front foot (on the right) shows how the toes spread with the animal's weight. The clawless opposable thumb is clearly visible on the hind foot (on the left).

barely more than embryos. Sightless, hairless, less than half an inch long, and weighing 1/200 of an ounce (it takes approximately thirty-five hundred newborn opossums to make a pound), the young have no hind legs or tail, and their internal organs, including their heart, are in the very early stages of development. These tiny creatures spend their first minutes in the world crawling blindly up the mother's hairy belly and into her marsupial pouch, where they spend the next sixty days nursing and maturing into adolescents.

FIGURE 4.5
Under certain conditions, the opossum's front feet leave a star-shaped track, as seen here in snow. This is the result of the splayed foot shown above.

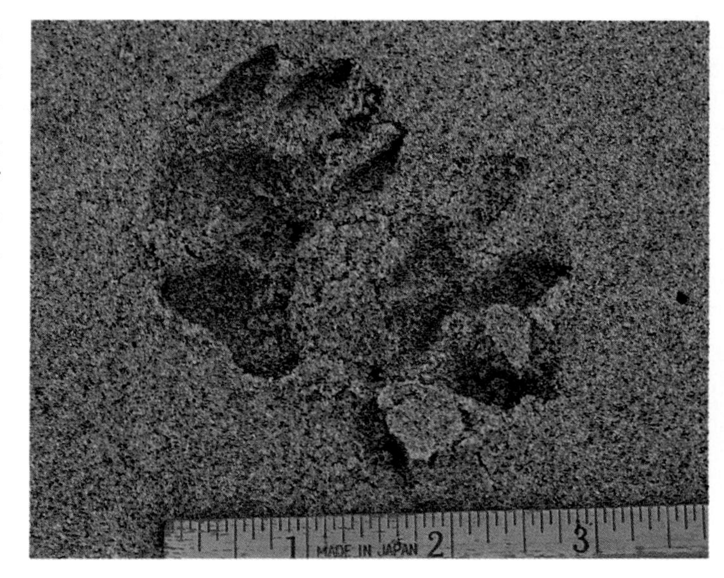

FIGURE 4.6
The opossum's hind track (left) is partially superimposed over the front track (right). Note that four toes of the hind track point in one direction, the opposable thumb in another.

Opossums seem to be immune to the venom of certain snakes, including copperheads, timber rattlesnakes, and water moccasins, sometimes withstanding up to sixty times the dose lethal for other mammals. They may provide a more tolerable antidote to rattlesnake poison than the serums currently derived from horses and sheep, which often trigger severe allergic reactions in humans.

TRACKS AND TRAIL PATTERNS. The opossum's feet are more primatelike than even the raccoon's. Opossums have five toes on the front foot (Figure 4.2) and five on the back (Figure 4.3), and the fifth toe on each hind foot is opposable like a thumb—it projects toward the inside of the track and can stretch right back in the opposite direction from the front toes. A look at Figure 4.4 shows how the toes spread out when the animal puts its weight on its front foot. This leaves a star-shaped track (Figure 4.5), with the toes going out in five different

FIGURE 4.7
This alternating walking pattern shows the hind tracks partially superimposed over the front tracks, in an indirect-registering trail.In coarse, wet snow, trail widths may increase to 5½".

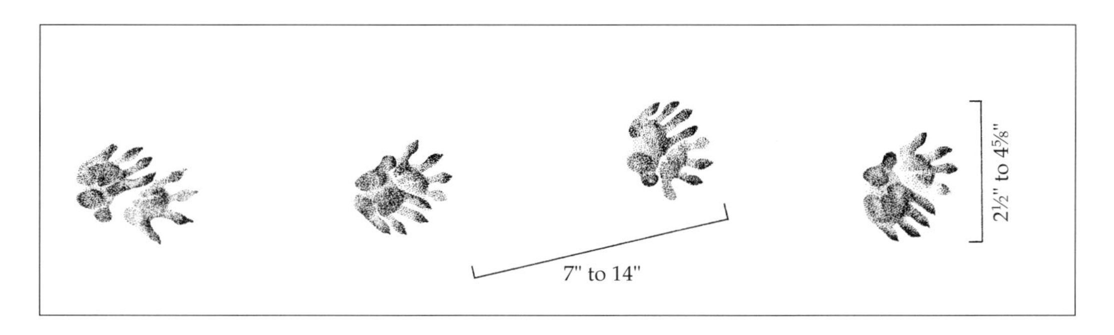

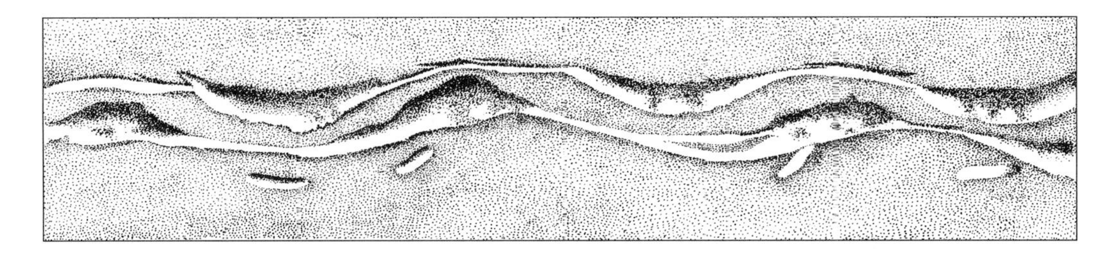

directions. Only the opossum leaves such a track. Also, its hind foot is strangely shaped and leaves a unique track (Figure 4.6) because it has the extended thumb that reaches way off to the side and sometimes also back to the rear. This thumb has no nail, another distinguishing feature of opossum tracks.

FIGURE 4.8
The opossum's alternating walking pattern looks quite different in 3" of soft snow. The zigzag parallel lines represent tracks and foot drag. The four marks to the lower side of the trail are from the animal's tail slap.

Like many quadrupeds, when opossums walk, their rear feet fall next to or partially on top of their front tracks, leaving what is called a double or indirect register. In Figure 4.7, you can see the typical walking gait, with the front foot making the starlike track and the hind foot stepping on top of part of the front track. The front track is usually about 1½" to 2⅛" long by 1¾" to 2⅜" wide, and the rear track is 1¾" to 2¾" long by 1¾" to 2⅞" wide. Measure the width of the hind track from the tip of the outside toe to the tip of the thumb; measure the length from the heel pad to the end of the nail of the longest toe. These measurements are not always reliable, however, because of the dexterity in the foot and the position of the thumb.

Figure 4.8 shows the same walking gait, this time in three inches of snow. There's some foot drag, which is common in this depth of snow. Also, if you look closely, you can see where the opossum's tail slapped the snow as

FIGURE 4.9
Opossum scat is very hard to identify because of the many variations in its ever-changing diet.

it waddled along. I've never seen an opossum leave a long tail drag, but it sometimes leaves a very subtle tail mark that clearly identifies the animal. The trail width for an opossum is 2½" to 4⅝", and the usual stride is 7" to 14".

SIGN: ***Dens.*** Opossums do not dig burrows, preferring to take over dens abandoned by other animals. Sometimes they will even move into a den still being occupied by the original owner. One complex burrow system in Michigan was found to be simultaneously occupied by an opossum, a striped skunk, a woodchuck, and a raccoon. They are nocturnal and not territorial. They wander around at night foraging for food, covering anywhere from half an acre to sixty acres, and den up at dawn just about anyplace—in a basement window cell, under a porch, or in any crevice they can find.

SIGN: ***Scat.*** Because opossums are such omnivorous eaters, their scat is widely variable and, unfortunately, often indistinguishable from that of many other animals (Figure 4.9).

CHAPTER 5: RACCOON

Procyon lotor

A raccoon in its natural wetland habitat.

Raccoon
Procyon lotor

RACCOONS ARE THE only members of the Procyonidae family found in the north temperate zone. They range throughout the continental United States, except for portions of the Rocky Mountains; through southern Canada, from Nova Scotia to British Columbia; and in the prairies as far north as Churchill, Manitoba, and across to Fort McMurray, Alberta. Other procyonids, such as the ringtail and coatis, have stayed in the American Southwest, and the kinkajou is found only in South and Central America. Although some writers place raccoons close to the bear family—their tracks are plantigrade, or flat-footed, like the bear's—their direct ancestors, as the root word *cyon* suggests, are actually the canids. The word *raccoon* is from the Algonquian *arakunem*, which means "least like a fox."

The *lotor* in the raccoon's Latin name means "washer" and refers to the popular misconception that raccoons like to wash their food in water before eating it. Raccoons in the wild do not "wash" their food. Studies have shown that only raccoons in captivity dunk their dinners before eating, and they'll do so whether the food is wet or dry, clean or dirty. According to wildlife biologist Jim Cardoza, this "dabbling" is a fixed behavioral motor pattern that facilitates finding food in the aquatic environment. In captivity, food is provided, but the dabbling search pattern is innate and is expressed by what we see as washing.

At any rate, the act of dunking does illustrate the fact that raccoons' forefeet are extremely dexterous. With their long, humanlike fingers, raccoons can untie knots, unscrew the tops of jars, and turn door handles. A friend of mine once watched a particularly determined raccoon trying to get into his compost bin. He had "raccoon-proofed" the bin with a lid equipped with a latch and had pushed a wooden peg through the loop. The raccoon calmly climbed up the side of the bin, pulled out the pin with its fingers, lifted the latch, raised the lid, and climbed into the bin. I also heard of a fellow who let a group of raccoons march in and out of his house at will. They had learned how to open the door to the house and even the door to the refrigerator. One particular raccoon would often come in when the man was playing Bach on his stereo. As soon as the record was over, the raccoon would leave.

Raccoons thrive in populated areas. A recent study shows that the highest density of raccoons in New York

FIGURE 5.1
Because of the raccoon's ability to adapt to a variety of habitats, these babies are as likely to grow up in a suburb or even a large city as they are in the country.

State is in New York City. I once did some tracking programs for a school in Marblehead, Massachusetts, a coastal suburban community with a lot of houses and hardly any woods, and I was amazed at the density of the raccoon population in the area. Unlike foxes and coyotes, which will follow ready-made human trails or whatever is easiest to follow (power lines, for example), raccoons seem to prefer making their own trail systems. In Marblehead, the raccoon trails were well worn through the woods and along the edges of wetlands. One of the trails followed a stream that passed through a culvert under a dirt road. Rather than go through the culvert, the raccoons climbed up the embankment and crossed the road, leaving a trail about 2½′ wide. There were hundreds of telltale tracks. We erased them all, but when we returned the next morning, the ground was filled with more tracks. I could not believe how many raccoons had passed along that trail in just one night.

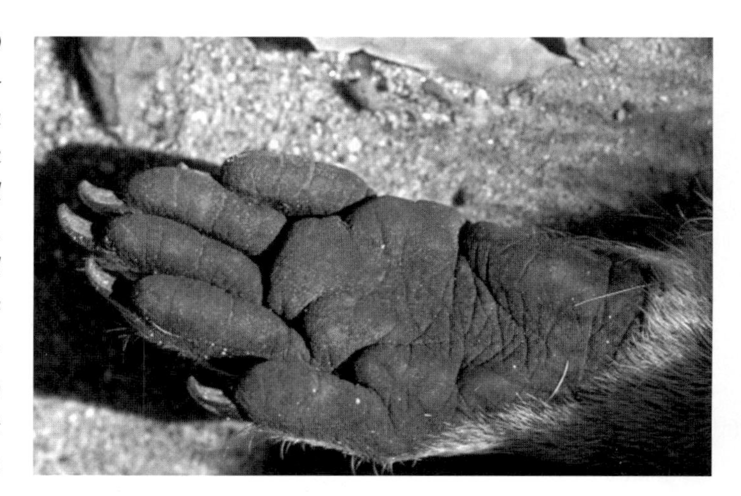

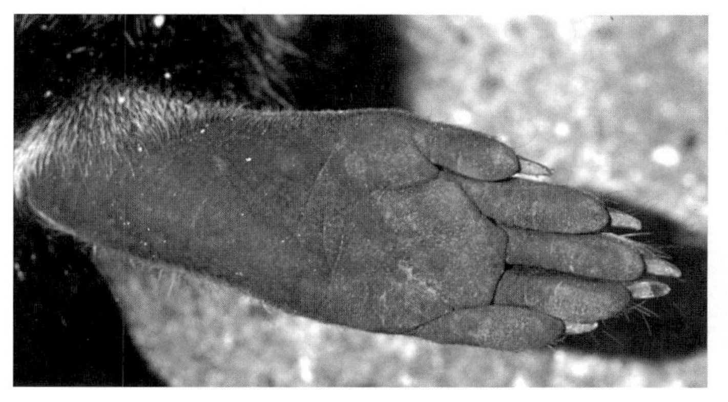

FIGURE 5.2 *(top)* *Raccoons have five well-developed, elongated toes on their front and hind feet, both of which also are naked and have C-shaped palm pads. The front foot has two heel pads, one of which is barely discernible.*

FIGURE 5.3 *(bot.)* *The heel area of the hind foot is smooth and longer than that of the front foot.*

TRACKS. A look at a raccoon's feet (Figure 5.2, front foot; Figure 5.3, hind foot) clearly shows how such delicate feats as untying knots are possible. Both the front and hind feet have five very well developed, elongated toes, and although the "thumb" is not actually opposable, it is long and supple enough to allow the raccoon to grasp things such as clam shells and door latches. Each foot also has large, C-shaped palm pads that make a curved impression arching back and away from the toes. The heel pads don't usually register in tracks of the front feet, but since the rear feet carry most of the animal's weight, as in primates, the rear tracks are often longer and deeper than the front in a soft substrate, and the heel pads are more likely to register. The nails of both the front and rear feet show up in tracks as small dots ahead of the toes, rather than deep impressions, as in canines.

Because raccoons' feet have such dexterity, their tracks vary widely. In soft substrate, the tracks can resemble those of a small human hand, the five "fingers" clearly showing

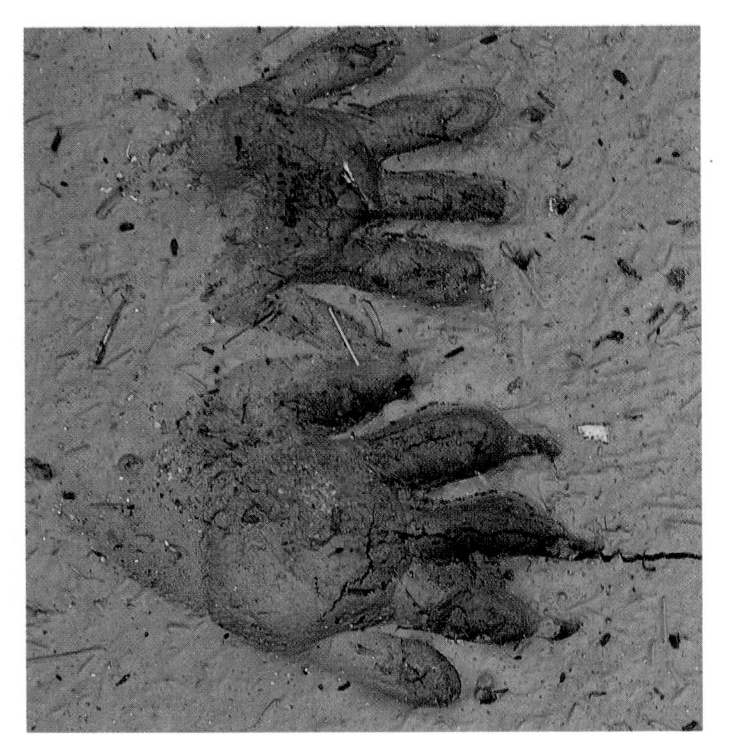

FIGURE 5.4
These raccoon tracks show all five toes and the C-shaped palm pads. The small dots made by the nails of the front foot (top track) are reliable identifying features. The hind track is on the bottom.

all the way from the tips to the palm pads. The toes of the front feet may be spread out, while those of the hind feet tend to be more in alignment with the direction of the track. At times, they, too, have a tendency to spread out. Variations in this pattern are indicated in the measurements. Front tracks can be 2″ to 3″ long and 1⅞″ to 2½″ wide. Rear tracks are 2⅜″ to 3¾″ long, depending on how much of the heel pad is registering, and 2⅜″ to 2½″ wide. Figure 5.4 shows a variation in which we see just a hint of the heel pad of the rear foot at the bottom of the picture; note that the nails of the front foot, at the top of the picture, barely show up as dots. Other variations in the raccoon's tracks are illustrated in Figures 5.5 and 5.6.

FIGURE 5.5 *(left)*
The toes are positioned close together in this hind track, which registered in mud. Note the small dots left by the nails.

FIGURE 5.6 *(right)*
Also imprinted in mud, these tracks were made by a raccoon's hind (on left) and front (on right) feet.

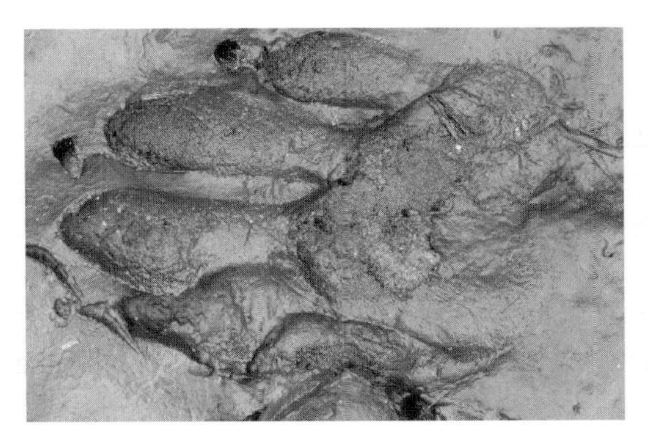

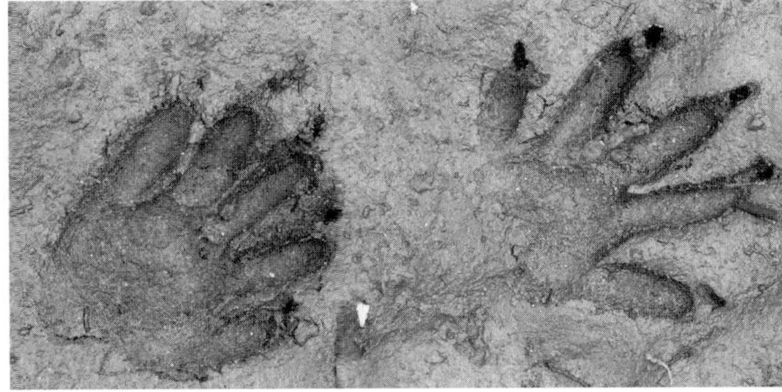

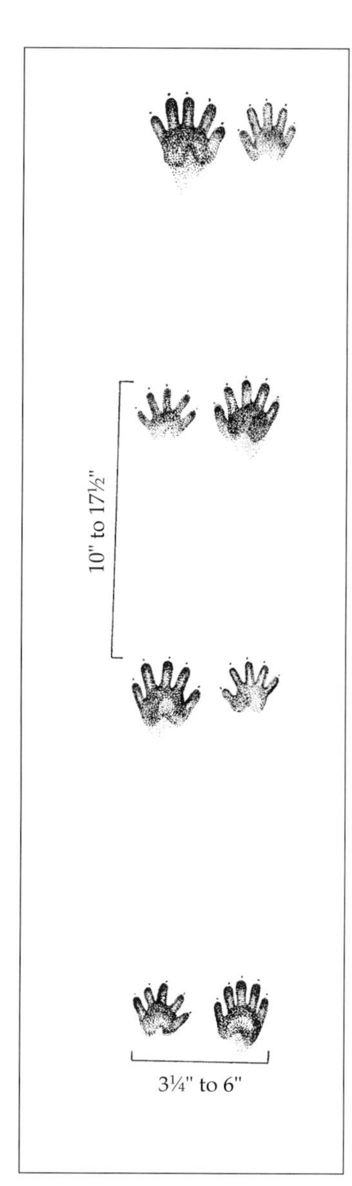

FIGURE 5.7
This common 2-2 walking pattern of the raccoon is unique to the animal. Front and hind feet alternate sides in each group of two tracks. Starting from the bottom, the larger hind track is on the right, then it switches to the left in the next group, to the right, and, at top, to the left.

TRAIL PATTERNS. Because of the many variations in the shape and size of individual tracks, the raccoon's walking pattern is often a better way to identify this animal. The raccoon has three distinct walking gaits. Its usual walking pattern is shown in Figure 5.7. The tracks are side by side in a distinct 2-2 pattern; no other animal walks this way. If you look closely, you'll see that the front tracks are smaller than the rear and that the pattern is small on the left and large on the right; then it changes to large on the left and small on the right. So the front and rear feet alternate. Also, the hind feet are heavier than the front, so the hind tracks are more pronounced and detailed and the hind feet register more deeply than the front. If you draw lines across this pattern, through each set of two tracks, the lines will be parallel. Such parallel lines are a sure indication that these are raccoon tracks.

Something to be aware of is that as this set of tracks ages, the lighter front tracks will begin to disappear first. Eventually, only the rear tracks will remain, which could mislead you to believe that the tracks belong to a different animal. I've seen a worn raccoon trail, with only the hind tracks visible, mistaken for the wide, alternating trail of a bobcat.

Figure 5.8 demonstrates a second walking pattern. Again, the front and rear feet alternate in the pattern, but in this case, one foot is slightly ahead of the other. If you draw a line across each set of tracks, these lines will form a series of alternating diagonals. No other animal, to my knowledge, leaves this type of pattern.

Figure 5.9 shows the third walking gait. I call this variation the diagonal, or alternating, walking gait. It is the result of direct or double registering.

The raccoon's stride in the first two walking gaits is 10″ to 17½″ or longer, and in the third 8″ to 14″. The trail width for all three is 3¼″ to 6″. The diagonal walking pattern demonstrated in Figure 5.9 is very similar to that of the fisher's walking pattern. The fisher's strides are 7″ to 14″, and its trail width is 3″ to 5½″. As you can see, there is a tremendous amount of overlap between these two animals. Furthermore, the tracks of the raccoon and the fisher sometimes look the same. The best thing to do is to track the animal for some distance to see whether it changes its walking pattern. If it changes into the pattern shown in Figure 5.7 or Figure 5.8, it's a raccoon. If the pattern more

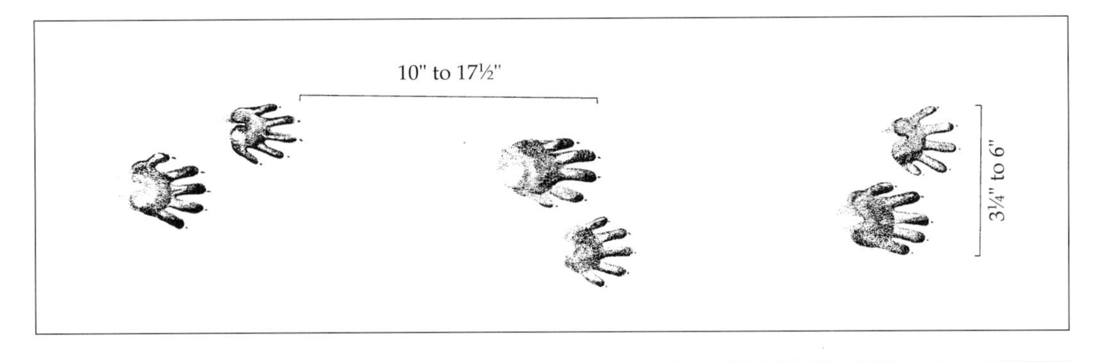

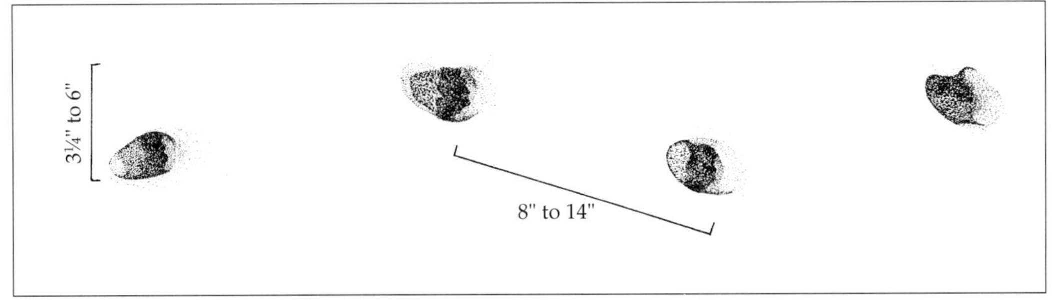

closely resembles the 2-2 or 3-4 pattern of the fisher (see Figures 3.36 and 3.39 in the fisher section), it's a fisher. The raccoon's alternating pattern also may be confused with the opossum's walking pattern. Look for the opossum's tail slap in snow.

The raccoon's bounding gait (Figure 5.10) is typical of that of most quadrupeds and is similar to that of chipmunks, red and gray squirrels, rabbits, or snowshoe hares: two front feet, one almost but not quite in front of the other, and the hind feet side by side but also with one slightly

FIGURE 5.8 *(top)*
In this 2-2 walking pattern, the angle of the track pairs alternates. Only rarely will the angle of the raccoon's tracks be parallel like that of the fisher. Even when the angle is parallel, the raccoon's front and hind tracks still alternate sides in each track pair.

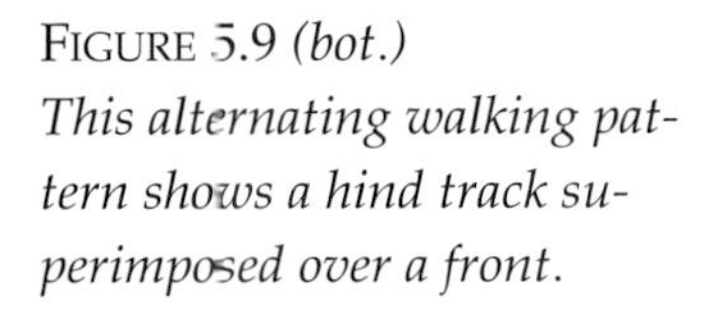

FIGURE 5.9 *(bot.)*
This alternating walking pattern shows a hind track superimposed over a front.

FIGURE 5.10
When raccoons bound or start to run, the sequence of tracks is (from right to left) front-front-hind-hind.

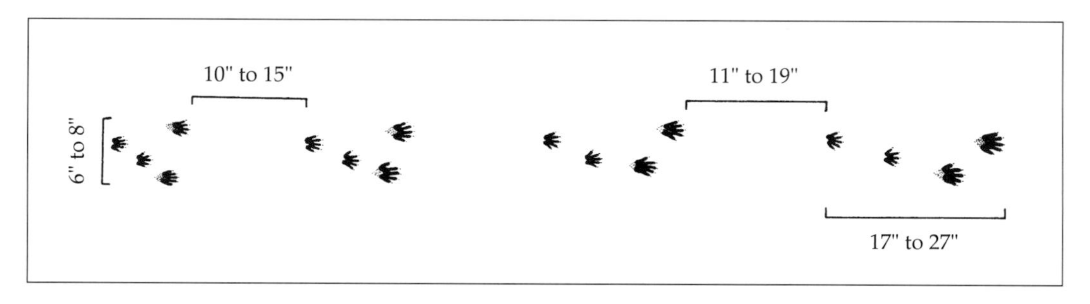

FIGURE 5.11
The two sets of tracks on the left illustrate the raccoon's bounding gait; the two sets on the right are gallops. In each group of four tracks, the sequence is front-front-hind-hind (from left to right), and the configuration changes as the animal increases its speed.

ahead of the other. This is the gait the raccoon uses when it has just started to run, is not really serious about running, or is taking a few bounds toward something. When it begins to run in earnest, it will stretch out into a C-gallop. When tracks are in the bounding pattern, the stride (the distance between each set of four tracks) is usually 10″ to about 15″, except in extreme bounds, and the trail width is 6″ to 8″. As the animal picks up speed, the trail width narrows to resemble a straight line (Figure 5.11).

SIGN: *Dens.* Raccoons use many different kinds of shelter for a variety of purposes. I've seen them use a large, abandoned bird's nest for refuge. Hollow trees, with cavities around eleven to fourteen inches wide, are commonly used for denning. Other possible denning sites include rock crevices and caves, abandoned buildings and culverts, wood or brush piles, and abandoned fox dens. Look for isolated raccoon hairs in areas where you suspect they might be denning. The hairs have long waves and are banded black, with lighter bands of gray or yellowish gray and black.

Raccoons are not deep hibernators. In the fall, they put on huge stores of fat, which they use over the winter. They tend to den up only during very cold spells. You'll often see them foraging for food, digging in forest seeps, or traveling along or in waterways.

SIGN: *Digs.* Raccoons are omnivorous opportunists, taking advantage of whatever food is available to them in their environment. They are known to eat hundreds of different plant and animal species, usually plants more often than animals. Corn growers are aware of the raccoon's fondness for sweet corn. Raccoons have a habit of knocking down the whole stalk to get to the ears of corn. All kinds of fruit also seem to appeal to them, as do nuts, grains, and seeds. Raccoons like to eat crayfish and will hunt small ro-

FIGURE 5.12
Raccoons, like skunks, often dig for ground bees. When investigating such a dig, be mindful that the bees might still be present—and angry.

dents and cottontails, some birds and their eggs, amphibians, turtle eggs, and fish. Raccoons often leave digs made in pursuit of yellow jackets, or ground bees (Figure 5.12). Because skunks do the same, it is very difficult to distinguish between the two animals' digs (look for hair and tracks). Like skunks, raccoons seem to be able to tolerate bee stings. My friend Bill Byrne is of the opinion that raccoons have a very high tolerance for pain. He has seen them climb an electric fence to get at a field of corn, then climb back out when they were finished.

SIGN: ***Scat.*** There is a definite danger in handling raccoon scat. In some areas, these animals carry a parasitic roundworm *(Baylisascaris procyonis)* that lives in the intestinal tract. The eggs containing the larvae are dispersed in the scat when the raccoon defecates. If these eggs are inhaled or ingested, they could cause serious harm to animals or humans. This is of such concern that many wildlife rehabilitators—people who take in sick or injured animals—will not accept raccoons anymore. Precautions also should be taken with children, where hand-to-mouth transmission is possible.

Although raccoons will defecate randomly, more often they will develop latrines. As mentioned earlier, raccoons travel along specific corridors and pathways. They often choose a large tree somewhere along the corridor as a latrine and defecate beside it. Other areas chosen for latrines may be on top of a stone wall, on a large horizontal limb, or in the fork of a tree close to the ground. Also look on fallen logs or under rock outcroppings. Latrines indi-

FIGURE 5.13
This raccoon latrine was found on a stone wall that the animals were using as a pathway. It contains a variety of typical scat samples. Raccoon scat usually measures between ½" to 1" in diameter, though it may occasionally reach 1³⁄₁₆".

cate high-traffic areas and usually that denning or resting areas are nearby. The larger the latrine, the more intensive the activity in that area. Figure 5.13 illustrates a raccoon latrine. Note the different types of droppings, suggesting that this latrine has been used throughout the year.

Scat may be reddish to yellow, black, or many shades of brown. It may be made up entirely of insects (often bees), grains, fruit (mostly seeds), or a combination of these. Black droppings often have an earthy appearance. Raccoon scat may have blunt ends and will break off bluntly. When dry, it crumbles easily. In shape, raccoon droppings range from tubular to ploplike, the latter usually when the animal's diet consists mainly of fruit and seeds.

Being fully aware of the concerns about handling raccoon scat, I exercise great caution when I do examine it by wearing gloves and using a stick when crumbling it. Black or brown earthy-looking droppings often have a very pungent, penetrating smell—an easy way to identify raccoon scat. Scat composed of grain smells like cereal, and that composed of fruit smells like the fruit. When smelling scat, there is a remote possibility of inhaling microscopic airborne eggs, particularly if there is a lot of scat and you are in a confined area. Again, be extremely careful.

CHAPTER 6: DOG FAMILY

Canidae

Young red fox kit outside its den.

Red Fox
Vulpes vulpes

LATE SPRING IS THE best time to observe red foxes. They are found throughout most of North America, except for parts of several western states (Washington, Oregon, California, Nevada, Arizona, New Mexico, and western Texas) and some arctic regions. A red fox may have several dens in its home range, but it doesn't always use them, often preferring to sleep in lays nearby. In spring, when the young are born, the foxes stick to their dens, and it's the perfect time of year to try to watch the kits playing outside. It's also a fine time to be in the woods: The leaves are just beginning to burst forth but are not yet so thick that they obstruct your view of the animals—or their view of you. It's a season of renewal and promise, a good time to remember that we share this planet with a great number of other creatures and that we are all partners in nature.

Foxes deserve their reputation for being secretive, wily, and cautious. They are the ghosts of the forest, and it's entirely probable that you'll be watched by a fox that will go entirely unnoticed by you. The tables, however, can be turned, and when they are, the pleasure of catching a glimpse of a fox family's private life is more than ample reward for the hours of patience and practice necessary to bring it about.

I remember one spring outing in particular because it led to one of my most haunting experiences with the red fox. It began on a whim. I left the house with my camera, planning to do some photography but with nothing specific in mind, and I decided to visit an abandoned woodchuck hole on the side of a southwest-facing hill not far from where I live. Some friends had been there a few weeks before, and they'd seen a fox in the vicinity, so I thought it might be a good idea to check it out.

Instead of heading directly for the den, I approached it like a fox. I made a wide arc, carefully working my way toward it from downwind. It was a beautiful, cloudless morning, and I could feel the sun's rays penetrating my jacket. I moved slowly and quietly, stopping often, until gradually, as I neared the den, I spent more time motionless than in motion. I knew that every time I moved, I separated myself from the forest and increased my chances of being seen and heard. As I remained still in the forest, I became still within myself. I blended with the forest. While

FIGURE 6.1
Red foxes are not always as red as the one pictured here, scratching itself. There are several color phases (black, silver, cross-phase) that are characteristic of individual animals and not the result of seasonal changes.

standing there, taking it all in, I gradually became aware of an odd-shaped shadow moving against a gray stone wall—a baby fox.

It ran along the wall and disappeared. I followed, and as I approached the den, I spotted two young foxes, their coats already beginning to turn from tan to red, which meant they were at least nine weeks old. A mound of yellow dirt told me where the den was, and I noticed a few small bones and feathers scattered around the base of it. I was no more than fourteen feet from the kits. One of them sat on its haunches and watched me curiously; the other lay sprawled out in a splash of sunlight, seemingly without a care in the world.

I began to circle the den, moving so slowly that sometimes I would remain motionless for an hour at a time, or until my legs went numb. I noticed two well-worn runs made by the adults in their hunting forays, and I kept an eye out for their return, but by midafternoon they still hadn't shown up. I was then within nine feet of the kits, and still they hadn't grown alarmed. Their routine consisted of repeated trips into and out of the den, interspersed with long stretches of just lying in the sun. Occasionally one would show some curiosity about me and begin to walk in my direction, but then caution would prevail, and it would stop and stare at me, tilting its head to one side. Gazing into its eyes was like stepping into a pool of innocence. If I so much as blinked, it would scurry back to the den, then poke its head out for another inquisitive peek.

Late afternoon set in, and I began to feel a little guilty for having taken up so much of their time. I waited until the kits were inside the den, then stole away as quietly as I could. I was overflowing with excitement as I hurried back to the house. When I told Paulette about my discovery, she wanted to see them, too. Although it was late—nearly 5:30 in the afternoon—and I felt I had disturbed the den enough for one day, I thought we could stay at a fair distance and watch them with binoculars.

When we arrived at the den site, the kits were inside, but as if on cue, they bounded out. We were about thirty feet away, and they seemed unaware of our presence. Suddenly they erupted into a frenzy of yipping and darting about, and it seemed that there were now more than the two I had watched during the day. We watched in silence, spellbound by the theater of nature taking place before us.

Eventually, a large adult fox, most likely the male, trotted along the main run toward the den, carrying something in its mouth. When the fox reached the spot in the run where I had crossed it that morning, it about-faced immediately and retreated some forty feet back along the trail; then it about-faced again and came toward us. This set the kits into an even greater frenzy, but the adult came no farther along the run than it had the last time. It did its about-face trick once more, retreated some distance, and then climbed a rock to get a better view of the terrain. I made the mistake of raising my camera, and the fox took off in great leaps. The kits scampered back into the den.

Paulette and I grinned at each other, glowing with the exhilaration of the scene we had witnessed. Then we headed home, leisurely walking along a path thickly carpeted with pine needles. Rounding a bend, we could see ahead to where the path formed a dark tunnel through the overhanging pines and hemlocks—one of Paulette's favorite spots—and there we saw yet another red fox, heading directly toward us. I touched Paulette's arm lightly, and we froze. The fox continued walking slowly in our direction, stopping once in a while, taking a few more steps, stopping again. We couldn't quite figure out what it was up to, but when it came to within 150 feet of us, it just lay down in the middle of the path and curled up. We were both shocked. What was this fox doing lying down in a well-traveled path?

We watched it for nearly half an hour. It didn't get up, though every so often it would stir, shifting its weight from one side to another, as if trying to get comfortable. We decided to stalk up on it while the daylight was still with us. We inched our way to within twenty feet of it, and it still made no attempt to rise. Could it be asleep? We crept to within ten feet, then seven. Then we stopped and waited. I was becoming a little concerned, but it finally began to move again. We watched, breathless, not knowing what it would do. Then, with what seemed to be a supreme effort, the fox rose slowly to its feet. It was a sad sight: horribly skinny, its fur dull and tattered, its legs almost too weak to hold it up. We could easily make out its entire skeleton beneath its dusty coat. At first it failed to notice us, and when it did, it veered drunkenly off into the woods, stumbling over deadwood and crashing into undergrowth. It took three tries before it managed to clear a stone wall not three feet high. Paulette stayed behind while I trailed it to a small cave that porcupines had occupied, and I called to Paulette to get some fruit. The fox was obviously starving to death. When Paulette returned with the fruit, we left it at the mouth of the den and continued home.

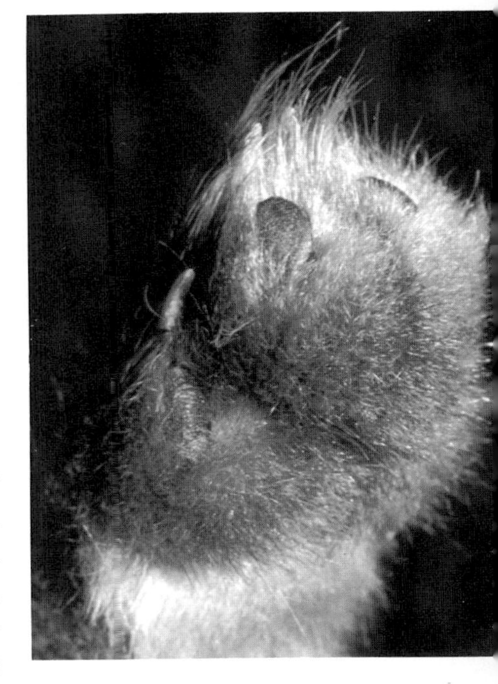

FIGURE 5.2
The front foot of the fox has five toes (only four show in the track) and a heel pad. The nails are nonretractable and blunt, and the bottoms of the feet are very hairy.

Daylight was draining from the forest as we picked our way along the path. Spring seemed to have left it, too. We had encountered the full circle of life in a single day—the young foxes in the morning, the acrobatic adult in the afternoon, and the dying old one in the evening—and we had been moved from joy to sadness within a few hours. We knew there could be no life without death, that the decaying forest litter beneath our feet was also the stuff from which new life would emerge, but it was a hard lesson. I promised Paulette that I would check on the sick fox in the morning. When I got to the cave the next day, it was still there, but it had not eaten the fruit. It would not eat again.

I think foxes are such interesting animals because they seem to be a link between the canines and the felines. In fact, there was originally some dispute as to whether foxes should be classed taxonomically as dogs or cats. Cats are direct-registering animals, and foxes are direct-registering animals. Foxes have eyes similar to those of cats; their pupils dilate elliptically, up and down, rather than in a round fashion, as dogs' eyes do. Gray foxes (*Urocyon cinereoargenteus*, which means "silvery gray–tailed dog") even have semiretractable nails and can climb trees, the only canid that can do so, which is why they are sometimes called the tree fox. All foxes are vocal, with a distinctly canine yipping bark. Thoreau described foxes ranging "in moonlight nights . . . barking raggedly and demonically like forest dogs."

FIGURE 6.3
This red fox front track, which registered in mud, shows four toes, a heel pad, and a line, or bar, that runs across the pad. The bar is unique to the red fox and is a good identifying feature.

TRACKS. A red fox's front foot (Figure 6.2) is extremely hairy, with the pads hardly showing through the hair, which greatly affects the appearance of the track. If you take a close look at a fox track in soft mud (Figure 6.3), you can often see the hair imprints. Red fox tracks in mud usually measure 2 9/16″ long with nails (2⅛″ long without nails) by 1⅞″ wide. An average width in snow may be 2″, especially when the animal is direct registering. Red foxes across the continent range widely in size. I found a fox in Alaska with a front track length consistently 3″ long with nails—an unusually big animal, or at least one with big feet. I also have recorded front-foot measurements as small as 2⅛″ long with nails. Red fox track measurements are usually 2⅛″ to 2⅞″ long with nails by 1⅝″ to 2⅛″ wide for the front feet, and 1¾″ to 2½″ long by 1½″ to 1⅞″ wide for the hind feet.

It's sometimes difficult to distinguish fox tracks from those of a small domestic dog, but there's one aspect of fox tracks that readily separates the two: the presence of a "bar" in the fox's front track. Sometimes this bar forms a straight line running across the heel pad (Figure 6.3); other times it can be shaped like a boomerang (Figure 6.4). It is not found in any other canine track. Some tracking books show a bar across the heel pads of both the front and rear tracks; some illustrations even show a more pronounced

FIGURE 6.4
These red fox tracks in snow show the smaller hind track (left) and the front track (right) Notice that the bar in the front track is boomerang-shaped.

bar in the rear tracks than in the front. However, the hind tracks rarely show a bar at all. Figure 6.5 shows the part of the heel pad that forms the bar in the front track. The bulbous heel pad of the hind foot (Figure 6.6) is far less likely to make a bar, although sometimes it does (Figure 6.7). Note the hind heel pad in Figure 6.8, where the heel pad barely touches the mud, forming only a small dot. This is common for both the coyote and the fox, which distribute most of the weight carried by the hind feet to the toes. The weight is more evenly distributed in the front feet. Because of this, the bar in the direct register is that of the front feet, not the hind. Although the hind feet fall on top of the front tracks, the hind heel pads do not come down far enough to erase the front heel pads.

Also note that the front feet, which carry the heavier load, are larger than the hind, as in all canines, although

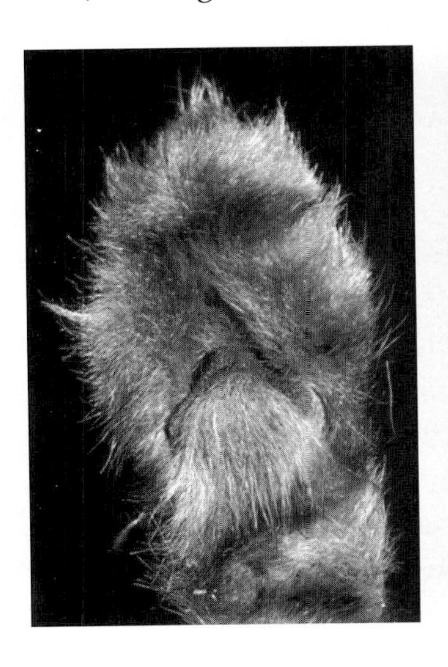

FIGURES 6.5 *(left) and* 6.6 *(right)*
The heel pad of the red fox's front foot (on left) has a boomerang or chevron shape that sometimes registers as a bar in its track. The hind foot (on right) has a rounder, bulbous pad that very rarely leaves a bar.

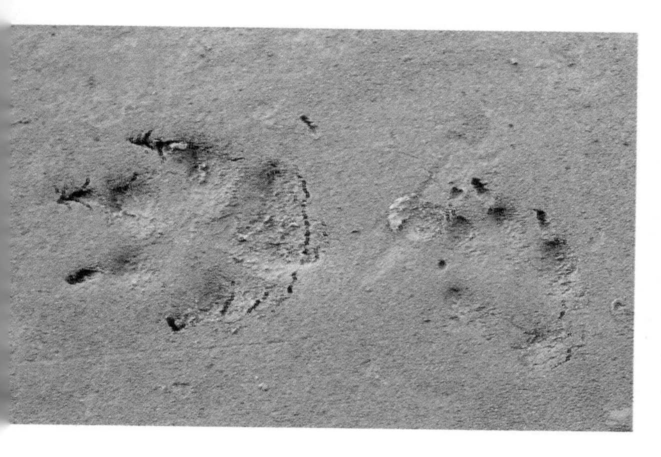

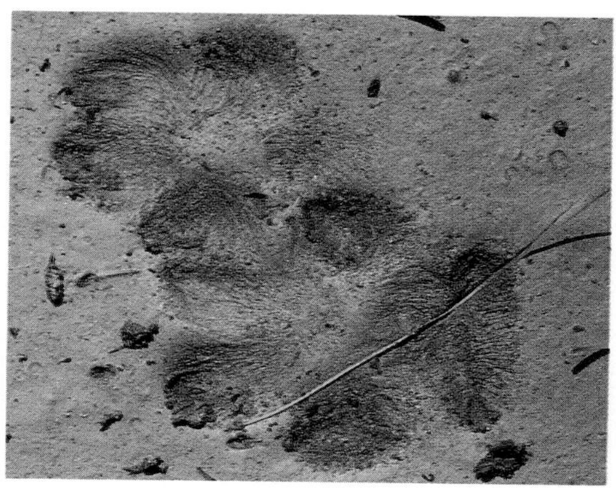

FIGURE 6.7 *(left)* *It is typical for a red fox's tracks in firm sand or other hard substrate to show lots of space between the toes. The smaller hind track (to the right) shows a thin, short bar at the heel, which is rare.*

FIGURE 6.8 *(right)* *The heel pad of the red fox's hind foot (track at top of photo) hardly registers in a firm substrate. Notice all the hair marks in both the front and hind tracks.*

fox and coyote tracks show a greater difference in size than do those of domestic dogs.

TRAIL PATTERNS. The red fox's trotting pattern (Figure 6.9) is easily identified by its regularity. A direct-registering animal, the fox trots almost in a straight line, with each mark representing two tracks—forefoot first, overprinted by the hind foot. Except in deep snow, a domestic dog is a double- or indirect-registering animal—the hind foot does not fall directly on top of the front track, but instead next to or partially on top of it (Figure 6.10). It rarely trots in a clean, straight line, but will wander and lope around as if it has no particular place to go. A fox usually moves as though it knows exactly where it is going.

The fox's average trotting stride is 14" to 16", measured from nail to nail or toe to toe, but it can vary from 13" to 18¾" depending on the size of the fox and its speed. Because of the typical straight-line pattern, the fox's trotting trail is very narrow, ranging from 2" to 3⅞". Other fox patterns, described below, are much wider.

There are variations in the fox's trotting pattern. When the fox shifts its rear end to one side, the pattern changes to a side trot with the hind track falling ahead and to the side of the front track (Figure 6.11). This 2-2 pattern is very similar to that of the weasel family. Domestic dogs, coyotes, and foxes all trot in this fashion: front-rear, front-rear, front-rear, front-rear. All the front tracks are on one side, and all the rear tracks are on the other. (The tracks of the weasel family do not line up this way.) If you have watched a dog running toward you, it looks as though it's running sideways, like an old jalopy with a bent chassis;

FIGURE 6.9
The red fox's trotting gait is usually a straight, precise, narrow line of tracks with strides between 13" and 18¾". Strides may occasionally reach 20½", at which point the fox's gait closely resembles a coyote's.

this is how this 2-2 pattern is made. Because coyotes, domestic dogs, and foxes all make a similar side trotting pattern, distinguishing between them can be difficult. Side trotting strides for the red fox are 14" to 21"; the group length is 4½" to 8½". As above, look for consistent patterns in wild canines and erratic patterns in domestic canines.

To measure track patterns correctly, it's important to know what the animal's gait is—whether it's trotting, loping, or galloping. Measuring the strides will not tell you how fast the animal is going; instead, the juxtaposition of the front and rear tracks is the best way to determine its speed. Once you know that, you will be able to apply your measurements to the correct trail pattern. The following sketches demonstrate the fox's various trail patterns as its speed increases.

In the direct-registering trotting pattern (Figure 6.9), each dot represents two tracks, with the hind foot coming down exactly on top of the track made by the front foot.

When the animal changes to a side trot, the hind foot comes down slightly ahead and to the side of the front foot. The faster the animal is going, the greater the overstep. Figure 6.11 shows increasing speed from a slow trot to a fast trot. When it's a slow trot, the rear track will be close to the front track, and when it's a fast trot, the rear track will be

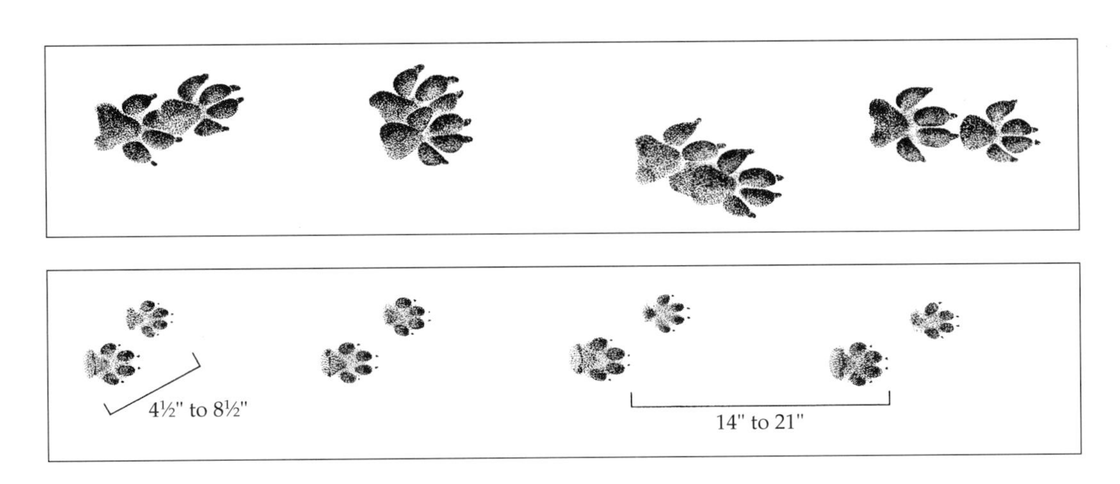

FIGURE 6.10 *(top)*
A domestic dog generally does not direct-register, and, compared to that of the red fox, its trail pattern looks sloppy.

FIGURE 6.11 *(bot.)*
In the red fox's side trotting pattern, the hind track (top row) falls farther ahead of the front track as the fox's speed increases.

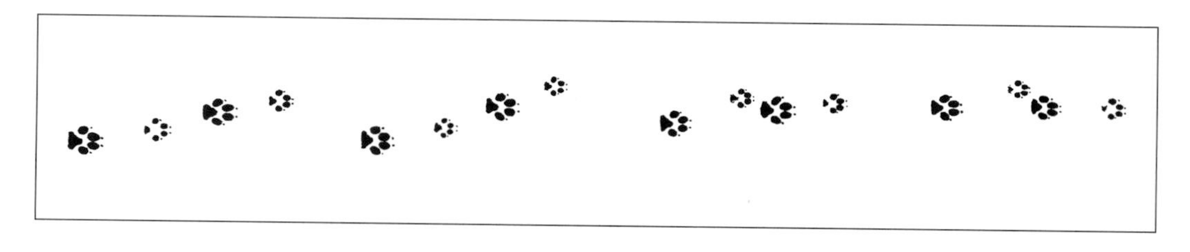

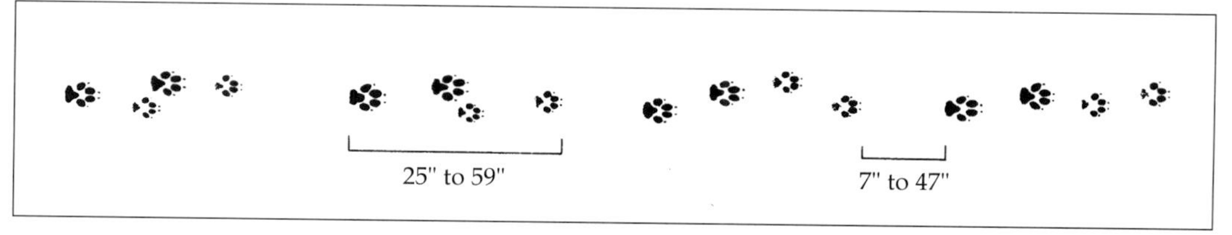

FIGURE 6.12 *(top)* *This loping (moderate running) pattern of the red fox shows an increase in speed from left to right. The sequence in groups of four is front-hind-front-hind. The smaller hind tracks move farther forward as the animal increases speed.*

FIGURE 6.13 *(bot.)* *This shows the red fox's running pattern increasing from a lope to a gallop (from left to right). The sequence in the first group of tracks, a fast lope, is front-hind-front-hind; in the others, front-front-hind-hind. As the speed increases further, the hind tracks move forward and the groups straighten out more.*

farther ahead. But the pattern is still front-rear-front-rear in a basic 2-2 pattern.

Figure 6.12 shows a lope with variations. In the lope, the two front feet hit the ground one after the other in rapid succession, followed by the two hind feet. The hind feet do not pass both front tracks, so the track pattern configuration is front-hind-front-hind.

Figure 6.13 shows the increase in speed transition from the lope to the C-gallop to the straight gallop. In any gallop, the hind feet pass the front, giving a front-front-rear-rear or large-large-small-small configuration. The coyote has similar track patterns; its galloping gait is also front-front-rear-rear. When the fox picks up speed in its gallop, it will gradually go into a straight line. Figure 6.13 shows the large-large-small-small pattern of the fox in a C-gallop, but it also shows the C-gallop losing its curve as the animal picks up speed. As animals move faster, their straddle decreases and the gait straightens out. Fox strides for gallops are 7″ to 47″, with groups 25″ to 59″.

In Figure 6.14, the alternating trot (center trail) is created when the fox's opposite diagonal legs move as a synchronized pair: the left front and right hind feet hit the ground at about the same time. You can see a bounce in the animal's movement when it trots. The two outside trails are gallops. The animal's body moves differently here: the two front feet hit the ground one right after the other, followed by the two hind feet one after the other or almost simultaneously. Because the speed of the animal is greater in the gallop than in the lope, the two hind tracks register ahead of both front tracks, creating a front-front-hind-hind configuration.

FIGURE 6.14
The red fox trail in the center shows an alternating direct-register trot, with the fox moving toward us. Note the set of tracks at the bottom left of the photo where the fox did not quite direct register. The two outside trails are C-gallops.

SIGN: *Dens.* Typical fox dens (Figure 6.15) have entrances as narrow as seven inches in diameter but more often are eight to nine inches. In spring, when the den is in use, you can expect to see a pile of loose dirt in front—unless, of course, the animals are using a rock crevice or cave. The kits can often be seen lying on the dirt mound. Evidence that the den is in use includes food remnants around the entrance: feathers, bone fragments, animal hair, and kit scat (¼″ to 5⁄16″ in diameter, usually containing hair).

Foxes often have more than one den and may use the same ones year after year. The dens are often found on south-facing slopes with good drainage and easy digging. Although cover is thought to be one of the most important

FIGURE 6.15
This typical red fox den, dug out of a sandy slope, has an entrance hole of about 7″ in diameter. Den sites might also include rock piles and small caves.

criteria for den selection, I've also seen them in open fields and beside busy highways.

Red foxes can dig their own dens, but they seem to prefer taking over those dug by other animals, such as badgers and woodchucks. In fact, foxes have been found using woodchuck dens with the woodchuck still occupying part of the tunnel network.

Den tunnel systems may be complex, often twenty to twenty-five feet long, and sometimes more than sixty feet long. The chamber for the kits may be lined with soft vegetation. I've often wondered whether such extensive tunneling was the work of the foxes or the system's original inhabitant.

SIGN: ***Lays.*** Lays are found near den entrances and are usually twelve inches in diameter (in snow). They also may be near a tree or other large object that serves as a windbreak and may be situated so as to catch the sun's warming rays.

SIGN: ***Scent Posts.*** You don't always have to see fox tracks to know that one is around; often you can smell it. The red fox's urine scent is strongest in January and February, but even at other times of the year, it is possible to detect the odor. In snow, look for yellow marks; foxes usually urinate along trails, on prominent objects either in the trail or beside it. The urine has a skunklike odor. You may have to get close to it to smell it, but if conditions are right, you can smell it up to thirty feet away. This is another good identification tool when you're following a small canine track in the snow and you're not sure of the species. Gray fox, domestic dog, and coyote urine has a very mild smell compared to that of the red fox.

Urine marks also can tell you whether the animal is male or female. If the snow shows that the animal squatted to urinate, it's a female; if it lifted its leg and clearly projected the urine, it's probably a male. Be aware, however, that females will lift their leg at a scent post to denote possession but will not project the urine. It is best *not* to have direct contact with fox or other wild animal urine (see pages 28-29).

SIGN: ***Scat.*** Like most canines, foxes tend to defecate right in the middle of a trail, often at the point where two trails cross. They seem to leave these scent posts in heavy traffic areas. Most wild canines have anal scent glands and are actually depositing scent on the scat. Some scientists believe that animals can identify other individuals from these scents. A likely place to find scat is on or near a prominent object, such as a stump, a humped ground root, or a frost-heaved rock. Also check places where there is a rise in the trail.

In summer, distinguishing the scat of different animals can be tricky because many so-called carnivores—the coyote, fox, fisher, marten, skunk, and raccoon—live almost entirely on fruits such as wild grapes, blueberries (Figure 6.16), raspberries, and wild cherries (Figure 6.17). During apple season, the scat of coyotes, foxes, raccoons, and other animals often is made up completely of apples; it will smell like fermented apples and have apple seeds and skin in it.

FIGURE 6.16
Fox scat in summer and fall may contain the seeds of whatever fruit is in season. This specimen is from a diet of blueberries.

When the berry season hits, these animals stop hunting and turn into "fruitarians," much to the relief of the rodent population. This type of scat also might lose its tubular shape, taking on all kinds of configurations. Add to this the fact that at this time of year, the young of these animals will be depositing scat of all sizes. Needless to say, berry-season scat can be hard to identify.

When winter comes, the animals are hunting again, and their scat can be easier to identify. Fox and coyote scat is almost always full of hair; domestic dog scat is usually all grains, because that's what domestic dog food is made of. Also, fox and coyote scat smells musky, not objectionable like some domestic dog scat. There can be quite a difference in smell between scat from domestic canines and that from wild canines, making scent an easy way to distinguish between the two types. I find it interesting that wild canines that get into garbage or are fed an unnatural diet leave scat that smells like that of domestic canines.

Distinguishing fox from coyote scat is trickier but not impossible. Look for size and composition. In winter, fox scat (Figure 6.18) usually consists entirely of hair from small rodents—voles, mice, squirrels, and chipmunks. There may be some small bones and even a few rodent incisors, but there are usually no large bone fragments; foxes don't have the jaw muscles to crunch large bones. The hair in coyote scat is usually from larger animals, such as snowshoe hares and deer. Large bone fragments present in coyote scat attest to their strong jaw muscles.

Hair and bones are circumstantial evidence. Diameter is the best distinguishing tool. In my own fieldwork, I

FIGURE 6.17
During berry season, scat can take on all kinds of shapes and sizes, and, consequently, it can be very difficult to identify. This fox scat is composed of wild cherries.

FIGURE 6.18
Winter and spring scat will be composed mostly of hair. It often has pointed ends and can vary in color between the two scat samples shown here.

have found that if wolves are not present, it is fairly safe to consider any canine scat ¾″ or more in diameter to be that of a coyote and scat ⅝″ or under to be that of a red or gray fox. Of course, any scat between ⅝″ and ¾″ will be ambiguous, so consider the other evidence noted previously.

In a coyote study I worked on in 1988 with Paul Lyons, wildlife biologist for the Metropolitan District Commission in Massachusetts, I found the diameter of eastern coyote scat to range from 9/16″ to 1⅜″, with an average of ⅞″. If I had used only the maximum diameter for each scat, my average would have been closer to 15/16″. In contrast, red fox scat averages ½″ to ⅝″ in diameter. A 1981 scat study conducted by Jeffrey Green and Jerran Flinders in Idaho found that the average diameter for coyote scat was ⅞″, with a range from 9/16″ to 1 5/16″; the average mean diameter for red fox scat was 9/16″ and the range 5/16″ to ¾″. They suggested that scat in their study area measuring more than 11/16″ be classified as coyote. Scat with a maximum diameter of ⅝″ to 11/16″ was considered not assignable to a particular species.

Another thing to watch for is color. With fox and coyote scat, black or dark brown indicates a fresh kill. The more blood and meat in the diet, the blacker the scat will be and the less hair it will have. When a deer is killed, the animals go for the organs first. Then, as they consume the carcass, their scat will show more and more hair and bones. Fresh scat also will have a mucuslike, shiny appearance.

Gray Fox
Urocyon cinereoargenteus

ALTHOUGH THE GRAY fox and the red fox are often confused, they are different genera. Both are about the same size, but the gray fox is more grizzled, having a grayish body and reddish patches on the back of the head and neck. It lacks the black "stockings" of the red fox. The most obvious difference is its tail, which sports a black streak down the back, ending in a black tip, as opposed to the white-tipped tail of the red fox. It weighs seven to fourteen pounds and can be up to forty-three inches long, counting the tail, which is bushy and eleven to sixteen inches long. Its height at the shoulders is about fourteen inches. The gray fox is found in most of the northern states, except the Midwest and parts of Maine, and only rarely in Canada (southern Manitoba, Ontario, and Quebec).

The gray fox has semiretractable nails and has retained the ability to climb trees, a trait that was confirmed for me one day while I was traveling on a dirt road in the Birch Hill Wildlife Area in Royalston, Massachusetts. As I was driving along, a gray fox appeared ahead in the road, running straight toward me. It veered suddenly to the right and, without slowing down, ran up a white pine to

FIGURE 6.19
The gray fox is most easily distinguished from the red by its black-tipped tail (the red's is white-tipped). In most areas, the gray fox has not adapted to people as well as the red fox, and thus it's usually found in more secluded rural areas.

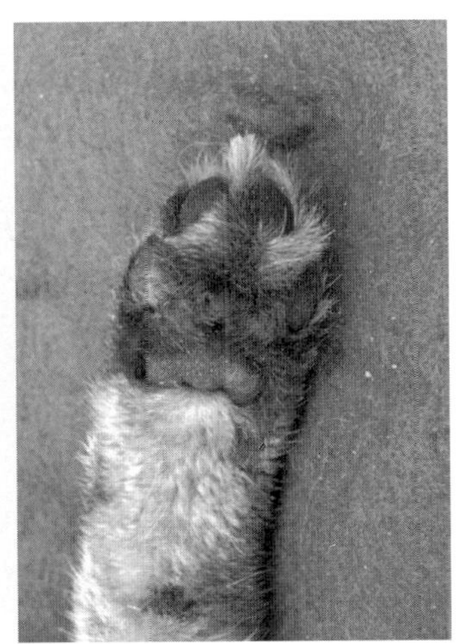

FIGURES 6.20 *(left)* and 6.21 *(right)*
The bottoms of the gray fox's feet are fairly hairy. Both the front (on left) and hind (on right) feet have four toes that register in the tracks, semiretractable nails, and a heel pad. Gray foxes in the north have larger feet than those in the south. Measurements here include both.

a height of ten or twelve feet. Then it immediately jumped down and quickly disappeared into the forest. I stopped the car and went over to the pine. A red squirrel was chattering excitedly from the top of the tree, and I wondered whether that was what the fox had been after. Gray foxes are also known to climb trees to avoid leaving a scent trail for enemies to follow. Some say it was this ability to evade capture by hounds that inspired frustrated fox hunters to import the earthbound red fox from England.

In most of its range, the gray fox seems to prefer deep, wooded areas and thick brush more than the red fox, but both species like to mix woods with open fields and brushy habitat rather than stick to large tracts of contiguous forest. I have often seen the trails of both foxes cross each other, and more than once I have tracked a gray fox that was scent-tracking a red fox.

TRACKS AND TRAIL PATTERNS. The gray fox's tracks are very different from those of the red fox. The gray's front tracks measure 1⅜″ to 2″ long by 1¼″ to 1⅞″ wide, whereas the red's front tracks are rarely under 2⅛″ long (all red fox measurements include nails). The gray fox's nails do not always register, especially those of the rear feet. Rear tracks measure 1¼″ to 1¾″ long by 1″ to 1¾″ wide without nails.

Another difference between the tracks can be seen in the heel pads. The gray fox's front heel pad (Figure 6.20) looks somewhat like a "winged ball," while the red's front heel pad (see Figure 6.5 in the red fox section) is more

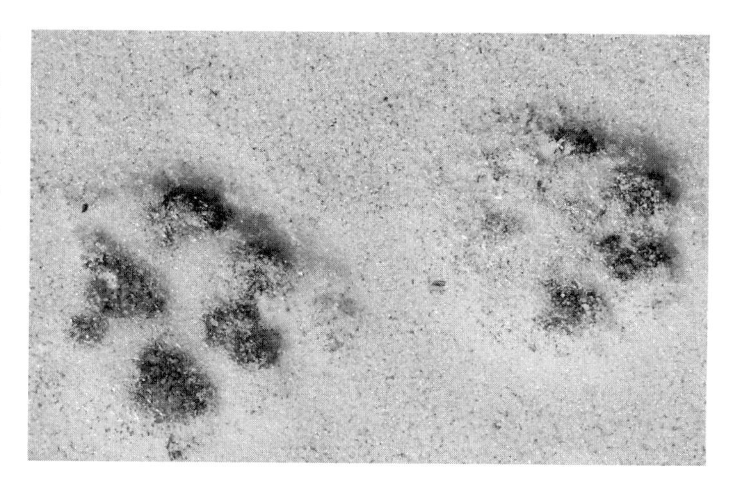

FIGURE 6.22
In these gray fox tracks in snow, the heel pad of the front foot (left) registers as a ball with small wings, while that of the hind foot (right) barely registers at all.

boomerang shaped. The red fox's rear heel pad (see Figure 6.6 in the red fox section) is not as well developed as that of the gray (Figure 6.21). The gray's winged ball sometimes shows up in its tracks (Figure 6.22), more often in the front tracks than in the rear, but not always (Figure 6.23).

Because the gray fox's nails are semiretractable and often do not register, its tracks can be mistaken for those of a cat. Its track size is between that of a domestic cat and that of a bobcat, and its trail pattern is very similar to a domestic cat's. Some biologists have suggested that the gray fox represents an evolutionary link between the canines and the felines. Gray fox tracks certainly display characteristics of both groups, as the overall appearance seems to say "cat" (Figure 6.22). A closer look, however, reveals a heel pad that is too small for a cat and a symmetry that is usually associated with canids.

FIGURE 6.23
Compare these front and hind gray fox tracks in mud to those in snow (above). The hind track is the smaller one on the left.

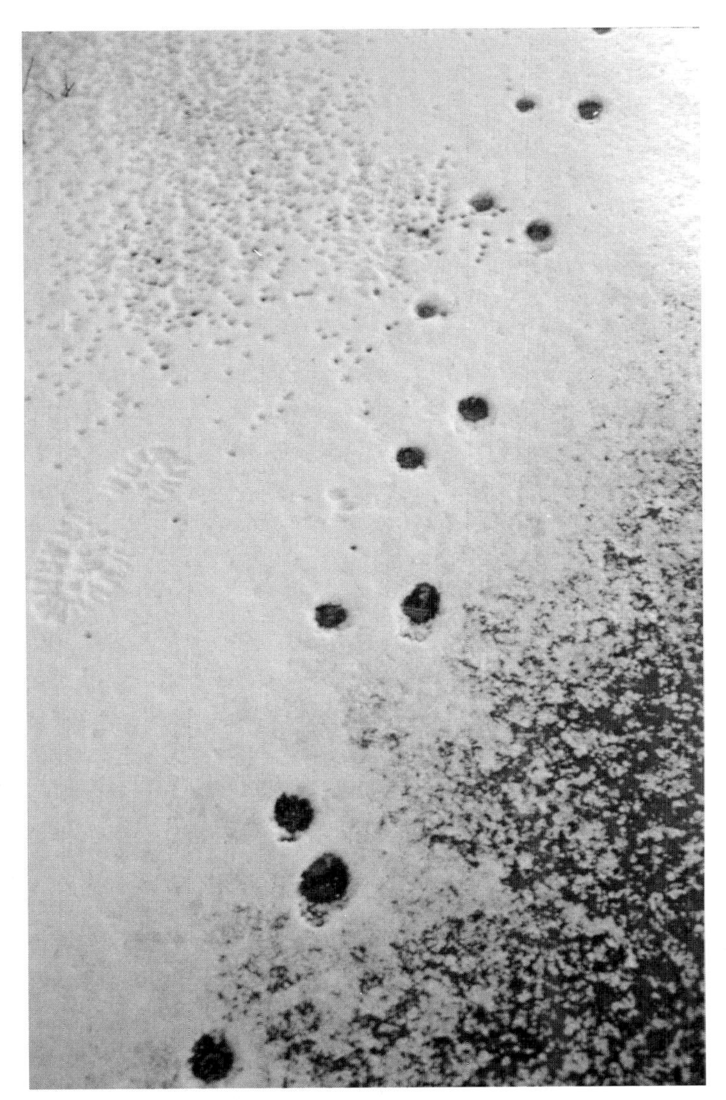

FIGURE 6.24
In this comparison of two canine trails in a dusting of snow, the gray fox's common trail on the left is wider and has shorter strides and smaller tracks than the red fox's trail on the right. However, under certain conditions, gray fox strides can reach up to 18". In those instances, the trail will look as straight and narrow as that of a red fox.

When I say that the gray fox's trail pattern is like a cat's, I mean that the domestic cat's strides and trail widths easily fall within the gray fox's range of 7½" to 16" for strides and 1¾" to 4" for trail widths. Although I've recorded a red fox's direct-registering stride at 10", that is very unusual. The red fox's trail width is usually narrower than the gray's, but, again, not always. As a general rule, when you compare red and gray fox trails (Figure 6.24), the gray (on the left) has a shorter stride and a wider trail than the red.

SIGN: *Dens.* A gray fox may enlarge a woodchuck den for its own use, but more often it will use a hollow tree, cave, rock crevice, or woodpile. Look for bone fragments, feathers, and other "table scraps" left around the entrance to one of these cavities.

SIGN: ***Scent Posts.*** Scent posts—on rocks, snow piles, small bushes, etc.—left by gray foxes have a different smell from those left by red foxes. Red fox urine has a strong, skunklike smell, whereas that of the gray is much milder. Remember that if both foxes have scented the same post, the red's scent will overpower the gray's. For example, if you're following what you think is a gray fox trail in newly fallen snow, you come to a scent post, and the smell is distinctly red fox, the red may have been there before the snow fell. If the smell is mild, it is gray fox urine. (Note: The urine of wildlife can be harmful, so take precautions when investigating it [see pages 28-29].)

SIGN: ***Scat.*** Gray fox scat is similar in size to red fox scat—½" to ⅝" in diameter. Since their diets are about the same, I haven't been able to find any way to distinguish between red fox and gray fox scat. Summer scat will contain whatever berries and fruits are in season; winter scat will be mostly hair.

Eastern Coyote
Canis latrans var.

Western Coyote
Canis latrans

Domestic Dog
Canis familiaris

THE CANINE FAMILY, which includes wolves, foxes, and coyotes, as well as dingos and jackals, is thought to have originated during the Pleistocene and to be descended from an even earlier family of arboreal mammals known as miacids. One branch of the canids—the wolves—migrated to Eurasia across the Bering land bridge and later returned to North America to drive their smaller cousins the coyotes south. When human activity eliminated wolves from the prairies, coyotes moved in. Coyotes are smaller than wolves, better adapted for running in open country, and less prey specific. Unlike wolves, which hunt in packs and live mainly on large game, coyotes are often solitary and can survive on anything from grass and grasshoppers to deer mice and deer.

There are many theories about the origin of the eastern coyote. One of the most widely accepted is that it has some introgression of wolf genes. These animals, in size between a wolf and a western coyote, then migrated east, expanding their range through Canada and into New York State in the 1920s, into New Hampshire and Maine in the 1930s, into Vermont in the 1940s, and into Massachusetts in the late 1950s. At first, some easterners thought they were coydogs—crosses between domestic dogs and true

FIGURE 6.25
The western coyote appears in many Native American cultures as both a trickster and a teacher. One legend states that "if all the creatures in the world were to die, the coyote would be the last one left," attesting to the animal's adaptability and intelligence.

coyotes—but we now know that when a coyote is crossed with a dog, its mating patterns change, making it a very unstable species.

Coyotes have a very small window during which they can mate, usually January and February, so the pups are born in the spring. During the birthing and early whelping season, the male cares for the female by bringing her food. Both the male and female care for the pups, occasionally assisted by an older sibling.

Domestic dogs may breed with coyotes, but the resulting hybrid has a delayed breeding season that results in the pups being born in midwinter. According to Jim Cardoza, wildlife biologist for Massachusetts and an expert in coyote concerns, "male hybrids, like male dogs, lack the parental care instincts typical of male coyotes," which means that single mothers left to care for four or five hungry pups at the harshest time of year make for a nonviable species. "These factors," says Cardoza, "militate against the survival of second-generation young." Wolf-coyote hybrids do not display this rephased breeding cycle, so the eastern coyote could be a cross between the wolf and the western coyote, and as such would be viable.

Another theory that I find interesting maintains that the eastern coyote is indigenous to the Northeast. Long ago, woodsmen there distinguished two kinds of wolves—large wolves and small wolves. They called the large wolf the eastern timber wolf and the small one the brush wolf, a term that survives in folklore but has no scientific credibility. Recent efforts to find archaeological evidence of wolves in Maine, however, have not yet confirmed that there were ever any timber wolves in that state. This suggests the possibility that what used to be called the brush wolf was in fact the eastern coyote and that these animals have always been there, surviving in small pocket populations in the eastern states and/or Canada and expanding their ranges when bounties on their hides disappeared and farmland reverted to forests. Paleontologists have found coyote remains in cave deposits and aboriginal village sites along the East Coast dating from the end of the last Ice Age to about one thousand years ago. This seems to add evidence to the theory that coyotes were in that part of the continent long before they were supposed to have migrated there from the West by way of Canada.

FIGURE 6.26
The eastern coyote is on average about ten pounds heavier than the western coyote and in some ways resembles wolves more than its western cousin.

In the West, when the western coyote inherited the wolf's territory, it also inherited its reputation. Likewise, by association, the eastern coyote has inherited this reputation. As a result, all coyotes, whether western or eastern, were hunted down as mercilessly as was the wolf. When the wolf was systematically eradicated from farmland in the Midwest, the coyote was lumped in with it. In 1946, seventeen western states began a coyote eradication program involving bounties, poison, and traps. In Arizona, $157,600 was spent in one year to kill coyotes thought to have caused $42,200 worth of damage. In the first year of the program, 294,400 coyotes were killed. In 1971, the U.S. government spent $8 million on the program. The money and effort were wasted, however; after twenty-eight years, coyote numbers had actually increased. In 1974, 295,400 coyotes were killed—1,000 more than were killed in the first year. Not only had the coyote expanded its range, but it also had proliferated. Today it is estimated that 87,000 coyotes are killed annually (a total of 4 million to date), but the coyote continues to do so well that some biologists claim that there are more coyotes than foxes in North America. No wonder Native American legends have "Coyote the Trickster" as the last animal left on earth when the rest of us have disappeared.

To put people's fears about coyotes into perspective, I've done some research on coyote and domestic dog attacks on humans. The best estimates assert that, in recorded history, there have been 20 to 30 coyote attacks on humans that resulted in injuries. A 1994 telephone survey conducted by the Centers for Disease Control and Prevention in Atlanta, Georgia estimated that 4.7 million people are bitten by dogs annually. In addition, the Humane Society of the United States reported that more than 300 people have been killed by domestic dogs in the U.S. since 1979.

Since the coyote doesn't seem to be going away, it's probably best if those who bemoan its presence simply learn to live with it, for recent studies show that coyotes are not the threat to domestic farm animals that they once were assumed to be. Researchers in Grand Teton National Park, in northwestern Wyoming, studied two coyote packs for a period of four years in the late 1970s and found that in summer, the coyotes lived entirely on small rodents—pocket gophers, field mice, and Uinta ground squirrels—and in winter almost solely on carrion from ungulates—deer, moose, and elk—that had died from natural causes or been killed by human hunters.

A more recent study conducted in the Lower Fraser Valley in British Columbia, undertaken specifically to determine whether local farmers' complaints about coyotes killing their sheep were well founded, discovered that small rodents constituted 70.2% of the coyotes' diet, rabbits 8.2%, and raccoons, opossums, muskrats, deer, plants, and insects most of the rest. Domestic sheep constituted only 0.2% of their total diet, and much of that could have been carrion from discarded sheep carcasses or stillborn lambs.

A photographer friend, Bill Byrne, took a picture of a coyote walking through a pasture very close to a supremely unworried-looking cow. The next picture shows the cow getting up and going over to the coyote, which has adopted a very submissive stance and even has its eyes closed as the cow approaches. Bill also has watched coyotes running between cows, catching field mice, and showing absolutely no interest in the cows. Now, I'm not saying that coyotes never kill livestock; they have, and they will. My point is that the portrait of the killer coyote is an exaggeration.

Both western and eastern coyotes have a long, bushy tail that hangs straight down, without the curl associated with a domestic dog's tail. Coyotes keep their tail down while running; wolves tend to hold theirs straight out, like foxes. The eastern coyote has longer hair and longer, bigger ears than the western. It's also a fuller, more muscular-looking animal. Eastern coyotes, according to a paper written by Gary Goff et al. and issued by the Department of Natural Resources at Cornell University, average thirty-five pounds for adult males and thirty-two pounds for adult females, or six and a half to nine pounds more than the average for western coyotes. This publication notes that few eastern coyotes exceed fifty pounds, but I've heard of one large male caught in Maine that weighed over seventy pounds. In fact, the Tweed wolves, weighing forty to forty-five pounds, and the Algonquin wolves, weighing sixty to eighty-five pounds, do not weigh much more than some eastern coyotes. So in terms of size, it could be difficult to distinguish between some eastern timber wolves and eastern coyotes from a distance.

Eastern coyotes tend to pack more than western coyotes, but not as much as wolves. When they do pack, wolves and coyotes exhibit similar social characteristics. There is usually only one mating pair in a pack, and all other members are offspring of that pair and do not mate themselves. There also may be one or more "orbit" or transient animals—older offspring of the mating pair that are not quite accepted into the family group but are tolerated, allowed to hunt and live in the same home range.

Here we should distinguish between home range and territory. A coyote's *home range* is the area it covers in a single year of hunting or roaming around; it may be a vast area, usually between two and a half and twenty-six square miles, depending on its gender and the availability of prey. The summer range is usually less extensive than the fall and winter ranges because of the abundance of food and the limited mobility of the young pups. A coyote's *territory* is a much smaller chunk of real estate—very private property that the coyote marks with scent posts and scat and defends against intruders, including other coyotes. A coyote's home range may overlap the home range of another

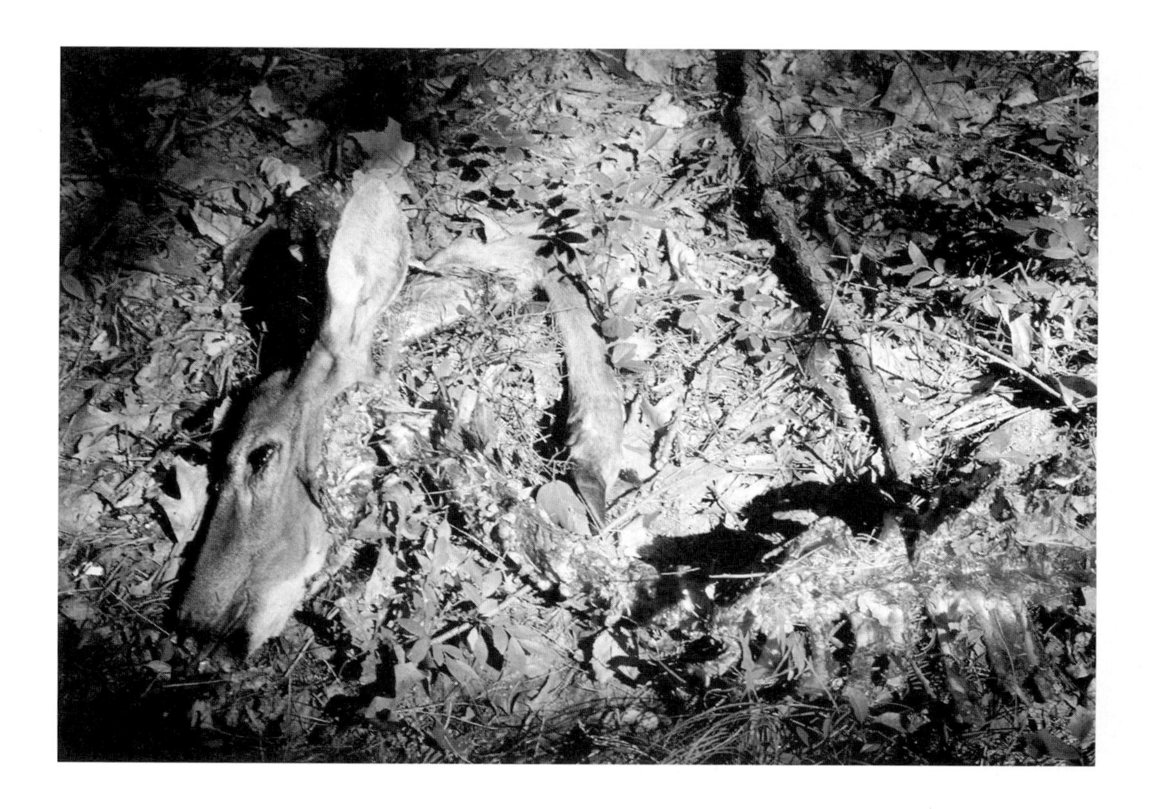

FIGURE 6.27
Coyotes have scavenged this white-tailed deer carcass. Note the broken ribs and that the head, leg ends, and spinal column are the last to be consumed. Eventually, the hair and stomach contents will be all that remains.

coyote; it's unlikely that its territory will. An orbit animal may share a mated pair's home range, but only the immediate family—the pups and previous offspring—will be allowed into their territory. When the female is having her pups, the dominant, or alpha, male will catch food for her (which she then eats and regurgitates for her young), as will other male members of the immediate pack. The orbit males will not do this. Orbit animals fall at the bottom of the hierarchy. Only when the immediate pack animals, in order of rank, have finished feeding on a carcass will the orbit animals come in and take what's left.

I've witnessed this pecking order in operation. I was working on a deer research project for the Metropolitan District Commission in Massachusetts. It was my job to track deer and determine what they were eating and how much of it was left. I was following a ravine up a hill away from the Quabbin Reservoir, when I caught some movement to the northwest out of the corner of my eye. I looked up, and there was a reddish coyote, looking in my direction. I was sure it couldn't see me, for only my head was above the top of the ravine, but as I watched, it ran off into the forest. As it disappeared, I thought I heard a few yips from the northwest. When I looked back, I saw a similarly colored coyote, also moving away, stopping for a second to look back in my direction.

When I looked to the west, I saw another coyote, a German shepherd–colored animal, slinking low to the ground, its tail low, moving south. I looked farther to the south, and there was a fourth coyote, moving north toward the other. These two met behind a tree and then continued north together, stopping to feed on a deer carcass. With my 9× binoculars, from a distance of about 150 feet, I watched them gorging for a few minutes. I could look right into their eyes, which slanted in toward each other and were a striking yellow. It seemed as though I was looking deep into the past.

After a while, I began to feel surrounded by coyotes. I could see movement in the forest, and shapes. I was pretty sure they were coyotes. The shapes came from the south and disappeared to the northeast. I watched the feeding pair for another five minutes, and then they walked off to the northwest. As soon as they were gone, another coyote came up from the south and started to eat. I decided to try to get closer. I ducked down behind the embankment, edging sideways until I had a tree between myself and the carcass. Using the tree as a blind, I moved very slowly about twenty-five feet closer to the feeding animal. After ten minutes, this coyote sauntered off into the woods with a deer leg in its mouth. That's when I realized that I had been watching a pecking order.

When no more coyotes appeared, I went to take a closer look at the deer carcass (Figure 6.27). I don't know whether the coyotes had killed this animal, but the broken rib bones indicated that it was, indeed, coyotes that were scavenging the carcass. Neither bobcats nor foxes would have broken the rib bones in this manner. Domestic dogs do not eat much on a carcass; they'll play with it and carry some of it off, but they're more used to eating cereal at home. Coyotes will pick the bones clean. They usually start at the hindquarters where the skin is thin and there is easy access to good portions of meat. They eat until only the head, the legs from the knees down, and the rumen (stomach contents) are left. Eventually, they will carry off even the head and the lower legs, leaving only the rumen and some hair at the kill site.

I had to take samples from a femur, the rumen, and the jaw for analysis at the biologist's lab. In this case, we found that the femur had plenty of fat in it, which indicated that this deer had probably not been in danger of starving. But the teeth were ground down to the gums, in some cases even past the gum line, so this was a very old deer. It may have slowed down, and the coyotes might have sensed that and taken it. That's where predators fit in. Animals live their best years and usually die quickly, without suffering for years on end. If there are predators around, injured, old, and weakened animals are taken out of the system, leaving healthy animals to keep the population viable.

TRACKS. Western coyote tracks are shown in Figure 6.28. The front foot of the coyote is larger and rounder than the rear. Like all canines, it has five toes on the front foot; the first toe—a dewclaw—is raised higher on the leg and usually does not register in the track, although I *have* seen it in a high-speed gallop. There are four toes on the coyote's rear foot.

The track sizes of various species of canids, including the nails, are as follows:

Animal	*Front Track*	*Rear Track*
Timber wolf	3⅞″ to 5½″ long by 2⅜″ to 5″ wide	3⅛″ to 4¾″ long by 2¼″ to 4¼″ wide
Red wolf	3″ to 4⅛″ long by 2³⁄₁₆″ to 3″ wide	
Western coyote	2⅜″ to 3⅛″ long by 1⅝″ to 2⅜″ wide	2¼″ to 2¾″ long by 1½″ to 2″ wide
Eastern coyote	2⅞″ to 3½″ long by 1⅞″ to 2½″ wide	2½″ to 3″ long by 1⅝″ to 2⅛″ wide
Red fox	2⅛″ to 2⅞″ long by 1⅝″ to 2⅛″ wide	1¾″ to 2½″ long by 1½″ to 1⅞″ wide

Eastern coyote tracks (Figure 6.29) are oval. The front nails often are close together and sometimes slant toward

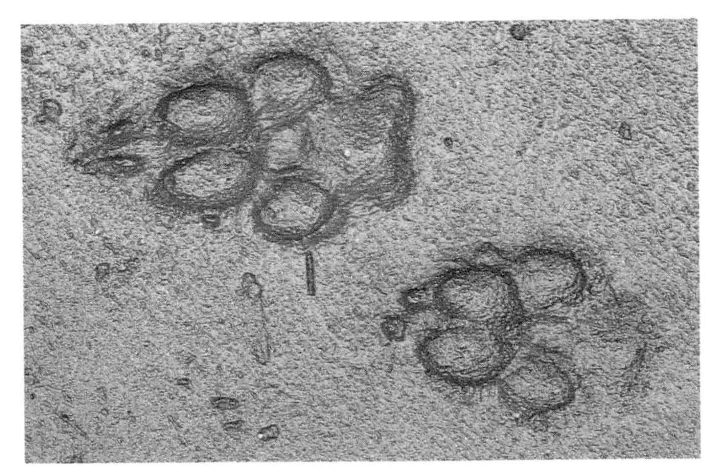

FIGURE 6.28
These front (upper left) and hind tracks of the western coyote have registered in mud. Notice the elongated, oval shape of the front track and how the heel of the hind track hardly shows.

each other. In contrast, the front nails of domestic dogs tend to spread out more and sometimes slant away from each other. This is especially noticeable in the walking or trotting pattern. The side nails of coyotes have a tendency not to register; those of domestic dogs usually do. The heel pad of the coyote is a good distance away from the front toes; that of the domestic dog is usually close to the front toes. A dog's track (Figure 6.30) is rounder, more robust-looking, and more splayed. The inner toes, sometimes called the front toes, spread off in different directions and do not point straight ahead. The outer toes also spread off to the side, like wings, and the nails seem to be pointing off to the side. Sometimes the heel pad of the eastern coyote will have two projections to the rear and to the outside of the pad (see Figure 6.29).

In the front and rear tracks of a coyote (Figure 6.31), you can see that the front heel comes down firmly, while the rear

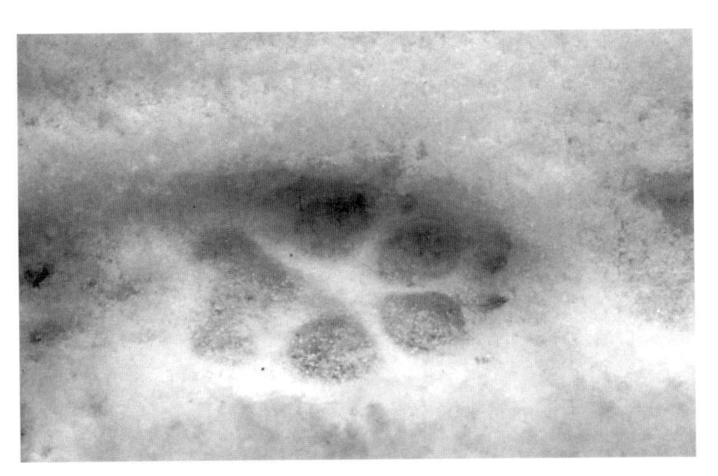

FIGURE 6.29
In this eastern coyote's front track, the front toes are parallel, with the nails held close together. The side toes seem to point straight forward and are tucked in close behind the front toes, creating the oval shape characteristic of coyotes.

Figure 6.30
Typical domestic dog tracks have an overall round appearance, especially evident in the front track (bot. right), where the four toes splay outward in different directions. The slightly smaller hind track is to the top left.

makes only a small mark. Not as much weight comes down on the rear heel. This also separates coyote tracks from those of the domestic dog. The domestic dog usually comes down more firmly on the heel, and the rear heel pad makes a much more pronounced indentation.

Figure 6.31 shows the tracks of two coyotes, a male and a female, trotting together. The front and rear tracks at the bottom are from the male, and the front and rear tracks at the top are from the smaller female. Note the rear tracks to the right. The front nails are very close together, and the front toes point either straight ahead or in toward each other. The side toes are tucked in behind the front toes, very close together, with the side nails not registering. The rear heel pads are just dots, and there was not much weight on

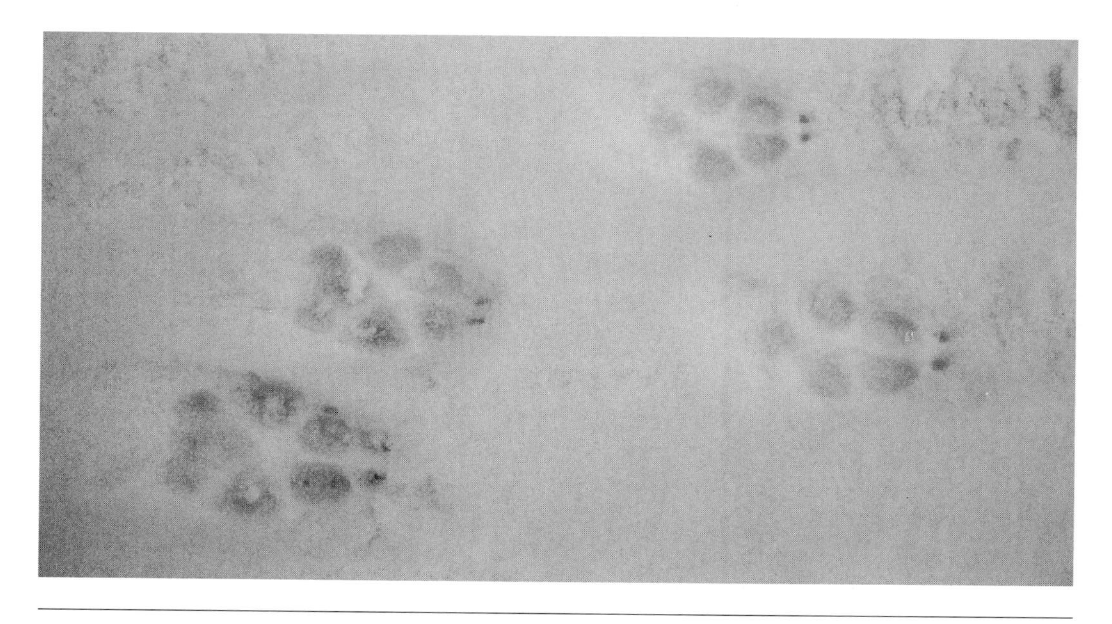

Figure 6.31
These are the tracks of a male and female eastern coyote trotting together. The two front tracks are to the left, the two hind tracks to the right. Looking closely, you can see that the heel pads of the hind feet show only as dots.

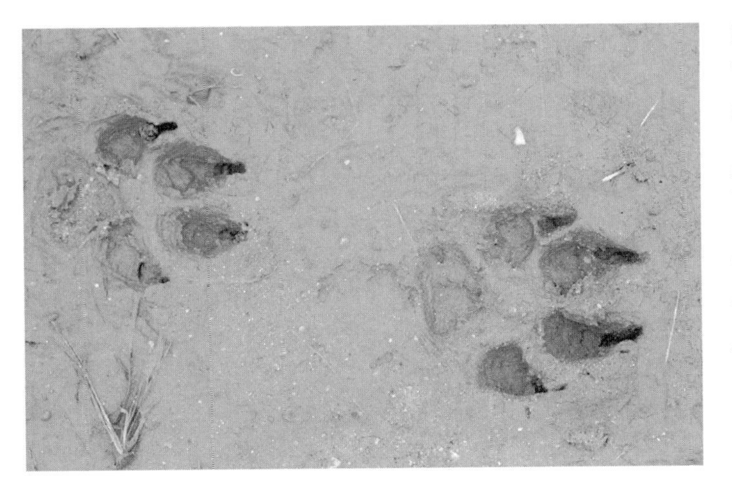

FIGURE 6.32
The hind (on left) and front tracks of the domestic dog.

FIGURE 6.33
When trotting, the eastern coyote usually leaves a fairly straight trail, with the hind foot directly registering on top of the front.

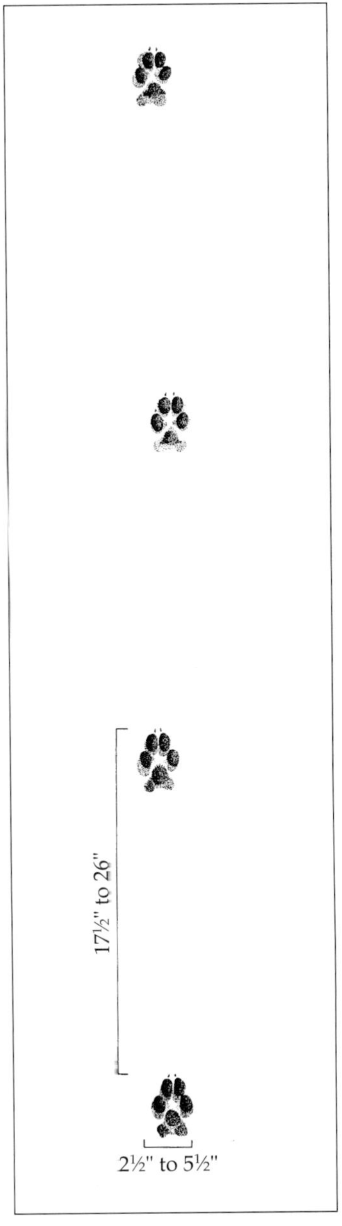

the rear heel. The front track shows good registration of the heel pads; the front nails hold close together, with some side nails not registering. These are coyote tracks: elongated, oval, and streamlined.

Compare these tracks to the tracks of a domestic dog (Figure 6.32). These are robust: round, outer pads pointing off to the sides, front nails spread way apart, heel pads close to the front toe pads. Both heel pads register strongly, making it hard to tell the front from the rear (the rear track is the smaller one on the left).

There are times when coyote tracks will resemble those of a dog. When a coyote is traveling in very coarse, granular snow, the snow may spread the coyote's toes so that its tracks look like dog tracks. Sand also will do this, especially if the coyote is in a high-speed gallop. Because of the high speed, the toes are spread—the two front toes digging in deeply and the side toes spread for traction. A plume extends out from the middle of the tracks, and the heel pad looks like a thick bar, similar to that in fox tracks.

TRAIL PATTERNS. The eastern coyote's trotting gait (Figure 6.33) is very similar to that of the red fox. Both are direct-registering animals, with one track directly on top of the other. The coyote's stride, however, is usually quite different from that of the red fox. The former is 17½" to 26", with an average of about 19" to 21", whereas the latter is anywhere from 13" to 18¾". There may be a little overlap at 17" to 18¾", but that's uncommon. Also compare the coyote's trail to that of the domestic dog (Figure 6.34). The dog, a double-registering animal, leaves a very sloppy trail.

FIGURE 6.34
Generally, the domestic dog does not direct-register, so its trail appears somewhat sloppy compared to that of the coyote.

In deep snow, one coyote may take the lead, and the others will follow very precisely in its tracks—so precisely that it may look as though there was only one coyote. I have never seen domestic dogs do this. Although coyotes don't always travel in this manner, when they do, it's a good way to distinguish their trails from dog trails, but remember that wolves and bobcats do this as well.

Figure 6.35 shows a coyote in a side trot: front-rear-front-rear. All the big front tracks are on one side, and all the small rear tracks are on the other. This is a very disciplined, precise trail, and it can be held for quite a distance. Sometimes a dog will fall into a disciplined trail like this, but not for very long.

For the side trot, eastern coyote strides range in length from 20½″ to 30″, with an average of 26″. Group lengths are 6½″ to 13¼″, with an average of 9″. Some Alaskan coyote tracks I measured also fell within these ranges. Comparing these measurements with those of a fox (the fox's stride ranges from 14″ to 21″, and its group lengths are 4½″ to 8½″) provides good criteria for distinguishing between eastern coyote and fox side trot patterns. Western coyote

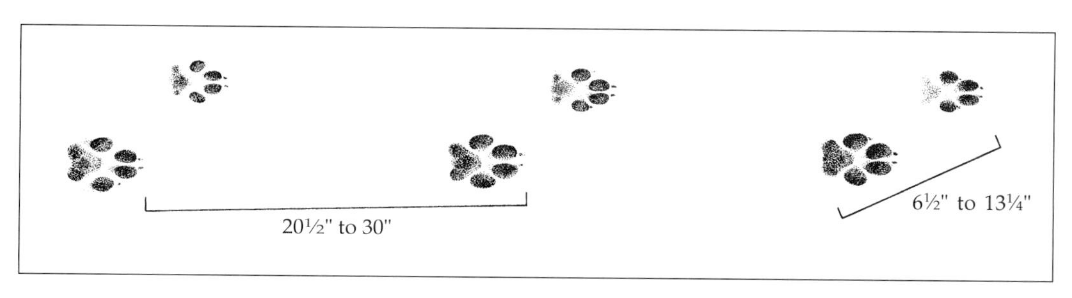

FIGURE 6.35 *(top)*
In this trail pattern of a coyote in a side trot, the larger front tracks are all on one side (at the bottom) and the hind tracks are above.

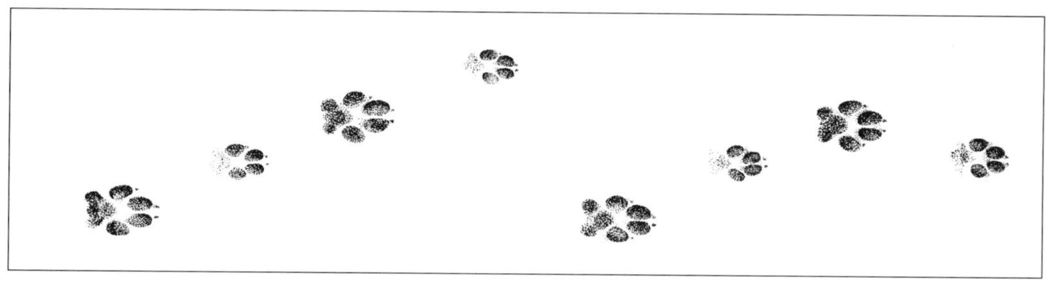

FIGURE 6.36 *(bot.)*
The coyote also travels in a slow lope. In these two variations, the sequence in each group of four is front-hind-front-hind.

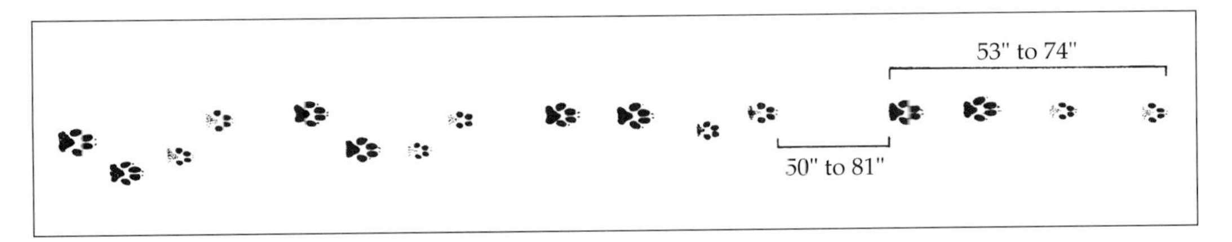

FIGURE 6.37
Starting from the left, each group of tracks shows an increase in speed. The track sequence, front-front-hind-hind, indicates a gallop. The faster the gallop, the straighter the pattern. Measurements are for a high-speed gallop only.

measurements are slightly smaller than those for eastern coyotes, with a little more overlap with fox measurements.

Figure 6.36 demonstrates tracks from a coyote in a lope. The tracks are very close together, but if you look at the individual tracks (in the groups of four), you can see they are front-rear-front-rear, front-rear-front-rear. This is a slow lope, only slightly faster than a trot, as the groups of four are very close together. As the animal gains more speed (Figure 6.37), the pattern can change to a C-gallop: front-front-rear-rear. When the coyote is in a full gallop (Figure 6.37), the front-front-rear-rear pattern holds but spreads out into a straight line, causing the trail to narrow as speed increases. The groups of four tracks in these high-speed gallops will range in length from 53″ to 74″. The strides between groups may be 50″ to 81″. Coyotes using these gaits can hit speeds of more than 40 miles per hour. The red fox's group lengths can be anywhere from 25″ in a slow gallop to 59″ in a high-speed gallop, with an average of around 37″. The strides between the groups of four can be anywhere from 7″ to 47″. Domestic dog trails may fit into any of the above dimensions. The main difference between wild and domestic canines is in the domestic dog's inconsistent, sloppy trail.

SIGN: *Dens.* Coyote dens can be found in areas where there are rocky ledges or steep or brush-covered slopes. Coyotes commonly move their pups from one den to another. A den also may have more than one entrance, plus a system of interconnecting underground tunnels. Coyotes sometimes use the same dens year after year, and as with foxes, they will use abandoned dens of other animals. Look for a mound of dirt outside the entrance hole and signs of activity (tracks, pieces of bone, fur, etc.).

SIGN: *Lays.* Eastern coyotes like to follow trails. When they have a choice between deciduous and coniferous forests, they seem to prefer the latter. When coyotes are moving, they often travel in single file, but when they reach the conifers, they usually spread out. When moving through snow-covered hardwood forests, they tend to remain in single file. This may be because the snow is deeper in hardwood forests and it's easier for them to walk in single file. It also could be a hunting strategy, since deer are more likely to make their bedding areas in coniferous habitats and spreading out is a more effective way of checking for them. I have often found coyotes checking deer beds.

It might be worth noting that pairs of coyotes, especially in snowshoe hare country, will split off from each other with thirty to two hundred feet between them, walk in parallel for some distance, come together for a short while, then split up again. No doubt this is a hunting strategy designed to catch hares.

Coyotes also bed down in conifers. One February, in one of my backwoods tracking programs in which we find the trail of a predator and follow it for the day, we came upon the tracks of a coyote and a fisher. We took a vote and decided to track the fisher. The trail led us up the side of a small mountain to an ear-shaped depression with a flat bottom, ending at a short precipice. In this bowl, we found ten coyote beds between fifteen to twenty-two and a half inches in diameter. A red fox had visited the area, apparently without any inhibitions about being in a coyote pack's bedroom. What I found really interesting about this bowl was that by standing on that precipice, I could see the whole valley below. Any approach to that hill would be detected very early; every sound in the valley would come right up the hill, and the wind would carry any scent up the hill as well. The coyotes had picked the most strategic area in which to hole up.

SIGN: ***Kill Sites.*** One winter I was tracking deer with my snowshoes on. The snow was deep, and it was very cold. I was following a heavy deer run down through some hardwoods, just where it entered some conifers. Suddenly, up ahead in the conifers, I saw something that froze me in my tracks. It was a beautiful silver coyote. I've seen only two in my life, and to me they're the most beautiful of the coyotes. This one was just standing there, looking at a deer, which in turn was looking at the coyote. I noticed that there was blood on the snow around the deer and that

FIGURE 6.38
This white-tailed doe was killed by an eastern coyote. As is evident here, coyotes usually enter a large carcass from the rear.

its right hind leg was broken, dangling at the joint. The two animals were less than ten feet apart, perfectly motionless. The whole forest stood still.

All of a sudden, the coyote attacked the deer, but instead of running, the deer charged the coyote—and the coyote ran! The coyote tried to run behind the deer, but the deer turned and kept the coyote on the defensive, running at it. Then the coyote tried to get behind the deer again. It was running in circles, and snow was flying everywhere. Suddenly, the coyote stopped and simply walked away, as if nothing had happened. The doe stood in silence, looking in the direction in which the coyote had disappeared.

About ten minutes passed. The doe and I did not move. My toes and fingers were getting numb. Fifteen minutes. No sign of the coyote. The doe slowly began to move. Although it had shown great agility before, it now looked weak and vulnerable. It moved maybe seventy-five feet, browsing a little as it went.

Suddenly, the coyote was back, moving slowly and low to the ground toward the doe. Once again, it burst forward. The deer turned quickly to meet attack with attack. The coyote turned away, circling to the doe's rear, but the deer also turned. They twisted about in a dance of life and death. After five minutes, the coyote stopped and again disappeared into the forest, but this time there was no question about the outcome. The coyote knew that the deer was in trouble. There was no sense in wasting energy. Its ancient wisdom told it to be patient, test the doe's strength from time to time,wait for the right moment.

It was getting darker and colder. Knowing the hike back out was long, I left, reluctantly. The next morning at sunrise I was back. Blood covered the snow not far from where I had witnessed the confrontation. A red-tailed hawk flew up and left the carcass as I approached. The coyote had taken the doe (Figure 6.38) and had entered the carcass from the rear, going for the soft organs first. A fetus had been pulled out and left beside the doe. It looked as though the doe's leg had been broken for some time, for the bone was worn at the tip. Native Americans believe that each animal has been sent to perform a certain task. Here, the predator had performed its function. The doe was in distress and losing life; the coyote was a blessing in disguise—the surgeon of nature, cutting out the suffering of life. And although the doe did not give birth to her fawn, maybe she helped sustain the coyote's pups. Her blood nourished a red-tailed hawk and many other creatures in the forest.

SIGN: *Scat.* Like most canine scat , coyote scat (Figures 6.39 and 6.40) is usually found in the middle of trails, sometimes at a high point, on a stone, or on some other raised object. It also may be deposited near these objects. Another good place to look is where trails cross. A trail sporting several coyote droppings usually indicates high coyote activity.

Coyote scat usually has a mild, musky odor, similar to that of a fox but unlike a domestic dog's. As you probably know, dog scat can stink to high heaven, making differentiation between coyote and dog scat quite easy. The smell of coyote urine is unlike the skunky odor of fox urine but similar to that of the domestic dog. Some say coyote urine smells stronger than dog urine, but I haven't been able to substantiate that.

Coyotes, as well as foxes and dogs, make elongated, narrow scratches with their front and hind paws after urinating or defecating. This may have something to do with scenting the ground with scent glands in their paws. There is some evidence that coyotes do more scratching closer to the center of their territory than on the periphery. These scrapes can sometimes be 3′ long and only 6″ wide.

Coyote scat may look very similar to fox or wolf scat. If these animals have been eating the organs from a fresh kill, the scat will be very dark, smooth, and wet-looking; be somewhat loose in consistency; and have little hair and few bone fragments. Summer scat may contain the same fare as fox and raccoon scat—fruit in season, if it is available. Scat composed of fruit may have unreliable diameters, so winter scat is probably the best bet for identification. Winter scat of wild canines contains mostly hair; if it has large pieces of bone, it probably is not from a fox. From here on, the diameter of the scat is the best criterion for identification.

Scat diameters for western coyotes are as follows: Arizona, ⅜″ to 1⅛″ (average, ¾″); Idaho, 9⁄16″ to 1 5⁄16″ (average, ⅞″). For eastern coyotes, scat diameters range from 9⁄16″ to 1⅜″ (average, ⅞″). Thus, some western coyote scat is slightly smaller than eastern coyote scat. When comparing coyote and fox scat, it is safe to say that most scat over ¾″ in diameter is coyote, not fox, with only a small margin of error. Scat over ⅞″ in diameter is coyote. Scat under ½″ or from ½″ to ⅝″ in diameter is probably fox, although some coyote droppings may fall in that range. In my experience,

FIGURE 6.39
The winter scat of coyotes is most often composed of hair and at times may contain bone fragments. As long as there are no wolves in the area, scat close to 1″ in diameter with large bone fragments, like this one, is unmistakably that of a coyote.

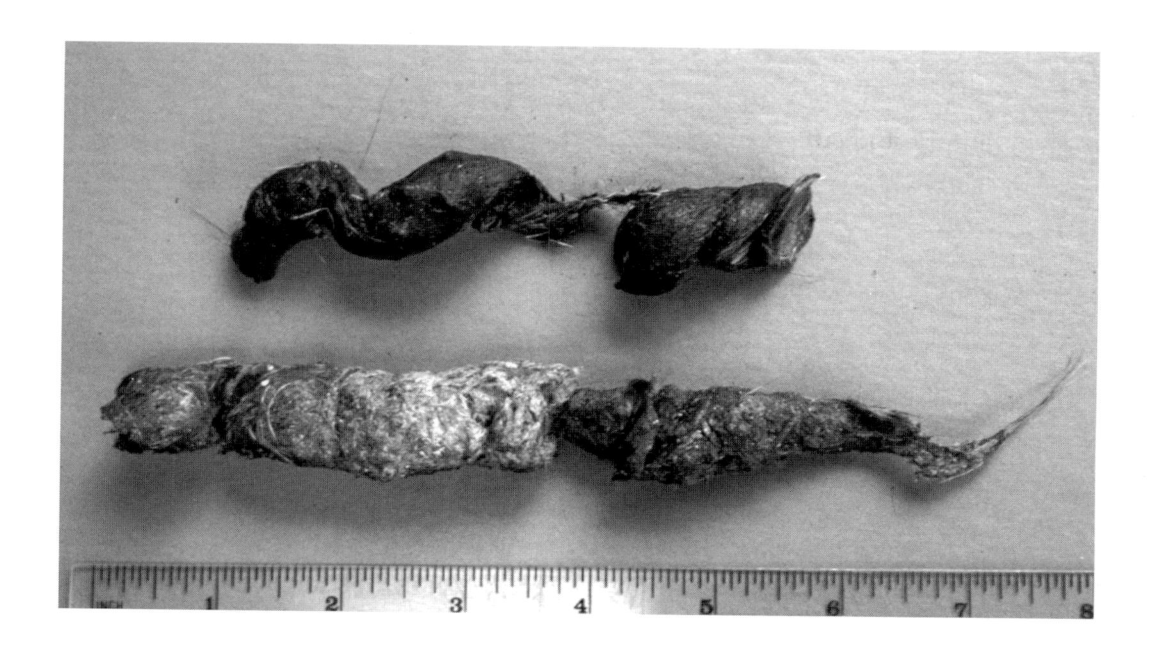

FIGURE 6.40
Coyote scat in the winter may look twisted (top) or have a tapered end (bottom).

very few coyote droppings are under 5/8″, maybe less than 10%. Scat from 5/8″ to 3/4″ in diameter is ambiguous without supporting evidence from tracks or large pieces of bone in the scat.

Wolf scat and coyote scat have considerable overlap in diameter. In a study conducted in 1979, John Weaver and Steven Fritts measured 616 western coyote scat samples from Wyoming and 290 wolf scat samples from Minnesota. They found that the coyote scat range was 1/4″ to 1 5/16″ in diameter, with an average of 13/16″, and the wolf range was 1/2″ to 1 7/8″ in diameter, with an average of 1 1/16″. Two-thirds of the wolf scat was under 1 3/16″, however, so scat between 9/16″ and 1 3/16″ could be either wolf or coyote. Scat over 1 1/4″ may safely be considered wolf, but this identifies only a small proportion of the animals.

Gray Wolf
Canis lupus

Red Wolf
Canis rufus

THE GRAY OR TIMBER WOLF is the largest wild dog in North America. An adult can weigh 60 to 130 pounds, although the largest North American gray wolf on record was a male from east central Alaska that weighed 175 pounds. The adult measures four and a half feet to six feet in length, including its straight, bushy tail; its height at the shoulders is twenty-six to thirty-eight inches.

In color, the wolf varies from white through creamy, tawny, rufous (reddish), gray, and black, with gray predominating (hence its name). The color seems to lighten in the northern extremes of its range—Alaskan and Canadian tundra wolves tend to be lighter than those farther south—though black individuals exist in the Arctic as well. No matter what the color, in some cases the animal's markings are so distinctive that researchers can recognize individuals easily.

Although wolves are built for running (successful carnivores have to cover a lot of territory to keep their larders full) and are capable of great endurance (they can easily travel twenty-five to thirty miles in a night), they are not fast runners. Their top speed is about 25 miles per hour, which they can maintain for only about twenty minutes, after which the animal usually has to lie down to recuperate. There are reports of wolves running 35 to 40 miles per hour when chased by vehicles on frozen lakes, but these speeds are extremely rare.

FIGURE 6.41
Because an increasing number of people understand the important role that predators play in the scheme of nature, the gray wolf may now have a better chance of survival.

The wolf is a very intelligent animal, with a large skull and a brain larger than that of the domestic dog. Its eyesight and hearing are keen but not as acute as its sense of smell, which is one hundred times better than man's. A wolf can detect another animal three hundred yards downwind, sometimes a great deal farther. World-renowned wolf biologist David Mech observed wolf and moose interaction in Isle Royale National Park. He reports that when wolves caught the scent of a moose, "the animals would suddenly stop and point stiffly upwind; then they would assemble, nose-to-nose, wag tails for ten to fifteen seconds, and veer straight upwind toward the moose. Once they did this when 1.5 miles downwind of a cow moose and twin calves."

The wolf was once the most wide-ranging mammal on the planet, with huge populations on every continent except Africa. In North America, wolves were found from Mexico north to the twentieth parallel. Since the arrival of Europeans, wolf-eradication programs have been very effective. In northern Montana alone, between 1883 and 1918, 80,000 wolves were poisoned, shot, or trapped under a state bounty system. Today the wolf has been eliminated from most of its former range in North America. Wolves still inhabit Canada and Alaska, but reliable population estimates are difficult to come by. Current statistics for the lower 48 states estimate approximately 2600 wolves. Recent reintroduction programs are helping wolves reestablish themselves in their former range in Idaho, Montana, and Yellowstone National Park.

Wolves hunt and live in packs. The pack social structure is both vital and intricate. Mech says, "By far the most important characteristic of the wolf as a species, what sets it apart in the animal world, is its elaborate social organization." Wolves are more pack oriented than coyotes. Although the lone wolf does exist (about 8% to 28% of wolves will stray from a pack from time to time), it is often simply in a solitary mood and will rejoin the pack in a few days. Since alpha (dominant) females are the only ones in a pack allowed to breed, nonalpha females that come into heat are sometimes chased away.

Wolves rely on packs to hunt animals larger than themselves. Unlike coyotes, which can run for hours in open country without tiring and are very opportunistic eaters, wolves tire quickly if traveling faster than a trot and

FIGURE 6.42
At the time of this writing, the red wolf was on the endangered species list. Until recently, the only animals left were in captivity. Changing attitudes and recent reintroduction programs are giving this species another chance.

resort to small game only when large prey is unavailable. Most studies of wolf prey species have found that 59% to 96% of a wolf's diet consists of animals the size of a beaver or larger. The most common prey animals are white-tailed deer, mule deer, moose, caribou, mountain sheep, and beavers.

Pack sizes vary widely, from two to fifteen, but most packs include seven or fewer wolves, all of them usually (but not always) the offspring of the alpha pair. The late Olaus Murie, who was one of the leading mammalogists in North America, suggested in 1944 that pack size might reflect the number of animals that could feed off a single carcass, but others speculate that pack size is determined more by social factors than practical necessity. Seven seems to be the optimum number of animals with which a wolf can socialize comfortably, rather than simply the maximum number that can fill their bellies from a nine-hundred-pound moose.

TRACKS. Like all canines, wolves walk on their toes, not on the full foot, and what we refer to as the heel pad is actually a palm pad. They have four toes on each hind foot and five on each front foot, but one toe in front (a dewclaw, which is higher up the leg) does not register. They also have blunt, nonretractable claws. Their tracks are very robust. Those I measured in Denali National Park in Alaska looked far more robust than those of coyotes. They also were much larger. The front tracks ranged from 4″ to 5¼″ long by 3″ to 4⅛″ wide (Figure 6.43). Rear tracks were 3¼″ to 4¾″ long by 2¾″ to 3⅜″ wide. Both tracks may be larger. Murie reports enormous tracks—6″ by 6″—and he usually didn't include nails in his measurements. There is a great deal of variability in track widths caused by the spreading of the toes, especially in the front foot in mud. I measured one track that was 5¼″ long by only 3⅝″ wide and another 4¾″ long by 4⅛″ wide.

The Algonquin wolf, so called because it has been studied chiefly in Algonquin Park in Ontario, is a smaller wolf than those found in Alaska, and it has smaller tracks: 4″ to 4¾″ long by 2½″ to 3¼″ wide for the front tracks and 3³⁄₁₆″ to 3¾″ long by 2⅜″ to 2¾″ wide for the rear tracks. Biologists think that these small wolves have some introgression of coyote genes.

The tracks of most gray wolves are much larger than those of coyotes. Overall measurements for them fall within the following parameters: 3⅞″ to 5½″ long by 2⅜″ to 5″ wide for the front track; 3⅛″ to 4¾″ long by 2¼″ to 4¼″ wide for the rear track.

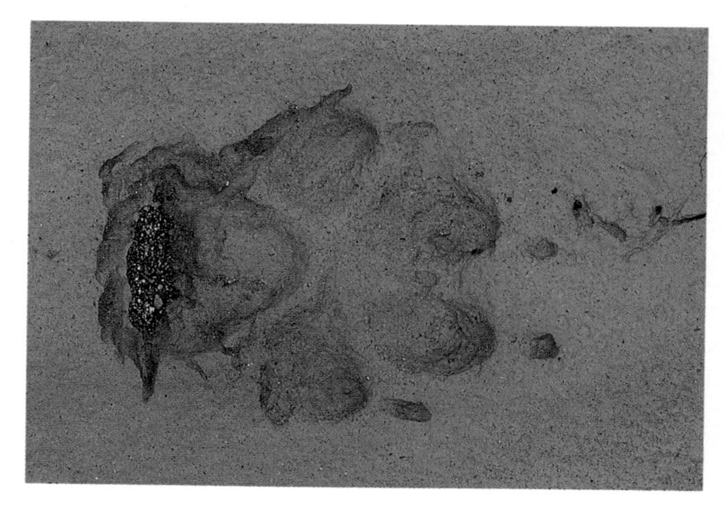

FIGURE 6.43
The front track of a gray wolf (Alaska) can be hard to distinguish from that of a large domestic dog; however, the two front toes in a wolf's track are often parallel, unlike those of a domestic dog.

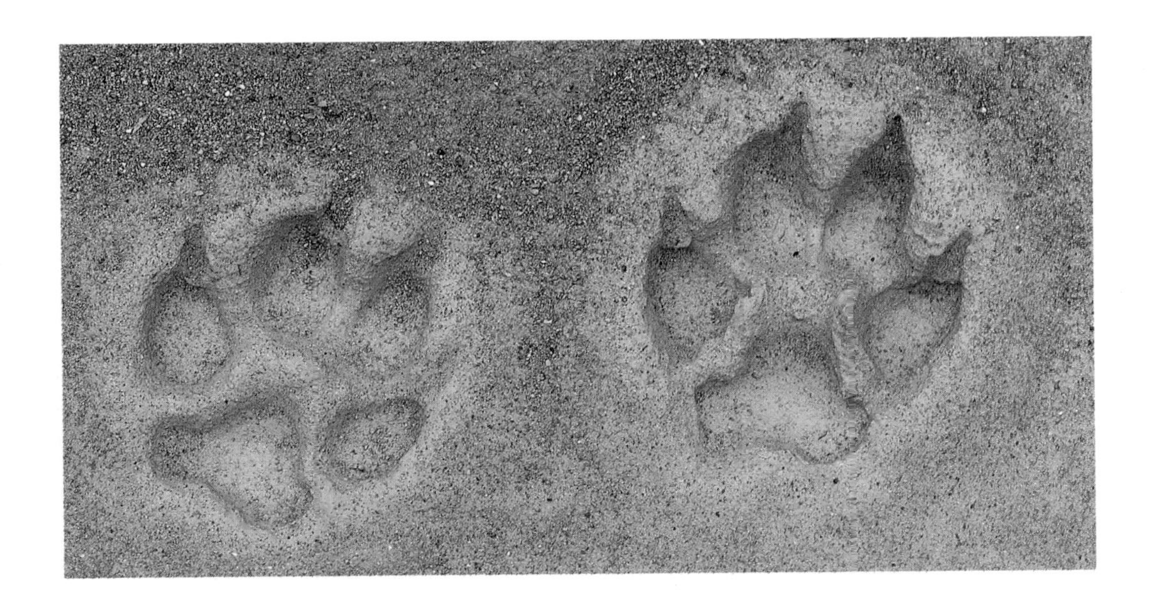

Wolf tracks resemble large domestic dog tracks more closely than they do coyote tracks, and the two are often impossible to tell apart. However, the inner toes of the front foot of a wolf tend to stay parallel and closer together than do those of a domestic dog (Figure 6.43)—except, of course, in a substrate that tends to spread the toes (Figure 6.44). Look for other evidence that will help differentiate between wolves and dogs. Wolf trails, for example, are much more businesslike than dog trails. Dogs tend to wander around, similar to weasels, investigating this and that, running off in one direction, only to stop abruptly and dash away in a new direction. Wolves seem to know exactly where they're going and why.

FIGURE 6.44
Substrate conditions that tend to spread the toes, such as soft mud, can make gray wolf tracks (shown here) difficult to tell apart from those of a large domestic dog.

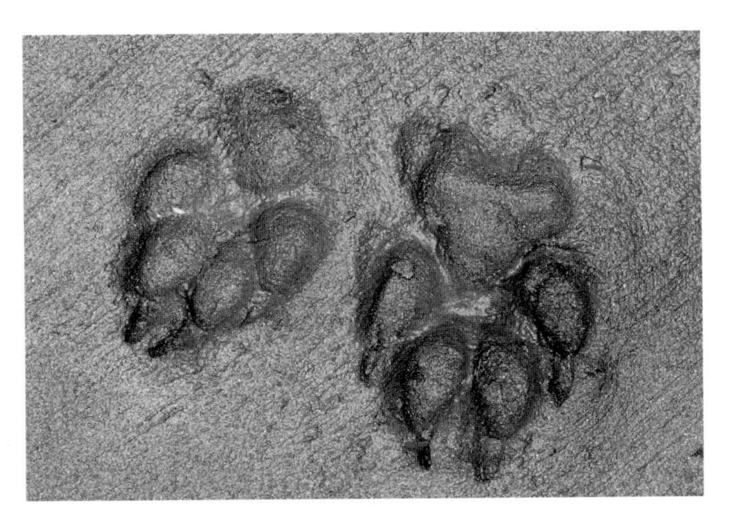

FIGURE 6.45
These are the hind (on left) and front tracks of a male red wolf, but they have the somewhat oval shape of coyote tracks. Female red wolf tracks could easily pass for those of a large coyote.

The red wolf is an extremely rare smaller species of wolf that may be a coyote-wolf hybrid. I had an opportunity to observe two red wolves, a male and a female, at Roger Williams Park in Rhode Island. Both, especially the smaller female, had a tendency to leave elongated tracks, like those of the coyote. In addition, although the male's tracks tended to splay a bit, like those of a gray wolf, they seemed to resemble the more characteristic oval shape of a large coyote. The female's tracks could easily have passed for those of a large coyote. Her front tracks were 3″ to $3\frac{9}{16}$″ long by $2\frac{3}{16}$″ to $2\frac{9}{16}$″ wide. Her rear tracks were $2\frac{7}{8}$″ long by 2″ wide. The male's tracks (Figure 6.45) were $3\frac{3}{4}$″ to $4\frac{1}{8}$″ long by $2\frac{1}{2}$″ to 3″ wide in the front and $3\frac{7}{16}$″ long by $2\frac{1}{4}$″ wide in the rear, easily within the parameters of those for a small wolf but exceeding those of a large coyote.

TRAIL PATTERNS. Although wolf tracks are noticeably larger than coyotes tracks, the two animals' trail patterns are very similar. Unlike the trotting patterns of domestic dogs, wolf patterns are usually direct registering, with the typical canine alternating step (Figure 6.46). The Algonquin wolf's trotting strides are commonly $20\frac{1}{2}$″ to $28\frac{1}{2}$″, and trail widths are 3″ to 7″. Strides for side trots (Figure 6.47) are 26″ to $34\frac{1}{2}$″, and group lengths are 7″ to $13\frac{1}{4}$″. Strides for gallops (Figure 6.48) are $14\frac{3}{4}$″ to 40″, with group lengths of 42″ to 63″. Larger measurements are possible for the Alaskan gray wolf. Note that when wolves are in a side trot, they tend to show some irregularity. The rear track may fall more in line with the front rather than be off to one side (Figure 6.47).

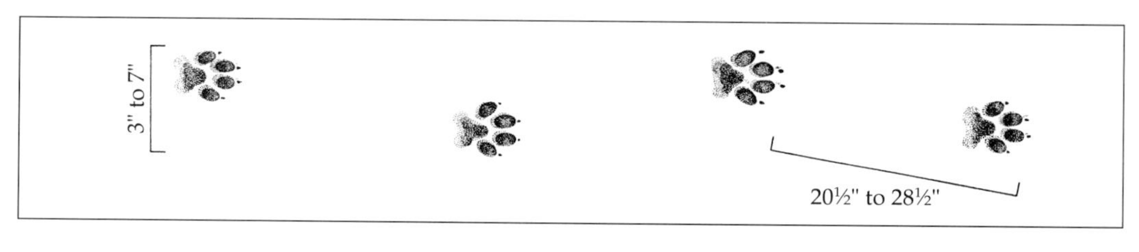

FIGURE 6.46 *(top) Wolves are mostly direct-registering animals in their alternating trotting gait. Their trail width is usually wider than that of coyotes, but the two animals' stride lengths overlap substantially.*

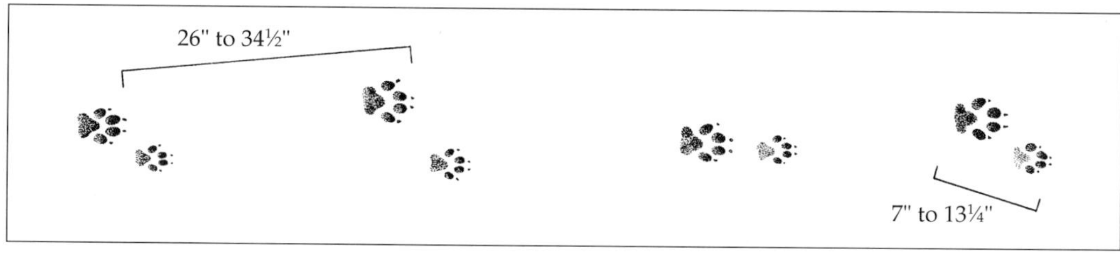

FIGURE 6.47 *(bot.) Gray wolves tend to leave irregular side trotting patterns. The configuration of the tracks and their spacing within each group vary considerably. The smaller hind track leads the front in each set of tracks.*

SIGN: *Dens.* Although wolves have been known to den in abandoned beaver lodges, hollow trees, and shallow rock caves, most dens are dug by pregnant females about three to four weeks before they give birth to three to seven pups. Dens are usually in light or sandy ground, preferably in the side of a sandy or gravelly embankment or ridge, in an elevated location usually near water. Look for many tracks and huge mounds in front of oval openings fifteen to twenty-five inches wide. The openings must be big enough for an adult male to enter easily, since providing food for the pups and the nursing mother is usually delegated to a beta male. A den may be in use until the pups are eight to ten weeks old, at which time it is abandoned. The same den may be used by a pack for many years.

SIGN: *Scent Posts.* There is much debate among wolf biologists on the significance of scent marking, but it seems fair to say that scent posts are to wolves what fire hydrants are to domestic dogs—not merely handy places to urinate, but individual territory markers or pack boundary posts, as well as a host of other things we don't fully understand, such as indicators of virility. The language of pheromones has yet to be translated. What is known is that males urinate on the ground as well as on objects (small bushes, tree trunks, rocks, blocks of ice, mounds of snow, etc.) in three ways, all of which may show up in the tracks: by lifting one hind leg and squirting on vertical objects, by squatting with four legs spread to squirt on the ground, or by crooking one hind leg under its body. Females scent-mark much less than males. They almost always squat but sometimes squat and lift a hind leg under the body. Both males and females scratch the ground after marking. Scratched areas usually appear as elongated scrapes. Sometimes each member of a pack will line up and urinate on a single object; other times only one animal, a sort of designated scent marker, will sign on behalf of the entire pack.

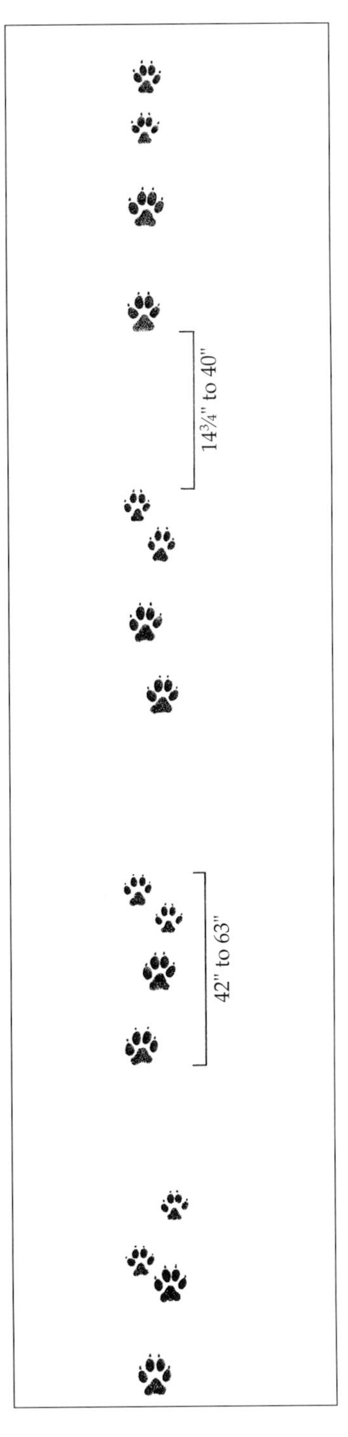

FIGURE 6.48
In these galloping patterns, each group of tracks shows an increase in speed, starting at the bottom.

SIGN: *Scat.* Wolf scat and coyote scat are similar in appearance. If the animal has been feeding on a fresh kill, eating the organs first, the scat will be very dark, smooth, and wet-looking; be full of blood meal; and have few bones and little hair. As it continues to consume the carcass, the scat will contain increasingly larger amounts of hair and bone fragments. Usually the hair will coat the bones, preventing them from scratching the intestines.

A study conducted by John Weaver and Steven Fritts showed that there is considerable overlap in the diameter of wolf and coyote scat. (See page 208 for the results of their study.)

Wolf scat, as is common for all canine scat, acts as a scent post and is usually deposited in the middle of a trail or at the junction of two trails, for exposure to the most animals. Wolf scat also may be found near their congregating areas.

CHAPTER 7: CAT FAMILY

Felidae

Close-up portrait of the elusive bobcat.

Bobcat
Felis rufus
Lynx
Felis canadensis

THE BOBCAT AND the lynx both belong to the Felidae family (from the Latin *feles*, which means "cat"). *Lynx* is Latin for "lamp" and refers to the family's highly specialized eyes, which have expanded irises that enable them to hunt very effectively at night. Their eyes also have reflectors that catch any light that isn't absorbed by the retina the first time. It's these reflectors that make cats' eyes light up when caught in a car's headlights. The word *bobcat* refers to the animal's short tail, which is about five to six and a half inches long. The bobcat's tail is striped, with a tip that is black on top and pale or white underneath. The lynx's tail is not striped and has a completely black tip. A male bobcat can weigh seventeen to fifty-seven pounds and a female nine to thirty-three pounds. Lynx are marginally lighter, weighing eleven to forty pounds.

Since there is so much overlap, the best way to tell the two apart, other than by their tracks, is by the habitat. Generally speaking, bobcats are a more southerly animal. Their range covers most of the United States, except the north central states and a few scattered pockets, and only the border regions of Canada. The lynx is a snowbelt cat whose range includes most of Canada and Alaska and only small portions of the northern United States and Rocky Moun-

FIGURE 7.1
This young bobcat kitten is about six weeks old. Bobcats are usually very secretive and rarely seen.

FIGURE 7.2
The lynx is a true phantom of the north woods. A snowbelt cat found mostly in Canada and Alaska, it has large, hairy paws that are well adapted for travel in deep snow.

tains. Where the two animals inhabit one area, as they do on Cape Breton Island, Nova Scotia, the bobcat sticks to the low-lying, open areas where there is less snow in winter, and the lynx is found higher up in the more heavily forested areas where there is more snow. Since prey availability is similar in both types of habitat, the difference in snow depth in winter is thought to be the main factor in explaining the distributions.

Bobcats are a very secretive animal; you rarely see them. Even finding a track may be difficult. They're out there, but they keep themselves well hidden. Their mottled coats blend in perfectly with their habitat, and even when they curl up and sleep, they are almost impossible to see. This elusiveness may explain why bobcats have never been completely pushed out of the New England forests, even though they have been hunted as mercilessly as other predators. In fact, the bobcat has increased its range since colonial times, which suggests that it is very adaptable and can withstand a lot of pressure.

Bobcats hunt by stealth, by ambush, and by slow, careful stalking. They prefer cottontails, snowshoe hares, and, increasingly in the North, fawns, but they are opportunistic hunters and will eat anything that comes along—beavers, muskrats, opossums, squirrels, mice, shrews and voles, insects, fish, and birds. Their usual hunting method is to sit patiently beside a rabbit run waiting for a rabbit to race by. Sometimes they will crouch in one spot, called a hunting bed, for hours on end, turning every so often to observe a new avenue of approach. In winter, they wait so long that bobcat hair will be frozen into the perimeters of the circle and the bed will have paw prints all around the edge. Lynx are more prey specific, concentrating on snowshoe hares when possible. They'll also eat other animals, such as rabbits, squirrels, mice, grouse, and ptarmigan, when hares are not abundant.

When stalking, a bobcat slinks along the ground, its belly almost touching the earth. It stops occasionally to check on its prey, then slinks again. Finally, as if it can stand the tension no longer, it rushes for the last ten yards or so, rapidly sinking its canines four or five times into the animal's neck and severing the spinal cord or, with a larger animal, seizing its throat and cutting the jugular. Bobcats are nocturnal, though not exclusively so, and when they kill larger animals such as white-tailed deer, they do so by

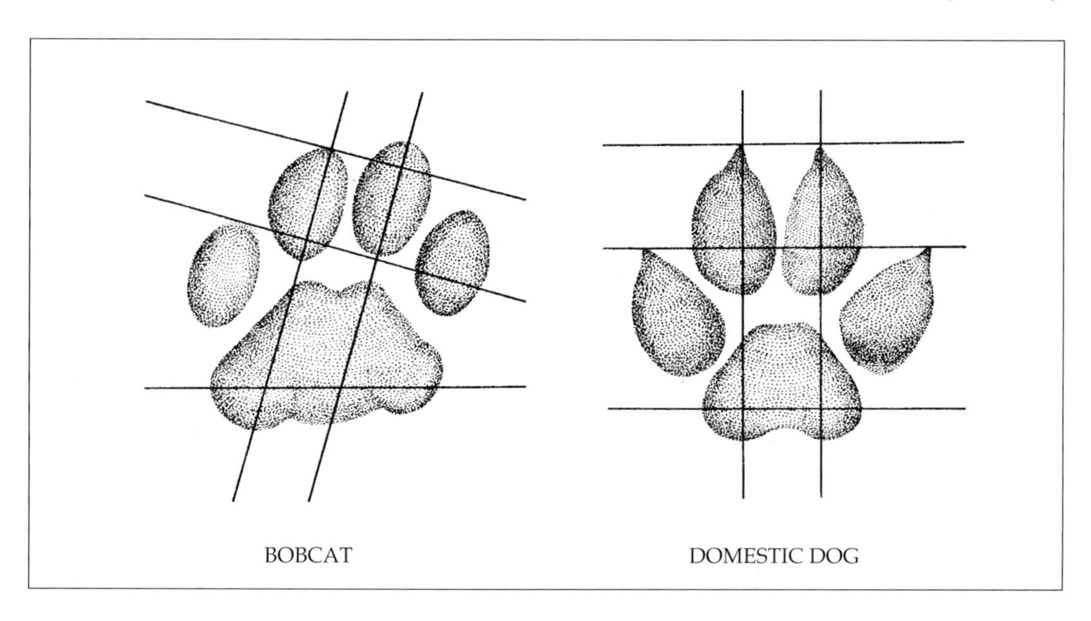

FIGURE 7.3
A comparison of cat and dog tracks highlights the asymmetrical shape of the cat's track. The toes point in a different direction from the heel pad, and the two inner (front) toes have one slightly ahead of the other, as with the two outer toes. In contrast, the dog track is more symmetrical.

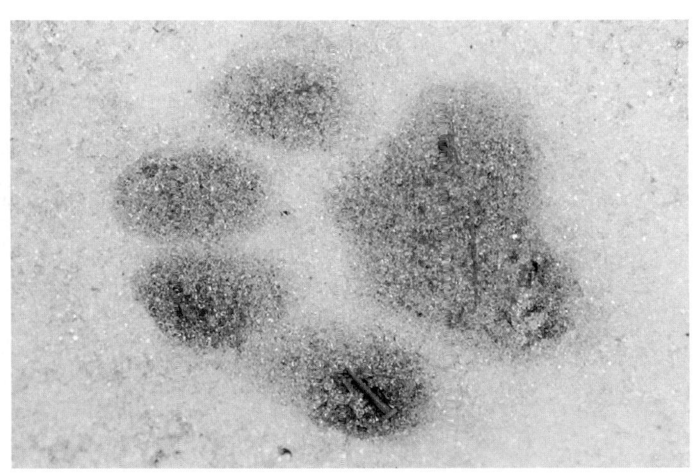

sneaking up on the animal while it's bedding. Less frequently, a bobcat will hunt from a tree. A friend of mine, a very talented woodsman named Bill Fournier, told me he once tracked a bobcat that went up a tree that had fallen over a deer run. The bobcat then jumped on and killed a young deer that passed under the tree. Sometimes the bobcat hunts by luck. I once came upon a place where a bobcat had killed a gray squirrel. As the tracks told the story, the bobcat and the squirrel just seemed to walk up to each other. The squirrel hopped right into the bobcat's path, and the bobcat killed the squirrel right there, without even breaking stride.

TRACKS. Distinguishing between feline and canine tracks can be a confusing task, but there are many differences to look for. The most obvious is the generally round appearance of a feline track compared to the oval shape of

FIGURE 7.4 *(top left)*
The bobcat has five toes on the front foot; the fifth is rudimentary, high on the foot, and does not register.

FIGURE 7.5 *(top right)*
This direct register of the hind track superimposed over the front looks like a perfect left front track of a bobcat.

FIGURE 7.6 *(bot. left)*
The hair on the underside of this lynx's front foot almost obscures the toe pads.

FIGURE 7.7 *(bot. right)*
This front track of the lynx in mud shows small toe pads.

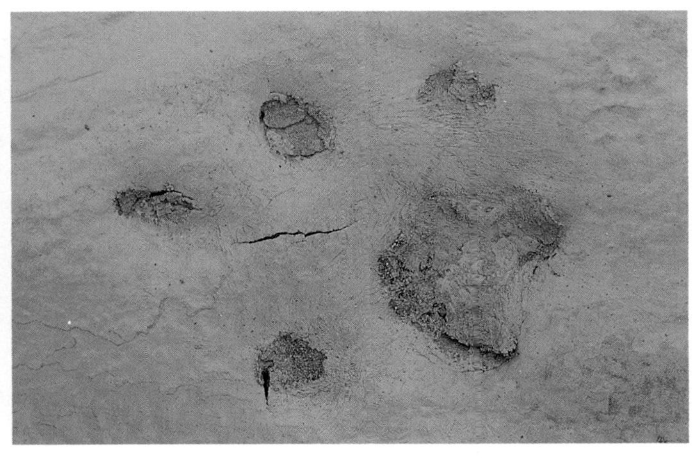

FIGURE 7.8 *(left)* *The lynx's hind foot has four toes and a single heel pad that are completely surrounded by thick fur.*

FIGURE 7.9 *(right)* *The hind track of the lynx is more symmetrical than the front. Here, the toes register as small pads, with the hair marks most noticeable around the heel pad.*

a canine's (which is especially pronounced in the coyote but much less so in the domestic dog). As bobcat tracks begin to age and lose definition, the roundness becomes even more exaggerated, and the tracks may fade into perfect circles. Cat tracks also may seem more asymmetrical than those of a dog. The cat's heel points in a slightly different direction than the toes (Figure 7.3), while the inner and outer toes are offset, one slightly ahead of the other, giving the overall impression that the track is somewhat off center. This is apparent in the front foot and track of the bobcat (Figures 7.4 and 7.5) and especially in the lynx's front foot and track (Figures 7.6 and 7.7). A cat's hind foot and track are more symmetrical than its front (Figures 7.8 and 7.9).

Under the right conditions, most cats' heel pads show a double lobe on the leading edge of the track, while those of canines do not. With the exception of the lynx, a cat's heel pads also tend to look oversized in proportion to the toe pads when compared to a dog's. An important difference I haven't seen mentioned elsewhere is the ridge that forms an arc between a cat's toes and heel pad (Figures 7.10, 7.11, and 7.12). Again, the lynx is an exception. In most cases, canines have a small pyramid or mound between the toes and heel pad (see Figures 6.28, 6.29, and 6.31 in the coyote section). This comparison—ridge versus pyramid—is especially helpful when the tracks are not clear. Another difference is the result of an important feature in the feline anatomy: its retractable claws. While the animal is running or walking, its claws are drawn up into its toe pads and thus seldom register in the tracks. Tracks of coyotes, dogs, and wolves usually show nail marks. Tracks of some canines on hard surfaces are an exception.

Once you've determined that the track is feline, the next step is to distinguish bobcat from domestic cat, lynx, and mountain lion. This can usually be accomplished by measuring the size of the tracks. Bobcat tracks are 1⅝" to 2½" long by 1½" to 2⅝" wide; domestic cat tracks are much smaller, 1" to 1⅝" long by 1" to 1¾" wide. A large house cat might make a track as large as that of a small bobcat, but that would be uncommon. Lynx and mountain lion tracks are quite a bit larger. Lynx tracks range from 3¼" to 3¾" long by 3" to 3⅜" wide (although if you measure along the outline of the hair impressions, the track may be 4½" to 5⅜" long by 3⅜" to 4⅛" wide). Mountain lion tracks are 3" to 4¼" long by 3¼" to 4¾" wide. These sizes are for front tracks; hind tracks will be slightly smaller and narrower, and the toes of the hind feet will not spread as much as will those of the front.

Lynx tracks overlap mountain lion tracks in size, but they look different. Lynx tracks show a lot more hair and smaller pads than mountain lion tracks, and since lynx weigh less and have such hairy paws, they tend to stay on top of the snow more. Occasionally, a large bobcat track can be confused with that of a small mountain lion. Track sizes may be exaggerated by the spreading of the toes or by substrate conditions. In this case, the heel pad measurements become an important tool in distinguishing between the two. Northern bobcat heel pads are 1⅛" to 1½" wide, while mountain lion's are 1⁹⁄₁₆" to 3" wide (for the

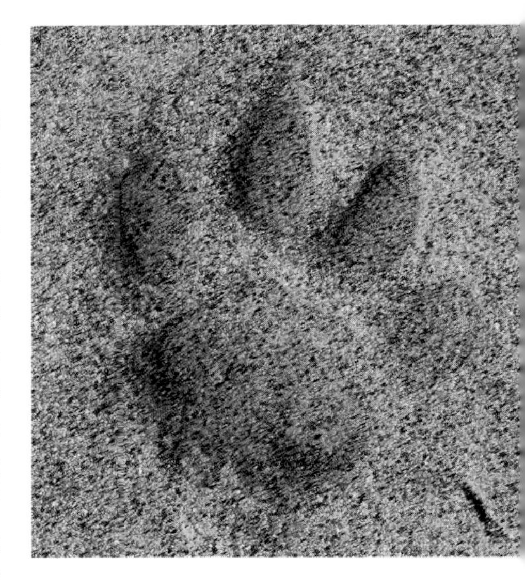

FIGURES 7.10 *(top) and* 7.11 *(bot. left) are bobcat tracks.* FIGURE 7.12 *(bot. right) is a house cat. These tracks are structurally similar, except for size. Measuring them is the best way to tell them apart. Since bobcats in the south are smaller than northern bobcats, their tracks may overlap with a large house cat, though this is uncommon. Measurements here include both northern and southern bobcats.*

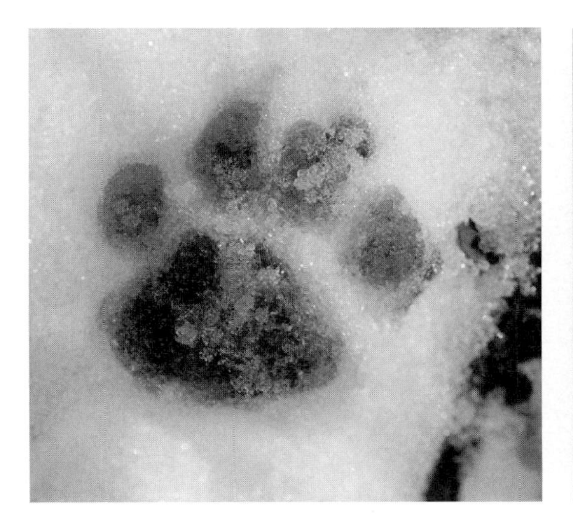

front foot). Mountain lion kittens may, however, have the same size heel pads as bobcats, so this tool works only with adults.

TRAIL PATTERNS. A bobcat's alternating gait in snow or while stalking may be direct registering; at other times, the cat will double-register. Either way, it makes a zigzag pattern—left-right-left-right (Figure 7.13). If there are several inches or more of snow, there will be substantial foot drag (Figure 7.14). The strides for bobcats in this alternating pattern can range from 11¼″ to 25″ or longer, but they are usually 14½″ to 16½″. Strides for lynx are 15″ to 31″. Domestic cats tend to have shorter strides—8″ to 14″, with a 9″ to 11″ average. Bobcat trail widths range

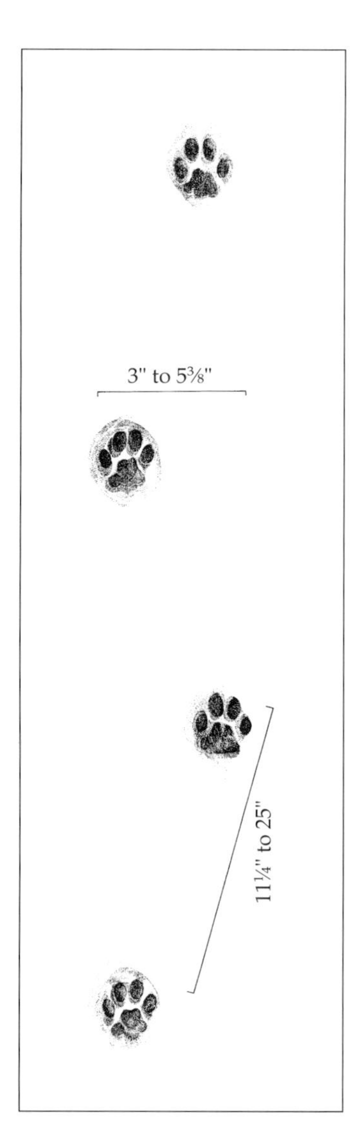

FIGURE 7.13 *(above)* *This shows the bobcat's typical alternating gait. When the bobcat's strides lengthen to 23″ or longer and trail widths narrow, the resulting pattern may be confused with the coyote's trotting pattern.*

FIGURE 7.14 *(right)* *Bobcats traveling in several inches of snow typically leave foot drag in the trail.*

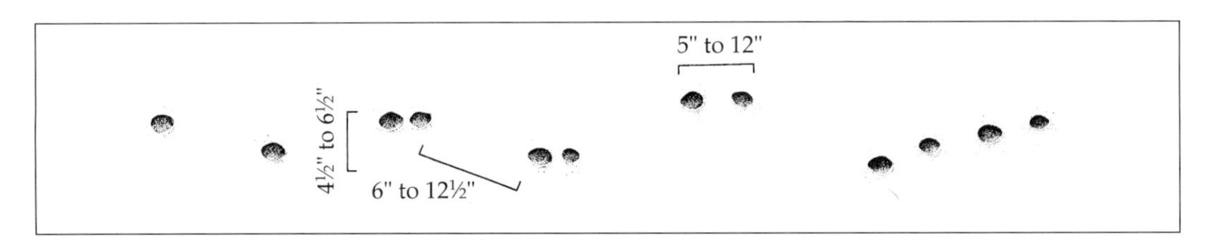

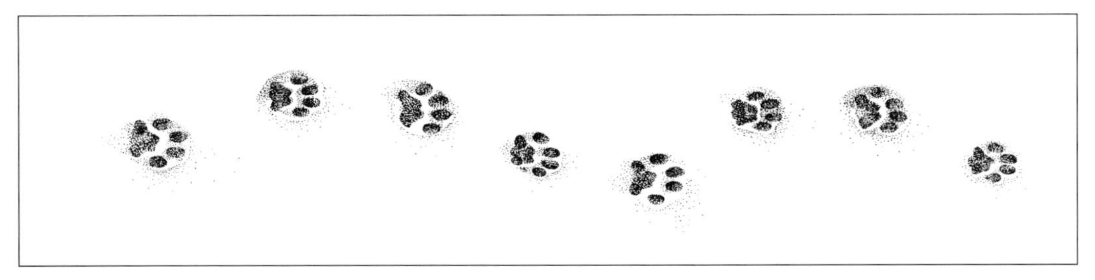

from 3″ to 5⅜″, while lynx trails are 6¼″ to 9″ wide and domestic cat trails 2⅜″ to 4⅞″.

As the bobcat picks up speed (Figure 7.15), the trail will have a 2-2 pattern, with one track almost directly in front of the other. This is a very distinctive pattern. Only bears, other cats, and, *very* rarely, canines will make a similar one. Trail widths are 4½″ to 6½″. You don't need toes or good track detail to determine species here; the pattern says it all. I believe this gait is a fast walk, with the hind foot overstepping the front in a front-hind-front-hind sequence. The stride is 6″ to 12½″, and the group length is 5″ to 12″. This pattern often turns into a lope (Figure 7.16). Another type of lope is shown in Figure 7.17. The sequence again is front-hind-front-hind, in a very wiggly and wide pattern. Cats also may leave bounding patterns that are extremely variable.

FIGURE 7.15 *(top)*
From left to right, this bobcat trail shows an increase in speed from a walk to a fast walk to a lope.

FIGURE 7.16 *(mid.)*
This running pattern is a continuation of the lope in Figure 7.15. The sequence of tracks, from left to right, is front-hind-front-hind.

FIGURE 7.17 *(bot.)*
This is another variation of a slow lope, with the same track sequence as above: front-hind-front-hind.

SIGN: ***Corridors.*** Bobcats travel in certain corridors, and once you know where these are, you can usually pick up a trail. The best way to find them is to walk along a back road until you come to a place where a bobcat has crossed and keep returning to that spot. If it's a corridor, tracks will consistently show up crossing in that area. The corridor may be 30′ to 40′ wide and will usually connect hunting areas with lays (places where the bobcat holes up). Sometimes these corridors will funnel down to a narrow strip of land dividing two wetland habitats. If the corridor is well traveled, you may find scrapes (Figure 7.18) and scat in the area.

SIGN: ***Lays and Dens.*** Bobcats sometimes rest during the day and have been known to seek refuge in bad weather. They often hole up under fallen logs, in rocks, or under overhangs—just about any place that is sheltered and provides some protection from the elements. All the bobcat lays that I have found have been on steep hillsides or ledges, in places where the cat had a good view. Ledges seem to be important in bobcat terrain, especially in the North.

My friend Ray Asselin and I tracked a bobcat one winter day and discovered two of its lays and one den. As we followed this cat, it led us up a very steep hill covered thickly with mountain laurels. Finally, we came to a small cave in the middle of the brush—a very inaccessible, secretive place with an opening only about nine inches in diameter. The cat had been going in and out of it and laying out in front. We even found cat hairs on the ground around the opening.

FIGURE 7.18
Scat deposited in a prepared scrape is a definite sign that it belongs to a feline—in this case, a bobcat.

FIGURE 7.19
Bobcats cover their kills with nearby debris, snow, or whatever is available, sometimes including the hair of their prey. Dirt was spread around this carcass to record the tracks of scavenging animals.

SIGN: ***Covered Kill.*** Like other felines, bobcats sometimes cover their kills (Figure 7.19). If they kill a deer or some other animal that they can't consume all at once, they'll cover it with snow or forest debris, scraping it with the forepaws. Often hair from the prey animal will be mixed in with the cover. Lynx are more inconsistent when caching their prey. The carcass may be barely covered or not at all. If you find a covered kill, check for fisher, mountain lion, or bear sign, because those animals also cover their kills.

SIGN: ***Scat and Urine.*** The diameter of bobcat scat is 7/16" to 7/8". The scat tends to be segmented and will contain a lot of rodent or snowshoe hare hair. Sometimes it is very tightly knit and hard to take apart. Ends are often blunt but may be tapered (Figure 7.20). When an animal has just eaten from a fresh kill, particularly the organ meat, the resulting scat will be very dark and smooth-looking, often appearing wet and loose. It will be composed mostly of blood meal, with very little hair.

Sometimes it is difficult to tell the difference between wild canine and wild feline scat, but as a general rule, bobcat scat tends to be more segmented and lacks the large bone fragments found in coyote or wolf scat. The drier the food source and the habitat, the more segmented the scat will be. It is common knowledge that cats cover their scat, but you can't always rely on that to distinguish it from canine scat. Our observations show that bobcats cover their scat less than half the time. When they are hunting or in another bobcat's home range, they are more likely to cover their scat; when they're on a trail or near their dens, they usually won't bother. I have found places where a bobcat has dug a rectangular scrape 12½" long by 6" wide and then defecated in it, with no attempt to cover it afterward.

Covered scat is not necessarily feline scat, as coyotes and domestic dogs will scratch a lot after urinating or defecating. Canine scrapes are usually long and narrow (sometimes 36" long by 6" wide). During the digging process, canines may inadvertently cover their scat. Cats, however, may show signs of purposely covering their scat by drawing in material from an arc-shaped area around the scat. Canines, unlike felines, will not prepare a site for defecation.

Uncovered scat is not necessarily canine scat. It may be part of a complex communication system bobcats seem to have set up involving scat, urine, and anal glands. Bobcats leave messages for other bobcats. Uncovered feces deposited on a prepared or scraped spot (Figure 7.18) near a den is one kind of message, as is urine sprayed on a rock

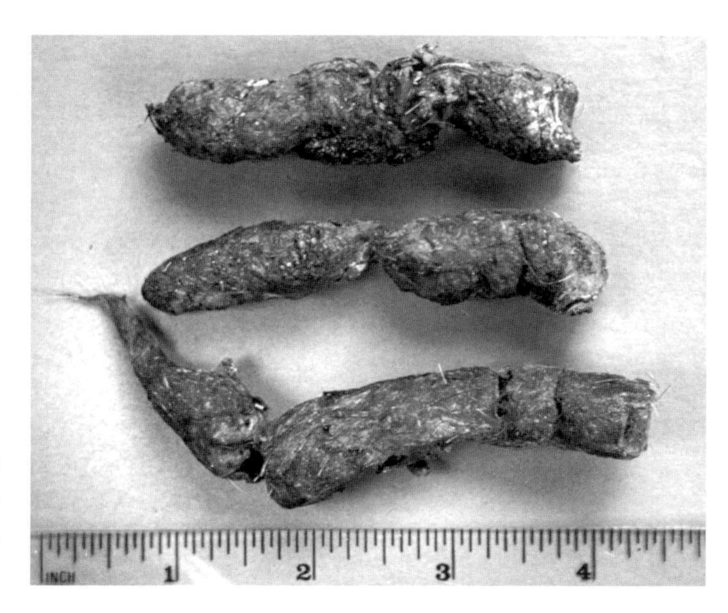

FIGURE 7.20
Bobcat scat is often tightly knit, full of hair, and dark in color. The ends are usually blunt but may be tapered.

or tree stump. Urine deposits are usually eight to nineteen inches above the ground.

After years of tracking bobcats, I've concluded they squirt backwards on certain objects more than others. The most common scent post is a short, decaying stump, usually not more than 6″ in diameter and 4½′ high. If the stump is leaning, they usually deposit urine on the underside where it is better shielded from the elements. Bobcats also construct scent posts by pawing leaves or other debris into a mound and spraying urine on it. These mounds are most often located at the base of an overhanging ledge, an outcrop, or a boulder, and are sometimes associated with lays. Mounds are typically 6″ high and 12½″ in diameter. When leaving the sheltered spot, they may spray urine onto the face of the rock. Since bobcats often travel along steep ledges and cliffs, rock edges in these areas are also typically used as scent posts.

What's most fascinating to me about all this urinating is that it provides an opportunity to track bobcats without any tracks. All you need to do is learn how to use your nose. Bobcat urine has the same distinctive odor as a house cat's. Once you know the types of objects bobcats urinate on, you should be able to find their urine deposits. After the cat has sprayed backwards, it usually walks straight away from the scent post. This is a clue to the direction in which the animal is traveling and to where to find the next scent post.

Lynx are similar to bobcats in that they usually urinate on stumps, bushes, and other objects. The frequency, however, seems to be far greater than a bobcat's. Lynx deposit scat at random. The kittens cover their scat (as do very young bobcat kittens), but adults generally do not.

Mountain Lion
Felis concolor

THE MOUNTAIN LION is known by a variety of names in different parts of North and South America: cougar, puma, panther, painter (a variant of panther), and catamount. In Brazil it's called *onca vermelha,* and in Spanish South America *léon*. The Aztecs called it *mitzli,* and the Ojibwa *mischipichin*. The variety of names is a good indication of the extent of its former range. Although now it is common only in the West (from British Columbia south into Mexico), in suitable habitats from sea level to elevations of ten thousand feet (hence the name *mountain lion*), it once spread across the continent.

Columbus described the mountain lion in 1502 when he cruised up the coast of Florida, where the animal is still found today. The Florida panther is gravely endangered, as an inbred population of fewer than fifty animals exists in the Everglades. Habitat loss, diminishing prey, and accumulated toxins in the environment seriously threaten this magnificent animal. Elsewhere, hunting and habitat loss have severely limited the mountain lion's range. Western populations are currently stable, and in some areas may be increasing slightly, but the eastern cougar *(Felis concolor couguar)* has been declared an endangered species. Various reports from the Appalachian Mountains, as well as from the Ozarks in Arkansas and Missouri, have come in. Specimens were taken as late as the 1960s and 1970s in northwestern Louisiana, southeastern Arkansas, and eastern Oklahoma, although there is not enough evidence to determine whether these were transient lions, natives, or lions released from captivity.

Although mountain lions are said to have been wiped out in the northeastern United States, sightings have been reported regularly in New England and New York. Biologists are reluctant to legitimize these reports without empirical evidence. Confirmation of such evidence, however, may be at hand. In April, 1994, Wayne Alexander and his sons tracked what they believed to be three mountain lions in the Northeast Kingdom of Vermont. A scat specimen they collected was sent to a forensic laboratory in Oregon. Analysis of the scat confirmed the presence of cougar hairs.

On April 4, 1997, my staff instructor, John McCarter, was scouting an area in central Massachusetts for an animal tracking class he was to teach for us. John discovered

what might be considered the find of a lifetime: four covered scats next to some covered beaver remains. The scats ranged from 7/8″ to 1 1/16″ in diameter. One scat was 11″ long with an accompanying segment 3 1/2″ long. It was obvious the scat had been covered by a cat, but John questioned

FIGURE 7.22
One of the most spectacular features of the mountain lion is its amazing size. Some can reach almost 8′ in length.

FIGURE 7.23
The mountain lion's front foot (on left) has five toes, but only four register in the track. One is high on the foot and is just about visible at the bottom of the photo. The hind foot (on right) has four toes. Note that the front foot is larger and more asymmetrical than the hind.

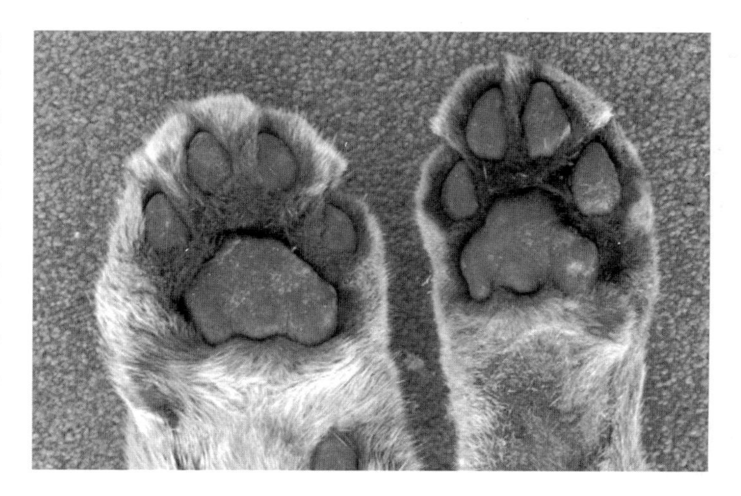

whether a bobcat could defecate such a large specimen.

We sent the scat to the laboratory at the Wildlife Conservation Society in New York for a DNA analysis. The tests concluded that the scat was from a mountain lion. Though state wildlife officials want to see more evidence before confirming the existence of this species in Massachusetts, I have no doubt there was a mountain lion in the area.

According to Susan Morse, a colleague who has researched cougar, bobcat, and lynx for twenty-two years, "In addition to the possibility that captive cougars are periodically being released throughout northeastern woodlands, recent confirmed reports of cougars in Quebec, Ontario, New Brunswick, Maine, and Vermont intrigue us with the possibility that occasional transient animals from source populations in western Canada are moving here." In my opinion, it's only a matter of time before enough evidence is collected to definitively confirm that, released animals or not, mountain lions are once again roaming the forests of the northeast.

Mountain lions need large areas for their home ranges. In the Northwest, a single male may require 175 square miles and a female between 76 and 131 square miles. Within that range, their preferred habitat is more or less that of the mule deer, ranging from open woodland (with large stands of oak), pine, juniper, and chaparral to dense coniferous forests. They like rocky cliffs and ledges for the security they provide. In the Northeast, the historical evidence suggests that their habitat was primarily that of the white-tailed deer: open areas with an abundance of edges (transitional zones), marshes, beaver meadows, swamps, and clearings in hardwood cover.

No one who sees a mountain lion in the wild is likely to forget it. It is an enormous and beautiful animal. Although its species name, *concolor,* literally means "one color," the mountain lion is in fact a shimmering blend of three or four earth tones. It's been described as several shades of tawny, reddish (or rufous) brown, and golden yellowish brown to dull gray. Its belly is buff, and its chest and throat are white. Brown-black stripes ring its muzzle, and the back of its ears and the tip of its tail are blackish brown. The animal's length varies from 6½ feet for females to almost 8 feet for males, not including its 2- to 3-foot tail. Mountain lions usually weigh 75 to 200 pounds, with an average of 100 pounds for females and 150 for males. The largest recorded was by Ernest Thompson Seton, who stated that a male cougar killed in Hillside, Arizona, in 1917 measured 8 feet 7¾ inches without the tail and weighed 276 pounds without the entrails.

The mountain lion's main source of protein is the mule deer and elk in the West and the white-tailed deer in the East. A study conducted in the West found that mule deer, porcupines, and grass constituted 86% to 100% of its diet. Mountain lions will eat porcupines wherever their ranges overlap, but they'll also prey on elk, moose, raccoons, foxes, beavers, mice, livestock, and even coyotes and grasshoppers whenever their preferred food is unavailable.

A mountain lion can outrun a deer, but only for a short distance. Its preferred method of attack is to slink along in a stalking stance, belly low to the ground, until it gets to within twenty to thirty feet of its prey, and then to pounce. Once it has made contact with its target, it grasps the shoulder and neck with its front claws, digs its hind claws into the animal's flanks, and bites the back of its neck with its scissorslike canines (look for canine marks one and one eighth to two inches apart), forcing apart the vertebrae and snapping the spinal cord.

Once a mountain lion has brought down an animal, it will drag the carcass off to a sheltered spot, usually in a shallow ravine or under some low brush, and gorge itself. It usually begins by opening the abdominal cavity just behind the rib cage and eating the organs—liver, heart, and lungs. This may help to distinguish a lion kill from that of

FIGURE 7.24
This left front track of a mountain lion exhibits many feline characteristics: It is asymmetrical and has a curved ridge between the heel and toe pads, and the heel pad looks large compared to the toes.

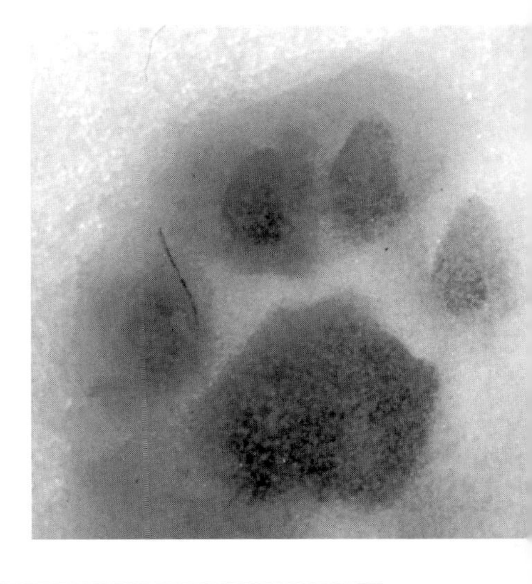

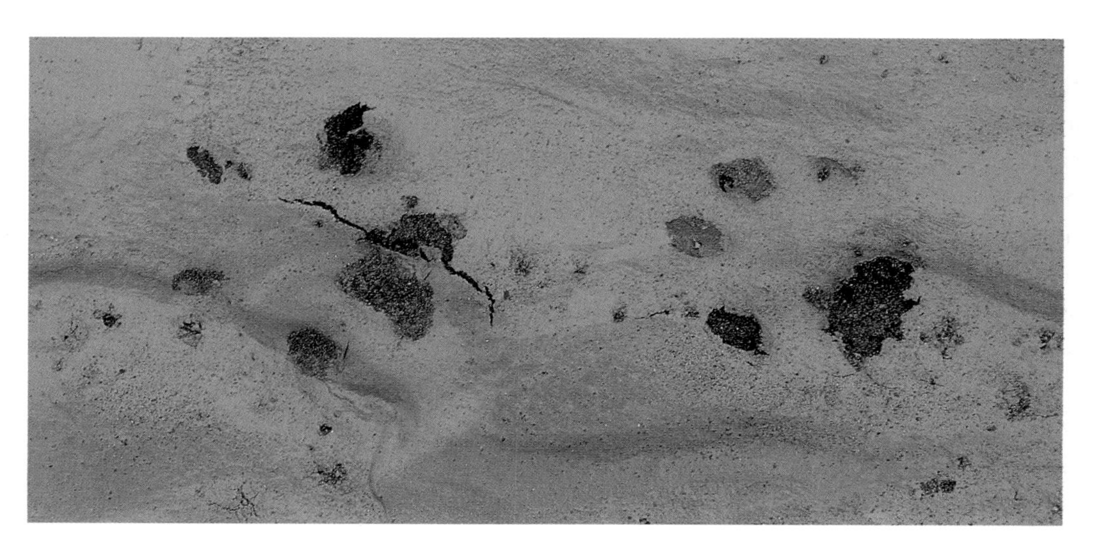

FIGURE 7.25
The lynx's front (on left) and hind (on right) tracks may measure as large as some mountain lion tracks. The hairiness of the lynx's foot and the space between the toes and heel pad when registered in mud help to distinguish lynx tracks from mountain lion tracks.

a wolf or coyote, which normally begins by entering under the carcass's tail. What the lion doesn't eat right away, it usually covers. It will return to a covered carcass for several days. A friend of mine watched a radio-collared female mountain lion in Alberta, Canada, remain with her moose kill for thirteen consecutive days, until it had consumed the entire carcass.

TRACKS. Like the tracks of all cats, mountain lion tracks show four toes on their front and hind feet. The front foot (Figure 7.23) actually has a fifth toe that is located higher up on the foot, but this never shows in the tracks. Each toe has a sharp, curved, retractable claw that usually doesn't register because it is drawn in when the animal is on the move. Also like other cats, the mountain lion has tracks that are often wider than they are long (Figure 7.24). The eastern cougar tracks recorded by R.D. Lawrence in 1954 in Ontario are typical: front tracks, 3¼" long by 3¾" wide; hind tracks, 3" long by 3⅜" wide. The tracks he saw in 1963 were bigger but still wider than long: front tracks, 3 11/16" long by 4" wide; hind tracks, 3⅛" long by 3 9/16" wide. My own suggested parameters are similar to Lawrence's: front tracks, 3" to 4¼" long by 3¼" to 4¾" wide. The heel pads showing in the tracks will look similar to those showing in a bobcat's tracks but will be much larger. Also note that the hind foot is slightly smaller than the forefoot, tends to spread less, and appears more symmetrical (see Figure 7.23).

In snow, the tracks may be slightly longer, exaggerated by thicker winter fur, slippage, and some foot drag, making them more difficult to distinguish from other cat tracks. Lynx tracks, for instance, are bigger than those of a bobcat and may overlap those of a mountain lion, especially in snow. Under certain conditions, lynx tracks may be even bigger than those of some mountain lions. Lynx tracks (Figure 7.25), however, show small pads with a lot of space between them, while those of a mountain lion show large pads with much less space between them. Don't forget that the lynx weighs less than the mountain lion and, with its hairy feet, tends to stay on top of the snow, while the mountain lion will sink in much deeper.

The best way to distinguish between mountain lion and bobcat tracks is to measure the width of the heel pad in the front track. Tom Smith, a feline specialist who traveled the world studying mountain lion tracks, has given me some measurements. An adult female pad will be 1⅝″ to 1⅞″ wide, and an adult male's will be 1 15/16″ to 2¼″ wide. Another source, Harley G. Shaw, a noted mountain lion biologist, measured lion pads in Arizona and California and found that adult pads were 1 9/16″ to 3″ wide. A bobcat's heel pads are 1⅛″ to 1½″ wide. This method will not help you distinguish between a young mountain lion and a bobcat, as their pads are too similar in size.

FIGURE 7.26
When the mountain lion is walking slowly, the sequence of tracks is hind-front-hind-front. In its regular walking gait, it often direct-registers but sometimes double-registers. As its pace increases to a faster walk, the hind foot oversteps the front. Depicted here is a gait in between a regular walk and a fast walk.

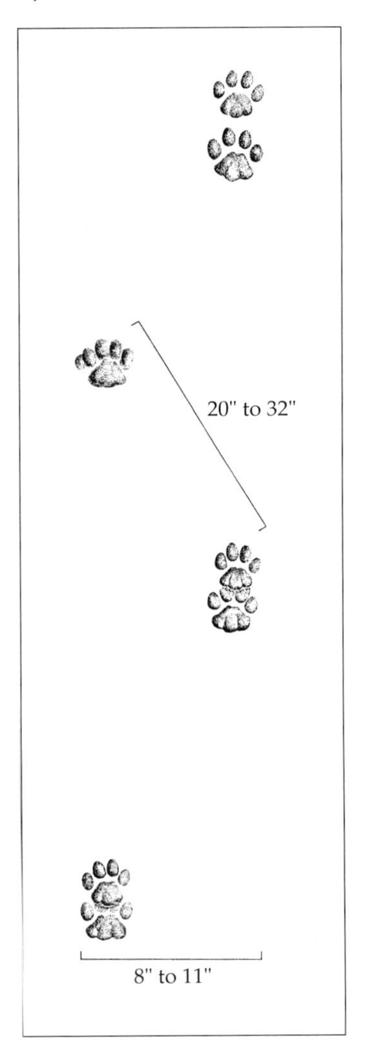

TRAIL PATTERNS. The mountain lion's stride in its alternating walking pattern is 20″ to 32″, and the trail width is 8″ to 11″ (Figure 7.26). Bobcat strides are usually much shorter—11¼″ to 25″ (average, 14½″ to 16½″)—and its trail is narrower—3″ to 5⅜″. Lynx strides are usually 15″ to 31″, and its trail width is 6¼″ to 9″. The overlap between the two animals is greater for stride than for trail width, so in the absence of good tracks, trail width is the best criterion for distinguishing between a lynx and a mountain lion.

Note also that in deep snow, the mountain lion's trail may show belly and tail drag marks. These marks easily distinguish a lion's trail from a lynx's or bobcat's.

SIGN: *Dens.* Female mountain lions select simple den sites in which to have their litters, usually with only enough cover to provide shelter from heavy rain and hot sun. In mountainous terrain, they'll search for a protected niche on the edge of a rocky cliff or outcropping; in scrubland, they'll hole up in dense thickets or under fallen logs. The dens may not have anything for bedding, although some observers have found moss and other vegetation, and one researcher reports a lining of the female lion's belly hair. Females may use the same den for several years.

Mountain lions usually have a litter every two years, giving birth to one to five kittens, with an average of two or three. Although births occur most often in midsummer, mountain lions can breed during any month. Kittens born in early spring have a better chance of survival. Weighing only about twelve ounces at birth, kittens are nursed for around three months. They are then weaned and introduced to meat. They may stay with the mother until late in their second winter, when they are encouraged to establish their own territories.

SIGN: ***Scrapes (Scratches).*** Mountain lions mark the boundaries of their home range with mounds of dirt, pine needles, and other forest litter soaked with urine and/or scat. Resident males make these scrapes (also called scratches) along paths, under trees, or along the edge of a cliff or ridge. Transient males and females without kittens also may make scrapes.

SIGN: ***Scat.*** Cat scat is usually copious and varies from masses to irregular tubular shapes to pellets (in dry areas). The scat will contain the hair of the animal's prey and sometimes bone fragments. It resembles wolf scat but has a greater tendency to segment. If the animal has just eaten from a fresh kill, particularly the liver, heart, or other organs, the scat will be very dark and smooth-looking, often appearing wet and loose. It will be composed mostly of blood meal, with very little hair.

Mountain lions sometimes cover their scat but more often leave it partly exposed as a scent post. (See the section on the bobcat and lynx for a discussion of the significance of covered and uncovered scat.) If the scat is covered, the scratch marks may indicate the animal's route, as mountain lions face the direction in which they are traveling as they scratch.

CHAPTER 8: BEAR FAMILY

Ursidae

Grizzly bear in tundra, Denali National Park, Alaska.

Black Bear

Ursus americanus

MEMBERS OF the black bear species also include the cinnamon bear of the western United States and Canada, the bluish glacier bear of Alaska, and the rare white Kermodes bear of northern and coastal British Columbia. Weighing between two hundred and six hundred pounds (females are smaller than males) and measuring up to six feet tall when standing on its hind legs (three to three and a half feet at the shoulders), the black bear still inhabits much of its ancestral territory—forested areas from the Arctic to Mexico. Home ranges for black bears can be extremely variable. Depending on age, location, habitat, and food supply, a male bear may need roughly 5 to 225 square miles for its home range; females sometimes require three to eight times less.

Bears have few predators. On occasion, a wolf may kill a bear or a large bear may kill a smaller one, but their chief enemies are human hunters, who take an estimated annual toll of twenty-five thousand to thirty thousand bears. Bears are omnivorous. They prefer vegetable matter to meat but at different times of the year will eat everything from berries and fruit to insect larvae, fish, carrion, and garbage. They've also been known to kill fawns and elk and caribou calves.

As nature writer Edward Hoagland notes in his essay "Bears, Bears, Bears," these animals "have been engineered to survive. Whereas wolves have their fabulous legs to carry them many miles between kills, and a pack organization so resilient that a trapped wolf released with an injured paw will be looked after by the others until it is able to hunt again, a bear's central solution to the riddle of how to endure is to den." In their extreme northern ranges, black bears go into deep winter hibernation, denning up for as long as six months of the year. During this period, bears do not eat, drink, or defecate. Females breed every second year. They give birth, usually to two or three cubs, during the denning season, in January or February. They nurse their young while in a comatose state, and when they emerge from their dens in the spring, they may have lost 20% to 30% of their fall body weight. Cubs stay with their mother for the first summer, den with her the next winter, and then are driven off during the second summer, freeing the mother to breed again that year.

During hibernation, a bear's heartbeat drops from about 40 beats per minute to 10, and its oxygen intake is

FIGURE 8.1
Despite the name, black bears are not always black like this one. Some may be cinnamon, "blue," or, very rarely, white.

reduced by one-half. Its metabolism also slows down, however, so its body temperature drops only a few degrees. This high body heat enables it to be roused quickly, so some caution must be exercised when coming upon a hibernating bear.

During the summer, most black bears would rather run than fight when encountering human beings. I discovered this myself once when I was camping in the Adirondacks. I knew that there were black bears in the area, but I was out on an island and had been told that there was little chance of bears swimming out to the island because there was plenty of forage on the mainland. One evening I'd set up my photography equipment in some alders, waiting for a fawn to come to a place where I knew it bedded down. I'd been waiting for a couple of hours when I heard some noises in the woods. I looked up, and there was a big black bear running very fast, straight at me, completely unaware of my existence. I thought, *If that bear finds out I'm here at the last minute, there's no telling how it will react.* So I yelled at the bear, "Bear, stop!" It didn't seem to hear me, so I yelled again, "Stop, bear, stop!" This time the bear immediately veered off to the left, stopped for a minute to look back to see what had made such a strange noise, and then disappeared into the woods. When I counted the steps from where I was sitting to where the bear had stopped, it was eighteen steps. I was pretty sure that a black bear here in the Northeast, without cubs, would not have been inclined to take a swipe at me, but I still felt that eighteen feet from a bear running full speed in my direction was too close for comfort. I never did get a photograph of that fawn.

FIGURES 8.2 *(left)* and 8.3 *(right)* *Black bears have five toes on both the front (on left) and hind (on right) feet. The large toe is on the outside of the foot. The front nails are long and curved; the hind nails are short. Both front and hind feet have palm and heel pads, but those of the hind feet are much more well developed.*

TRACKS. Each front foot of a black bear (Figure 8.2) has very long nails on all five toes. The small inner toe is to the side and rear of the other toes. Occasionally, this toe does not show up in the tracks. I've seen a few people mistake bear tracks for mountain lion tracks for this reason. The bear's nails are so long that sometimes they reg-

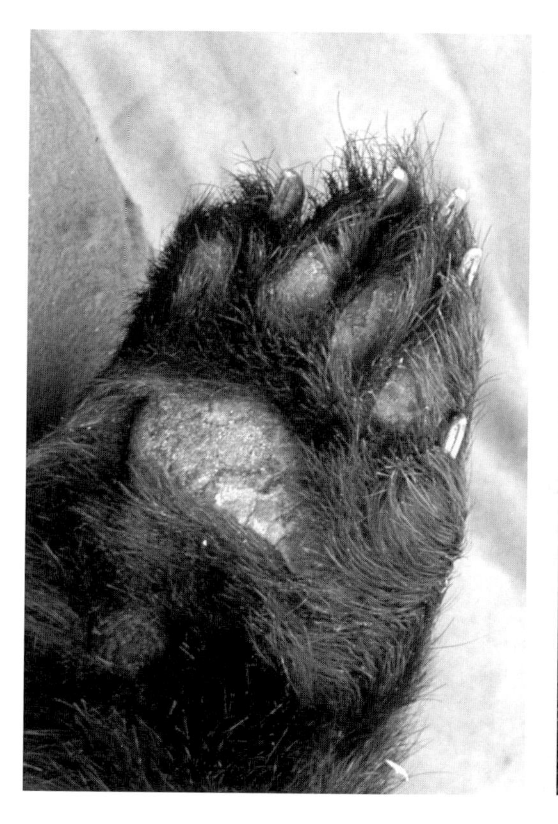

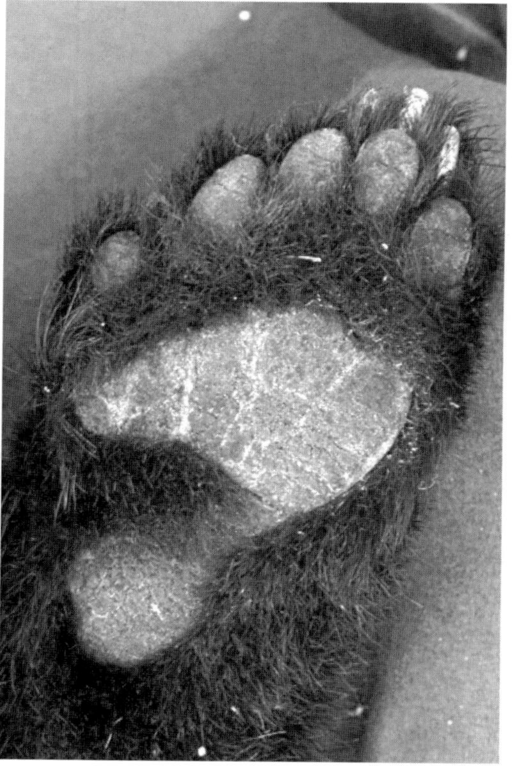

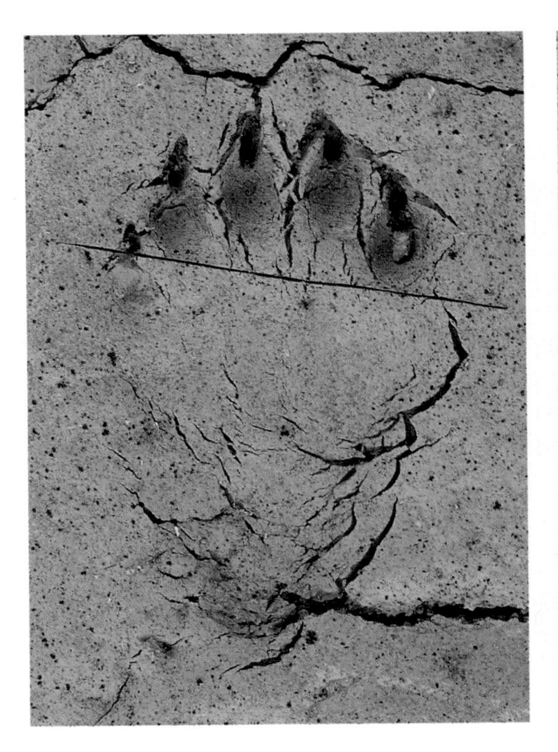

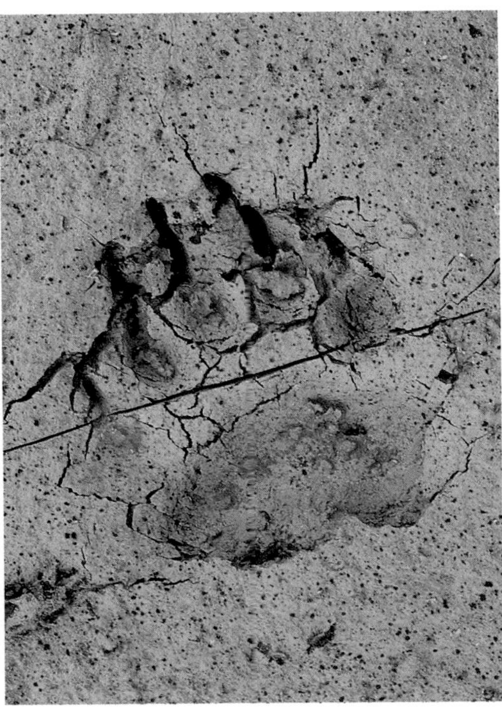

FIGURES 8.4 *(left)* *and* 8.5 *(right)*
All five toes register in this hind (on left) and front (on right) track of a black bear. In the hind track, the palm and heel pads are merged; in the front track, just the palm pad shows. In both, a line is drawn from the bottom of the large toe across the leading edge of the palm pad. The small toe falls mostly below the line, indicating the track is that of a black bear rather than a grizzly.

ister far ahead of the toes, causing them to be missed and thus adding to the confusion. The forefoot also has a palm pad and a ball or heel pad that sometimes shows up in the track.

The rear foot (Figure 8.3) has a palm pad, shorter nails than the forefoot, and a much more established heel pad, which usually shows up in the track (Figure 8.4). It has a tendency not to show, however, when the animal is in a slow walk.

The front track (Figure 8.5) often shows the bear's long nails cutting deeply into the substrate (in this case, mud). This is the right foot. The palm pad and the five toes are showing, though the little toe is not very discernible. Note that the bear's little toe is on the inside of the foot (opposite that of a human's foot). Note, too, that the palm pad is wider on the outside of the track, tapering as it goes toward the inside.

In areas where both black bears and grizzly bears are present, you'll probably want to be able to tell the difference between their tracks. This is not always a simple matter of size difference, because there is some overlap. For example, grizzly yearlings and adult black bears might have very similar sizes:

	Adult Black Bear	*Grizzly Yearling*
Forefeet	5″ to 6¼″ long by 3¾″ to 5½″ wide	5¾″ to 6½″ long by 4⅞″ to 5¾″ wide
Hind feet	6″ to 7¾″ long by 3½″ to 5½″ wide	9″ to 9⅝″ long by 5⅛″ to 5¾″ wide
	When the heel of the front foot registers, it can add 3″ to 4″. Black bear tracks can be larger. I've measured them to 6½″ wide.	When the heel of the front foot registers, it can add about 4″. Yearling tracks also can be larger.

Sizes for adult grizzlies commonly reach 8⅛″ long by 7⅛″ wide for the forefeet and 12″ long by 7″ wide for the hind feet.

Fortunately, there are other ways to differentiate between black bear and grizzly tracks. For example, in black bears (Figure 8.5), the spaces between the toes are usually greater than in grizzlies, depending on the substrate. In a soft substrate, a grizzly's toes look pinched together (see Figure 8.27 in the grizzly section). Also, a black bear's toes are arranged more in an arc than are those of a grizzly. A grizzly's toes run almost straight across the width of the foot. Draw a line from the bottom of the outer (or big) toe across the front of the palm pad. In a black bear, the smaller inner toe will be mostly below the line (Figures 8.4 and 8.5), whereas in a grizzly, the little toe will be mostly above the line (see Figure 8.28 in the grizzly section).

A third way to determine whether the track is black bear or grizzly is to measure the toe and nail lengths. Usually the nail length of a black bear is shorter than its toe length, and the nail length of a grizzly is as long as or longer than its toe length. I say "usually" because I have measured black bear and grizzly tracks in Alaska and found that 18% of the grizzly tracks I measured had ⅛″ shorter nail lengths than toe lengths. This means that if a bear track has a nail as long as or longer than the toe, there's a good chance it's a grizzly; if it has a nail shorter than the toe, it's probably a black bear, but if the nail is only slightly smaller than the toe, there's still about one chance in five that it's a grizzly. Note that black bear toe lengths

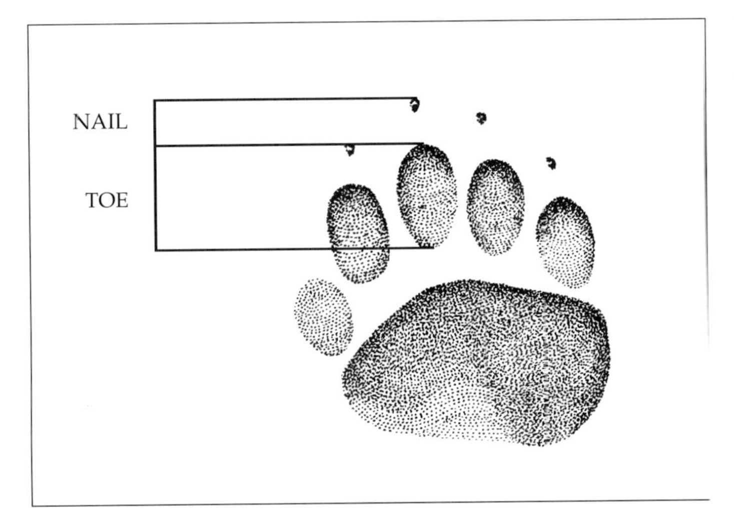

FIGURE 8.6
Measuring toe and nail lengths can help distinguish between black bear and grizzly tracks.

usually do not exceed 1¾″, and their nails usually do not exceed 1⅝″ (Figure 8.6).

	Black Bear	*Grizzly*
Forefeet toe length	1¼″ to 1¾″ long	1⅜″ to 2¼″ long
Forefeet nail length	¹³⁄₁₆″ to 1⅝″ long	1⅜″ to 2⅜″ long

TRAIL PATTERNS. The walking pattern in Figure 8.7 is the bear's alternating gait, sometimes double registering and other times direct registering. Notice that the tracks turn in, as though the bear were walking pigeon-toed. In this pattern, the stride is 17″ to 23″, and trail widths are 9½″ to 14½″. A more common walking pattern for the

FIGURE 8.7 *(top)*
This trail pattern shows the alternating walking pattern of the black bear and grizzly, which may be either a direct register, as shown here, or a double register. Strides are usually 17″ to 23″, but I've measured strides from 13″ to 20″ in 15 ½″ of snow.

FIGURE 8.8 *(bottom)*
Black bears and grizzlies leave a 2-2 pattern when in a fast walk. The hind foot oversteps the front in each group of two tracks.

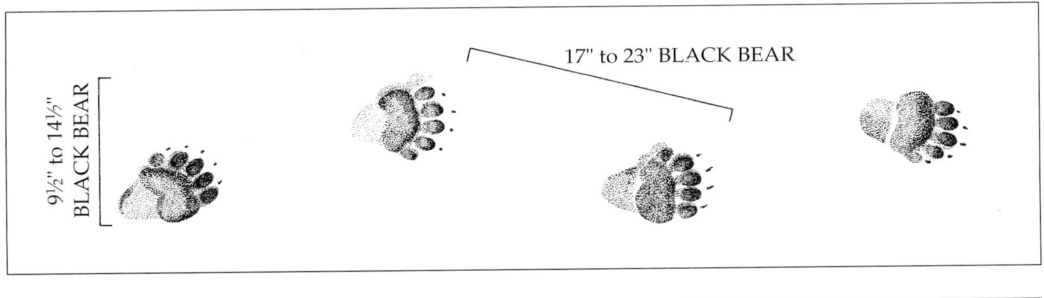

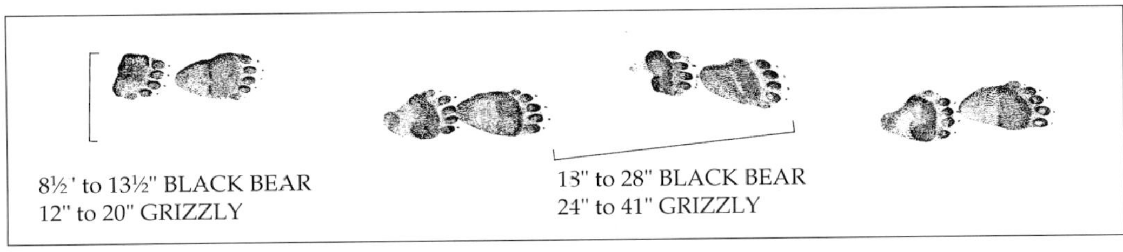

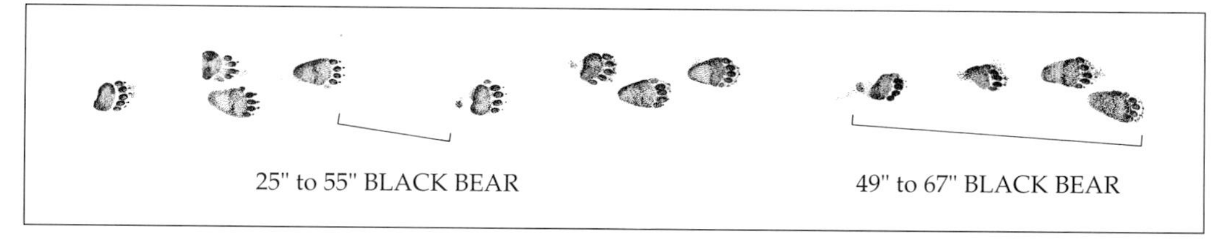

FIGURE 8.9
In these galloping patterns, each group of four tracks shows an increase in speed, from left to right. The track sequence in each group is front-front-hind-hind. These patterns are applicable to both black bears and grizzlies.

black bear (as well as for the grizzly) is the 2-2 pattern, with the hind foot overstepping the forefoot (Figure 8.8). I believe this is a faster gait than the alternating pattern. Black bear strides are usually 18″ to 28″, measuring from one rear track to the next rear track, and trail widths are 8½″ to 13½″. Grizzly strides in the 2-2 walking pattern are usually 24″ to 41″, and trail widths are 12″ to 20″.

Figure 8.9 demonstrates the running patterns of black bears and grizzlies, which are similar except for size. The sequence of tracks is front-front-hind-hind from left to right and shows a slow but continuous increase in speed of a black bear in Alaska. The strides for black bears usually range from 25″ to 55″, and each group of four tracks ranges from 49″ to 67″, with longer groups and strides possible.

Black bears make trails that look like simple, well-worn paths, often through wetland vegetation. Tundra trails may appear as two small trails side by side. Still another type will look like an alternating pattern of round depressions, the result of bears stepping in the tracks of bears before them, for many years.

SIGN: ***Dens.*** Black bears spend a lot of time in dens, which, in their northern ranges, may amount to half their lives. Unlike grizzlies, which sometimes construct elaborate dens, black bears opt for simple denning sites. For example, in my home state of Massachusetts, a black bear may den in a slight hollow on the forest floor with no overhanging cover.

Stephen Herrero, in his excellent book *Bear Attacks: Their Causes and Avoidance,* described a black bear building a bed for herself and her cubs after they were forced by rising water from their winter den in a culvert. According to Herrero, the sow had torn up saplings to make the den, which looked like an eagle's nest built on the ground.

Black bears may also choose to den inside brush piles, hollow logs, or rock crevices; under fallen trees; or in the sunken area beneath uprooted trees. They will even excavate a den under the roots of trees or any other convenient structure, including a snowbank.

In addition, black bears use summer bedding sites, usually shallow depressions in forest litter, and they may even be found sleeping in trees.

FIGURE 8.10
Black bears can do extensive digging, as seen in this beech forest in northern Maine.

SIGN: *Digs.* Although grizzlies are generally thought of as the diggers in the family, black bears also excavate for small animals, insect larvae, ants, and grubs. They'll overturn rocks and may even move, turn, or break up logs as they forage for food. Some of these digging areas can be quite extensive. While photographing black bears in Maine, I came across eight bears in a swale with a canopy of beech trees. A large female reared up on her hind legs and glared at me as her cubs climbed one of the beeches. The cubs soon came down, and they all moved on. When I examined the area (Figure 8.10), it appeared as though the bears had been digging for beechnuts buried under newly fallen leaves.

SIGN: *Bear Trees.* There are four types of black bear trees: bite trees, "whammy" trees, marking trees, and climbing trees. Bite trees (Figure 8.11) are usually 2" to 3" in diameter, with the bite marks located anywhere from ground level to 76 ½" high. In the northeast, bite marks have been observed on striped, red, and sugar maples, yellow birch, balsam fir, spruce, and hemlock trees, among others.

Sometimes bears not only bite these small trees, but

FIGURE 8.11 *(left)*
Bite trees are usually found in or near feeding areas. They often look as though the tree has been grazed by a 22-caliber bullet.

FIGURE 8.12 *(right)*
Abrasions on marking trees range from a height of 46" to 80" off the ground, with the main marking area between 60" and 70". Bear canine teeth marks are evident on this balsam fir marking tree.

beat them up, leaving a broken, mangled tree. These (as well as some larger trees 3″ to 7″ in diameter) I classify as "whammy" trees. In the northeast where I live, "whammy" trees are often hemlocks. The bear climbs the tree, bites at it, and breaks it off at the top, usually at a height between 11 and 25 feet.

Marking trees usually indicate high bear activity. With marking trees, the bear reaches up, and in various combinations, bites, claws, and rubs the tree, leaving scent and sometimes hair. Most marking is done in mid-summer, corresponding to the mating season, which adds support to the theory that the trees act as sign or scent posts. Some of these

FIGURE 8.13 *(above)* *Black bear claw marks are clearly evident on the trunk of this apple tree.*

FIGURE 8.14 *(right)* *This beech tree's smooth bark has recorded numerous claw marks left by black bears as they climbed the tree for food. You can estimate the size of the bear by measuring the claw marks. Locate five claw marks from the same paw and measure from the center of one outside toe across to the center of the opposite outside toe. Large bear claw marks measure 5″ to 7¼″, medium-sized bears 4″ to 4½″, and small bears and cubs, 2″ to 3⅞″.*

FIGURE 8.15 *(left)*
When black bears feed in trees, they often tear the small limbs to pieces, as shown by this apple tree.

FIGURE 8.16 *(right)*
Black bears also stack up broken branches in trees, making what are sometimes called "bear nests."

marking trees are used as such for decades, by generations of bears, creating large gouges in the trees. Marking trees are often located on bear trails or in feeding areas. Figure 8.12 shows a small marking tree, but this type of bear tree varies greatly, ranging in size from 3" to 3′ in diameter.

The fourth type of bear tree are those they climb repeatedly for food. Black bears are excellent climbers. Some biologists think this is a genetic throwback to the Pleistocene, when a bear's defense against large predators such as the giant canids was to scramble up a tree. Look for claw marks on apple (Figure 8.13) and beech (Figure 8.14) trees. The smooth bark of beech trees often record the passage of many bears far into the past. Claw marks may show up on aspens and the smooth bark of some conifers as well. Black bears seem to prefer trees that have a branch with a big bow in it, which acts as a kind of hammock in which they can recline while they forage for food. Other trees that tend to show a lot of bear activity are black cherry and mountain ash. Some hard and soft mast-producing trees, such as apple trees, can be damaged by repeated bear use. The apple tree in Figure 8.15 looks as though someone threw a hand grenade at it.

Sometimes ice damage can be mistaken for bear damage. Watch for a combination of debarking, claw marks, and broken branches to distinguish the work of bears.

Black bears also stack up broken branches in trees, forming what are sometimes called "bear nests" (Figure 8.16). You may find a bear sitting or lying in one of these nests, although it is not known whether bears make them specifically for this purpose. Look for bear nests especially in beech trees.

SIGN: *Kill Sites.* As with grizzlies, black bears will cover their kills with whatever material is available to protect them from scavengers. If you come upon any bear kill, you could be in extreme danger and should leave the area immediately.

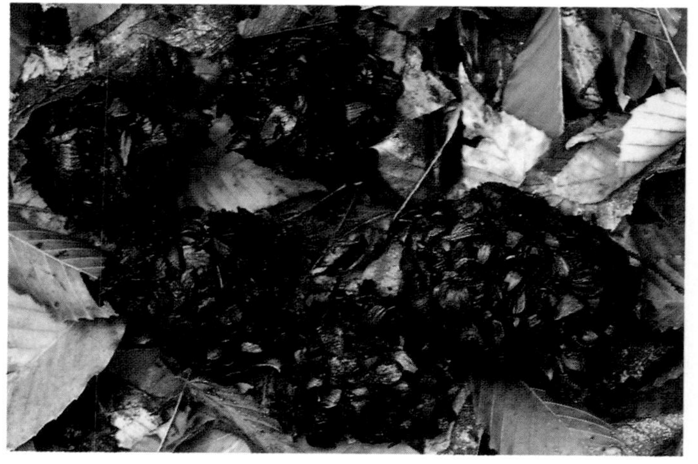

FIGURE 8.17 *(top)* *Bear scat is as variable as the bear's diet. Often, whatever it eats can be identified in the scat. This sample is from an early spring diet of grass.*

FIGURE 8.18 *(bot.)* *A diet of beechnuts in this scat can easily be recognized. When bears feed in beech forests, they often defecate under the canopy of a large conifer, such as a hemlock tree. Bears also mark these trees. Wherever bears feed, remember to look at and around large trees for signs of claw marks and scat.*

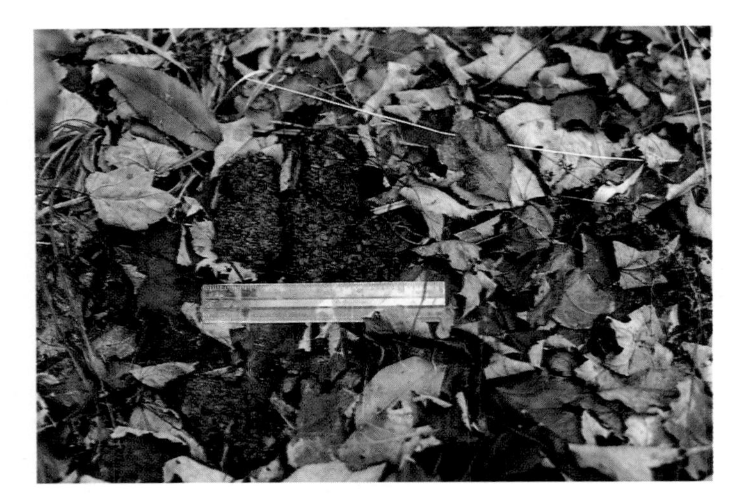

FIGURE 8.19 *(top)* *Scat from a diet of apples, shown here, will smell just like the fruit.*

FIGURE 8.20 *(bot.)* *Blueberries make up this scat.*

SIGN: ***Scat.*** Because bears are omnivorous, their scat can be variable. When black bears come out of hibernation in the spring, they eat a lot of grass, and their scat is black and full of grass fiber (Figure 8.17). As the seasons progress and different foods become available, the appearance of the scat will change. Beechnut scat (Figure 8.18) is easily distinguishable, as are apple scat (Figure 8.19) and blueberry scat (Figure 8.20). Figures 8.21, 8.22, and 8.23 show other possible variations as a result of different types of vegetation in the bear's diet. Scat composed mainly of fruit will smell like the fruit it contains. Scat left when the bear has been feeding on plant material will sometimes smell like the inside of a silo.

One way to tell whether bear scat is fresh is to turn it over. If the vegetation under it is still green, there may be

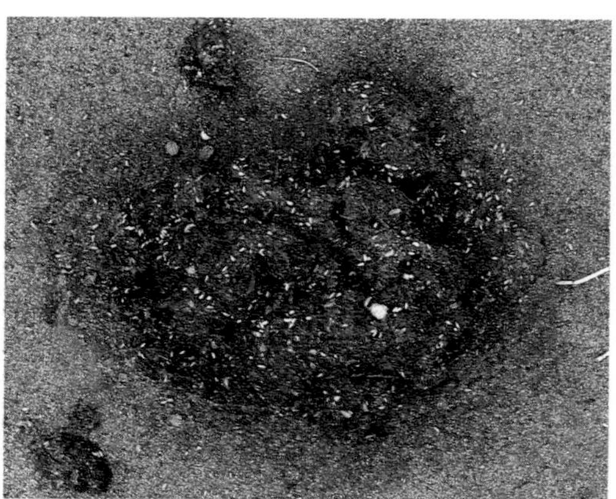

FIGURES 8.21 *(left)* *and* 8.22 *(right)* *These scat samples are variations from a diet of different types of vegetation.*

a bear in the area. Take sensible precautions. For instance, don't leave food out in the open. Black bears are usually shy, or at least wary, but where food is concerned, they also are crafty, inquisitive, stubborn, and, most important, unpredictable.

Adult black bear and grizzly scat may range from 1¼" to 2¾" in diameter, while black bear cub scat may measure as small as ¾" in diameter. I have measured some black bear scat at well over 2", and Stephen Herrero has measured grizzly scat under 2". I do not recommend using scat diameter to distinguish between the two species.

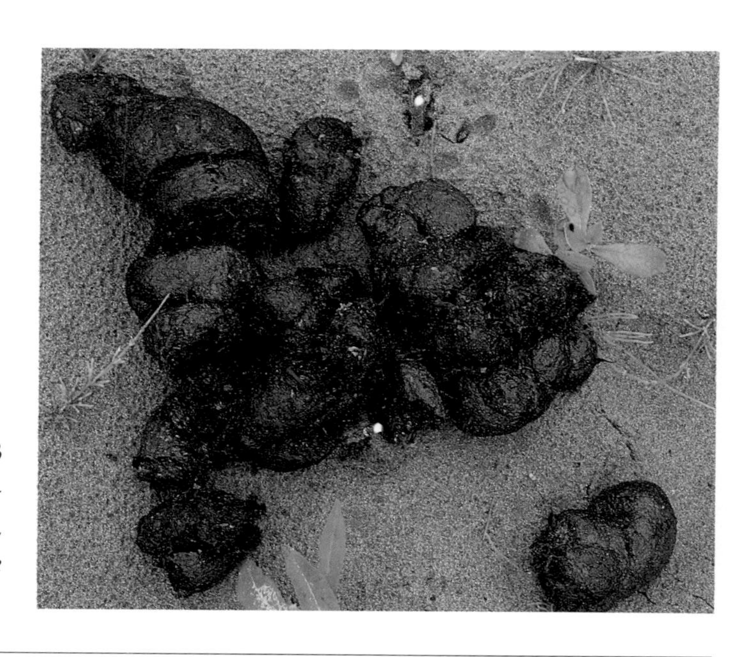

FIGURE 8.23
Scat composed of vegetation, like this specimen, sometimes smells like the contents of a silo.

Grizzly or Brown Bear

Ursus arctos

CONFUSION OVER grizzly bear names has existed since a list published in 1917 by American naturalist C.H. Merriam distinguished no fewer than eighty-seven species, but recent taxonomical streamlining has reduced the number to three subspecies: the brown or grizzly bear of continental North America *(Ursus arctos horribilis)*; the brown bear of the Alaskan islands of Kodiak, Afognak, and Shuyak *(Ursus arctos middendorffi)*; and the brown bear confined to the Alaskan peninsula *(Ursus arctos gyas)*.

The grizzly used to be common throughout western North America. Today, however, there are fewer than 1,000 south of the Canadian border, most of them in isolated pockets in national parks in Montana, Wyoming, and Idaho. A 1990 report by Christopher Servheen estimates that Alaska has 32,000 to 43,000 grizzlies and Canada about 22,000 more, but the grizzly is still considered an endangered species in the lower forty-eight states. Like the timber wolf, the grizzly was earmarked for extermination by early settlers, and it now occupies less than half its original range.

FIGURE 8.24
This inquisitive grizzly cub began walking toward a group of photographers in Denali National Park. We were extremely relieved when it changed its mind, since the protective sow was close by.

Individual grizzlies need enormous spaces in which to roam. Studies in Yellowstone National Park have found that during its life span, a male grizzly may need up to 1,004 square miles of home range. Annual ranges vary greatly depending on season, sex, age of the animal, and food availability. One female studied in 1963 was able to meet her nutritional requirements within a 27-square-mile area, while another required 106 square miles. In the Yukon, where there are about 7,000 grizzlies, home ranges are usually 28 to 33 square miles. In Alberta, which has fewer than 800 grizzlies, the range is 70 to 457 square miles. A female's home range is smaller than a male's, and a male's range may include those of three or four females.

Female grizzlies do not reach sexual maturity until they are four to eight years old, and they may breed only once every three years, usually giving birth to two cubs during hibernation. Grizzlies are the slowest reproducing carnivore in North America.

Grizzlies are truly magnificent creatures. They outweigh black bears by an average of 200 pounds, and a grizzly, when standing on its hind legs, can be seven and a half feet tall. (An adult male grizzly usually weighs around 600 pounds but can get up to 852 pounds. One Alaskan brown bear weighed in at 1,656 pounds.) The grizzly's massive foreleg muscles bunch at the shoulders to form a pronounced hump that is absent in the black bear and is a handy identifying feature. Another difference is in the grizzly's profile. Whereas the black bear's facial structure is flat, with its face descending in a more or less straight line from its forehead to the tip of its nose, the grizzly's profile is indented sharply at the eyes, giving it an almost canine appearance. A third difference is in claw size. Grizzly foreclaws are longer (Stephen Herrero gives claw lengths up to 3¼") and less curved than those of black bears.

Grizzlies are considered the world's largest land carnivores (excluding large male polar bears), but in fact they are omnivorous. In the words of nineteenth-century naturalist John Muir, talking about the Sierra brown bear, "to him, almost everything is food except granite. Every tree helps to feed him, every bush and herb, with fruits and flowers, leaves and bark; and all the animals he can catch—badgers, gophers, ground squirrels, lizards, snakes, etc., and ants, bees, wasps, old and young, together with their eggs and larvae and nests." Muir might have added beached whales, spawning salmon, bighorns, moose and elk calves, pine nuts, camas (a wildflower whose bulbs were also eaten by Native Americans), all the vacciniums (blueberries, cranberries, etc.), grass, and even cultivated corn to his list and still not exhausted the items in a grizzly's pantry. A study of the bear's diet reported in *Wild Mammals of North America* identified "well over 200 plant species whose seeds, fruits, foliage, flowers, stems, roots, tubers and root stocks are eaten by grizzlies within their North American range."

Grizzlies prefer to get their protein from meat—about 50% to 60% of their normal diet consists of animals—but will turn to vegetation when meat is not available, in which case they spend much more of their time foraging. The reduction of their range away from the fertile plains and coastal areas, where small mammals and fish are abundant, and into harsher regions such as mountains and tundra has been cited as a factor in grizzly decline. Eliminating the bison from the prairies (resulting from a campaign to eliminate Native Americans, as well as from the commercial trade in buffalo hides and meat to feed workers on the transcontinental railroad) also reduced the bison's other chief predators, the wolf and the grizzly.

As with the wolf, our haste to eliminate the grizzly from areas we want to inhabit may be based on a misconception. It is true grizzlies are very aggressive animals when provoked or perceive a threat (when you interfere with a sow and her cubs, for example, or seem to be trying to steal a grizzly's carrion cache). Sometimes they will warn you in advance, by making low coughing sounds or snapping their jaws, and under normal circumstances, they would rather pass you by than look at you. In Alaska's Denali National Park, I was photographing some scenic views off the side of a road when I noticed a grizzly sow and three cubs entering a ravine about a quarter of a mile away. Since I was standing at the spot where the ravine met the road, I figured I'd better get out of the way while still remaining close enough to get some shots of the bears if they crossed.

I knew that there was little chance of my outrunning them, if it came to that, as grizzlies can outrun a horse for

FIGURE 8.25
These left front grizzly tracks are from a sow and her yearling. The nails are long compared to the size of the toes. The sow's track, of course, is the larger of the two.

a short distance. Climbing a tree wasn't an option, since I was in tundra, where there are no trees. And that old idea that bears can't run downhill is "completely untrue," according to Stephen Herrero, who says he has "seen bears running downhill full tilt, and turn on a dime!" Some woodsmen recommend screaming at attacking bears, waving your arms, and maybe making a loud noise by banging rocks together. That might work, but it seems like pretty aggressive behavior to me, and probably would to a bear as well, especially a sow who may think she's protecting her young against a madman. Some people wear bells to warn bears of their approach, but I could never stand the idea of jingling and jangling through the woods. Instead, I choose to be as unobtrusive, nonconfrontational, and submissive as possible. If worse came to worst and an attack was imminent, I would probably try lying down and curling up into a fetal position to protect my head, neck, and vital organs. What happened after that would be in the hands of the grizzly.

I was still trying to decide what was a safe distance when I heard a vehicle coming down the road. The driver, seeing my predicament, pulled up near the ravine and shouted to me that his passenger door was open. I rushed for it, and once I had that escape route, I felt a great deal calmer about the grizzlies' approach. I stood by the open door with my camera poised and my hand relatively steady. Several minutes later, the sow emerged from the

FIGURE 8.26
These hind tracks are also a comparison in size between a grizzly sow (right) and her yearling (left). All five toes, palm pad, and heel pad show in both tracks.

FIGURE 8.27
These adult grizzly tracks in wet sand show the front track on the left and hind track on the right. All toes and pads show, except for the heel of the front foot. Unlike the black bear's toes, the grizzly's toes are very close together.

willows about thirty feet away. She didn't stop or look back at her cubs but continued across the road, the cubs close behind, without so much as a glance in our direction.

TRACKS. The tracks of a grizzly do not differ widely from those of a black bear, but there are several identifying differences. The front track of a black bear is 5″ to 6¼″ long by 3¾″ to 5½″ wide without the heel pad, which seldom registers. When it does show, it appears only as a round dot and adds 3″ to 4″ to the length of the track. A grizzly's heel pad can add more than 4″ to the track length. I measured some grizzly yearling front tracks in Cordova, Alaska (Figure 8.25 shows grizzly sow and yearling front tracks) at 5¾″ to 6½″ long by 4⅞″ to 5¾″ wide. As you can see, there is a lot of overlap between the tracks of a black bear and those of a grizzly yearling; it is impossible to distinguish between them simply on the basis of size. Adult grizzly front tracks can get up to 8⅛″ long without the heel by 7⅛″ wide. Rear tracks for yearlings and adults range from 9″ to 12″ long by 5″ to 7″ wide. In some areas, tracks may be even larger; inland grizzly tracks are slightly smaller.

The black bear's hind foot will leave a track 6″ to 7¾″ long by 3½″ to 5½″ wide, whereas my grizzly yearling rear tracks (Figure 8.26 shows sow and yearling hind tracks) were 9″ to 9⅝″ long by 5⅛″ to 5¾″ wide. Here, in the rear tracks, we can see the difference in size between the two animals, but this will not always be the case; criteria other than size must be used to distinguish between black bear and grizzly tracks.

FIGURE 8.28
One way to determine whether this is the front track of a grizzly or a black bear is to draw a line from the bottom of the large toe across the leading edge of the palm pad. If the small toe falls above the line, this indicates that it is a grizzly track.

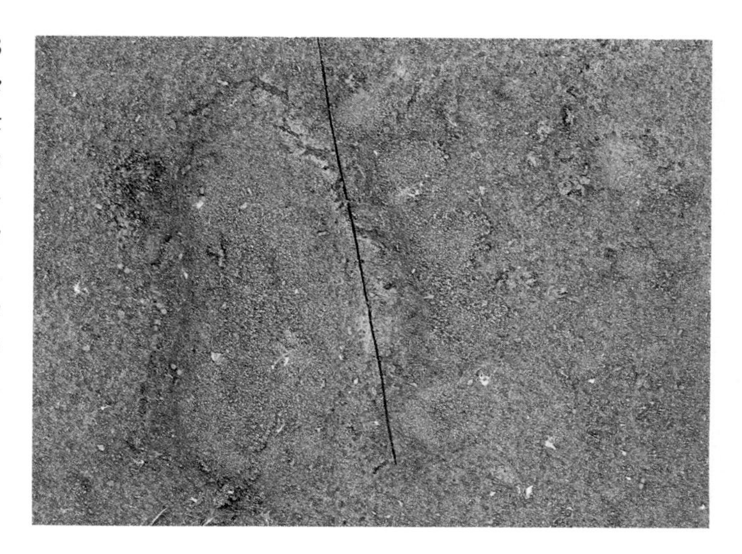

As a rule of thumb, the spaces between the toes in black bear tracks (see Figures 8.4 and 8.5 in the black bear section) are usually greater than those in grizzly tracks. Sometimes, depending on the substrate, grizzly toes look like they're squeezed together (Figure 8.27). Also, black bear toes are aligned in an arc across the width of the foot, while grizzly toes form more of a straight line. Draw a line from the bottom of the outer (or big) toe of the track straight across the leading edge of the palm pad. If it's a black bear track (see Figures 8.4 and 8.5 in the black bear section), the smaller inner toe will be below the line; if the line cuts through the toe, most of the toe should be below it. If it's a grizzly track (Figure 8.28), most of the little toe will be above the line.

Another way to determine whether you're dealing with black bear or grizzly tracks is by measuring toe and nail lengths (see page 243, Figure 8.6 for how to measure). Usually a black bear's nail is shorter than its toe; a grizzly's nail is as long as or longer than its toe (using measurements of the center toe only). Grizzly bear claws are commonly 1⅜″ to 2⅜″ long, although some may reach 3¼″ and Stephen Herrero has one 3⅞″—nearly twice as long as their toe pads, which are 1⅜″ to 2¼″. A black bear's toe is 1¼″ to 1¾″, and its nail is 13⁄16″ to 1⅝″. I have never seen a black bear track in which the nail was longer than the toe. However, when I gathered field data in Alaska, I found enough grizzlies with the nails shorter than the toes to make me cautious in this respect. Some of the grizzly tracks I measured—18% of them, in fact—had nails ⅛″ shorter than the toes.

Just by way of comparison, Figure 8.29 is the front track of a three-and-a-half-year-old male polar bear. Notice that no nails and only four toes (with lots of hair) show. There is also a lot of curvature in the palm pad. Figure 8.30 is the rear track of the same animal, almost obliterated by the animal's hair. Polar bears have much more hair on their feet than grizzlies, and the hair usually shows up in the tracks. Sizes of the above tracks, including the pads and fur, are 9½" long by 8" wide for the front and 11½" long by 8" wide for the rear. This polar bear had an alternating walking stride of 24" to 31" and a trail width of 13" to 17".

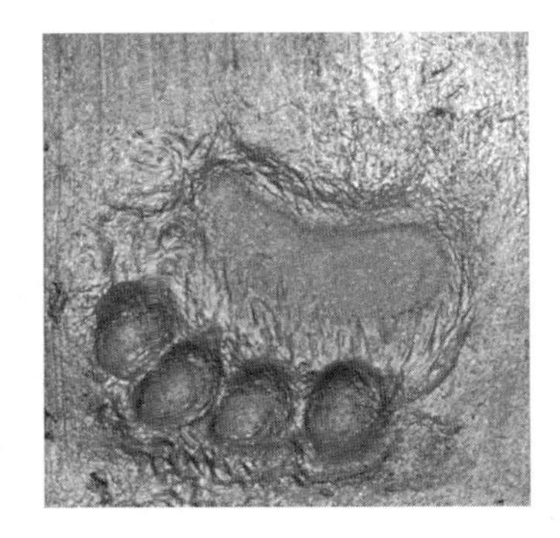

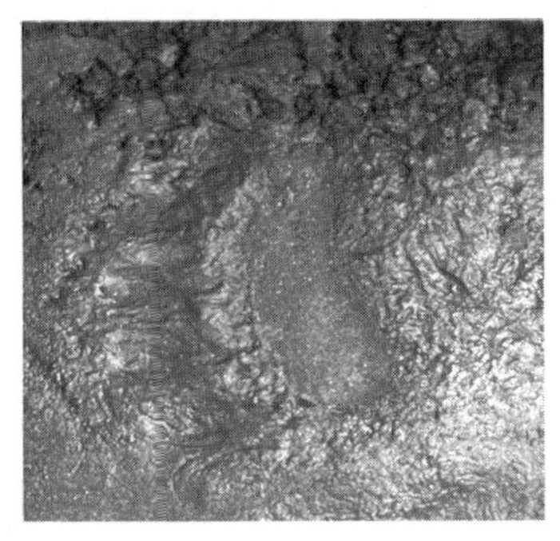

FIGURES 8.29 *(top)* *and* 8.30 *(bot.)*
Polar bears have a foot structure similar to that of other bears, but the most obvious difference is that their feet are heavily furred. Notice the hair marks in both of these tracks, especially the hind (bot.), where the heel area registers as all hair. Unless you have a really sharp eye, you will see only four toes in the front track (top) and will not find any nails.

TRAIL PATTERNS. The grizzly has two typical walking patterns, both of which greatly resemble black bear patterns except in their dimensions. The common 2-2—front-hind-front-hind—pattern (see Figure 8.8 in the black bear section) is used by both bears. The black bear's stride is 18" to 28", whereas the grizzly's is 24" to 41". The black bear's trail width is 8½" to 13½", and the grizzly's is 12" to 20", possibly narrower. As you can see, there can be some overlap in both stride and trail width.

A second walking pattern shows a direct-registering alternating pattern (see Figure 8.7 in the black bear section), which I think is evidence of a slower gait. The stride for black bears is only 17" to 23" (shorter than in the previous walking pattern), and the trail width is 9½" to 14½". Grizzly measurements are higher, with some overlap with those of the black bear.

Grizzly running patterns (lopes and gallops) are very similar to those of the black bear, although stride and group parameters usually are longer (see page 244, Figure 8.9).

SIGN: ***Trails.*** Grizzlies make three different types of trails. The first looks like a series of stepping-stones across an open field. Where they often go along one trail in an alternating walking pattern (direct or double registering), they will step in the same tracks over and over again, making a kind of zigzag pattern. In the North, Olaus Murie noted that where the bear's feet have tamped down the ground cover to expose bare soil, grass seeds will collect in the tracks, and the trail will appear as a series of grassy tufts or knolls across an expanse of tundra. Bears also make a double-track pattern, which appears as two

FIGURE 8.31
This grizzly trail heads straight through a wetland in the Copper River Delta area of Alaska.

lines or ruts in the ground, like the track made by a small wagon. The wide trail width of the grizzly makes it a difficult trail for a human being to follow; the two lines are just far enough apart to make walking awkward. Grizzlies also make trails through marshy areas (Figure 8.31). These trails are about 14" to 20" wide.

SIGN: *Dens.* Grizzlies may den up in a natural cave, but in most cases, they will dig their own dens, beginning anytime between early September and mid-November. They will enter their dens between mid-October and mid-November and emerge in late March or early May depending on weather conditions. In northern Alaska, they'll enter their dens a week or two earlier. Bears observed on Richards Island, in the Northwest Territories, entered their dens in late September to mid-October and emerged in late April to early May.

Den openings vary according to local factors, but the majority are oriented to the leeward of prevailing winter winds, to ensure a heavy accumulation of snow at the opening. The snow acts as insulation. Dens that seem to be oriented differently usually turn out to be cunningly placed in the direction of eddying winds that provide good snow accumulation. Such cunning increases with age, and there is some evidence that bears will return to particularly well placed dens year after year. Den entrances may be left bare or covered with brush.

The den's entrance tunnel may end at the chamber, or the chamber may be dug at a right angle to the tunnel. The floor and ceiling of the chamber are higher than those of the tunnel, and the chamber is longer than it is wide, with just enough room for the bear to stretch and turn. The den might be under the root system of a tree or beneath a large boulder or rocks, making for a stronger den to protect against midwinter or spring thaws. Dens investigated in Yellowstone National Park sometimes were lined with grass, roots, and boughs. This was found to be more common among older adults and females.

SIGN: *Digs.* Grizzlies are consummate diggers, always scraping for tasty tubers, insect larvae, and small mammals with their long foreclaws. If a bear has been after a ground squirrel, the dig will be characterized by large,

FIGURE 8.32
Grizzlies are well equipped for digging and seem to be fond of Eskimo potatoes, which they excavate for the edible roots.

gaping pits. Grizzly digs in plant patches will look less ambitious, with little pits and parts of plants scattered about. Figure 8.32 shows a dug-up Eskimo potato. Other signs of bear digs include overturned rocks, moved or torn-apart logs, or even overturned buffalo chips, where a bear has searched for beetles. They also will dig furiously into anthills.

SIGN: ***Bear Trees.*** Grizzlies select certain trees for scratching or rubbing posts, usually favoring a stout tree in a prominent place beside a trail. Although the function of these marking trees is not well understood, some speculate that they act as territorial markers informing passersby of the bear's height and prowess. The bear rises up on its hind legs and bites and scratches at the trunk. It also may rub its back against the tree to relieve itching. Teeth marks may be six feet or higher, while claw marks may reach well over ten feet high. A black bear's claw marks will not usually exceed a height of seven feet on a marking tree. Grizzlies also may strip the bark of pine, spruce, and fir trees to get at the sap. Repeated use can make for a pretty bedraggled-looking tree. Look for vertically shredded bark, deep claw marks, and bear hairs stuck to oozing resin.

SIGN: ***Kill Sites.*** If you come upon a high, loose mound of branches with some fresh earth or other forest debris heaped over it to conceal some carrion, get out of the area as quickly as possible. You've come upon a bear cache, and the bear will not be far away. Bears, especially

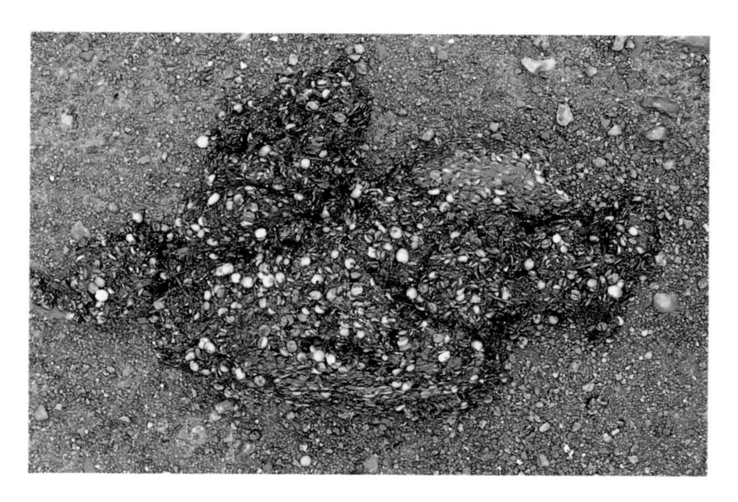

FIGURE 8.33
This grizzly scat contains cow parsnip and devil's club seeds and is only one variation of many types. Black bear and grizzly scat are indistinguishable. (See the black bear section for other scat examples.)

male grizzlies, will defend their store of instant protein almost as eagerly as a female will defend her cubs. They will make low coughing sounds or snap their jaws when they're very annoyed; mild irritation might be expressed by low grunts or growls. If you hear the latter, don't hang around waiting for the former.

SIGN: *Scat.* A bear's omnivorous diet makes for a large variety of scat. It may be elongated or a plop, with evidence of the food in question, such as apples or blueberries (see Figures 8.19 and 8.20, respectively, in the black bear section). Grass (see Figure 8.17 in the black bear section) will be present particularly in the spring. When the bear emerges from its den, grass is often one of the first food items it eats. Throughout the year, bears feed on various types of vegetation, resulting in scat similar to that shown in Figures 8.21, 8.22, and 8.23 in the black bear section. Sometimes the vegetation is mixed with whatever seeds are available (Figure 8.33). This scat was found in the Copper River Delta area of Alaska and contains cow parsnip and devil's club seeds. Scat containing large amounts of vegetation will often smell like a silo. If it contains fruit, seeds, or fish, the smell will usually resemble the contents. Fishy scat will often contain fish scales. If the scat is in the form of a plop, you can expect it to be around 11″ by 7″ in area and about 3″ high. Tubular scat from 1¼″ to 2¾″ in diameter and conforming to the descriptions given earlier will almost certainly be that of a bear. Diameter is not, however, a good criterion for distinguishing between black bears and grizzlies. In fact, there is no easy way to tell the difference.

CHAPTER 9: HOOFED ANIMALS

Ungulata

White-tailed doe bedded down in a grassy field, Adirondacks, New York.

White-tailed Deer

Odocoileus virginianus

THE WHITE-TAILED DEER (Figure 9.1) is one of the most beautiful animals in the forest. There is something ghostlike about it—in its silence and the way it suddenly appears before you and just as suddenly disappears. Its defense against predators is based on subterfuge. Too big to hide in burrows or underbrush, the deer conceals itself by stealth, immobility, and camouflage. Deer are swift and agile runners. They have been clocked at up to 36 miles per hour in open country and can bound over a seven-foot fence from a standing start. Some biologists believe that when they flash the white underside of their tail when they take off, they are, in effect, telling a predator that it's been spotted and it might as well give up the chase. Although the deer's natural enemies include coyotes, wolves, lynx, bobcats, and black bears, human hunters account for an average of two million deer deaths every year. Domestic dogs and road kills also take a heavy toll on deer populations. Still, the white-tailed deer remains one of the most abundant large animals in North America. Its range encompasses most of the contiguous forty-eight states (ex-

FIGURE 9.1
The white-tailed deer is admired for its grace and beauty. Its highly acute senses, speed, and agility have been honed by many predators. This summer photo shows the buck's antlers "in velvet."

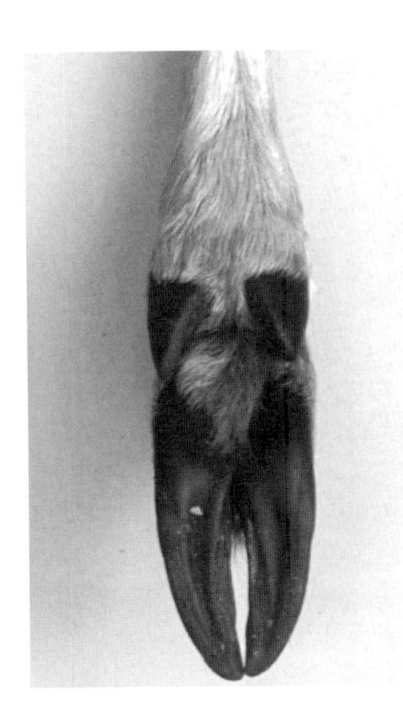

FIGURE 9.2 *(left)*
The deer's foot is a heart-shaped cloven hoof with two appendages called dewclaws, just above the hoof. Under certain conditions, the dewclaws will register as part of the track.

FIGURE 9.3 *(right)*
This very fresh white-tailed deer track shows a pronounced ridge in the center. The direction of travel is to the right.

cluding Utah, Nevada, and most of California) and parts of the southern Canadian provinces, except Prince Edward Island and Newfoundland to Labrador.

White-tailed deer mate in the fall. Females have a gestation period of 195 to 212 days, which means that fawns are born in spring or early summer. Females bear from one to four young, although quadruplets are very rare. If there is more than one fawn, the female will hide them in separate sheltered places and nurse them by making the rounds every three or four hours. If you're patient and observant, you can find them. They may be on a knoll, in tall grass, or up against a fallen log. Fawn tracks can be very small, sometimes just ½" in length, like dozens of tiny hearts pressed into the dark spring earth. Remember that clear, sharp edges are a sign of a fresh track, so if you find such a track, look around carefully.

TRACKS. A white-tailed deer's foot (Figure 9.2) consists of two crescent-shaped halves and two dewclaws (appendages behind and just up from the hoof itself). It leaves a heart-shaped track (Figure 9.3), with the bottom or sharp end of the heart pointing in the direction in which the deer is going. Notice the high ridge between the two halves of the track. In certain conditions—for instance, when snow covers the track—deer tracks can be confused with dog tracks, as the two hoof points resemble the inner

FIGURES 9.4 *(top)*, 9.5 *(mid.)*, and 9.6 *(bot.)* *The photographs on this and the following page demonstrate variations in white-tailed deer tracks. Factors such as the quality of the substrate, the speed of the animal, and whether the hoof is splayed or closed will greatly affect the appearance of the tracks. Figure 9.4 is a closed hind track over a closed front track. Figure 9.5 is a closed hind track partially over an opened front track in hard sand. Figure 9.6 is an opened hind track directly on top of an opened front track in soft mud.*

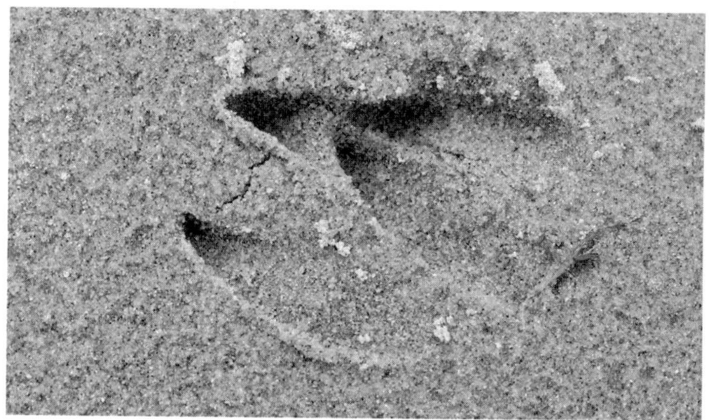

toes and/or nails of a canine track. Look for the center ridge; dog tracks do not have it. Deer also will sink farther into deep snow than will dogs.

Deer tracks measure 1¼″ to 3½″ long by 1⅜″ to 2⅞″ wide. This includes fawns and adults. There are extremes, as fawn tracks in spring may measure only ½″ long, while large deer tracks may be 4″.

TRAIL PATTERNS. Like most ungulates, deer leave an alternating walking pattern, sometimes in a direct register, with the hind track superimposed directly on the front track (Figure 9.4), but more often in a double or slightly off-register pattern (Figure 9.5). In Figure 9.5, the front hoof was spread open when it came down, even though the animal was walking. Some people say that the hoof will open only if the deer is a buck or is running at a high speed or walking in soft mud, when it requires extra

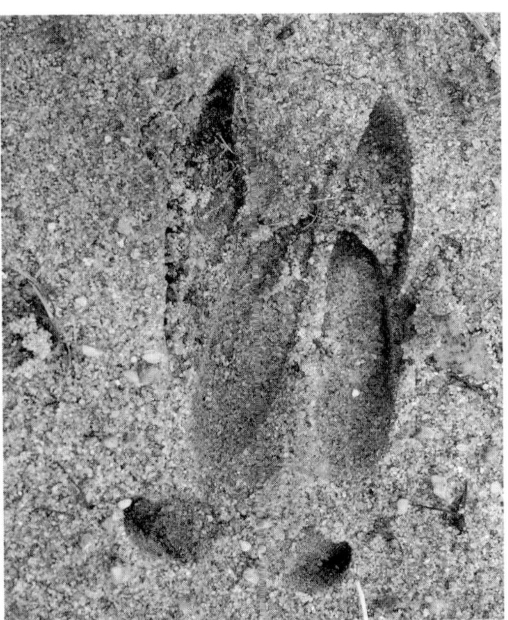

traction, but sometimes deer open their hoofs when stepping on a hard surface. There are no hard and fast rules. See Figures 9.6, 9.7, and 9.8 for other variations.

Deer walking strides (Figure 9.9) are 18″ to 26″, with strides of 14″ to 34″ possible, depending on the size of the animal and its walking speed. Common trail widths are 5″ to 9½″, with widths of 3½″ to 11½″ possible. The deer's wide trail will often be enough to distinguish it from a domestic dog's.

Deer also leave a direct-registering trot (another alternating pattern) similar to the walking pattern in Figure 9.9. However, in the trotting pattern, strides are longer, from 41″ to 56″, and trail widths narrower than in the walking pattern. Deer also may make interesting drag marks as they move through snow, sometimes leaving two lines made by the points of the hoof as it is lifted out of the snow and forward.

FIGURE 9.7 *(top left)*
Just the tips of the hoof register in these four deer tracks in very firm sand. The track to the upper right is a combination of two tracks.

FIGURE 9.8 *(top right)*
This shows a closed hind track over an opened front track. The dewclaws of the front foot have registered.

FIGURE 9.9 *(bot.)*
The white-tailed deer's alternating walk is a mostly double-registering pattern.

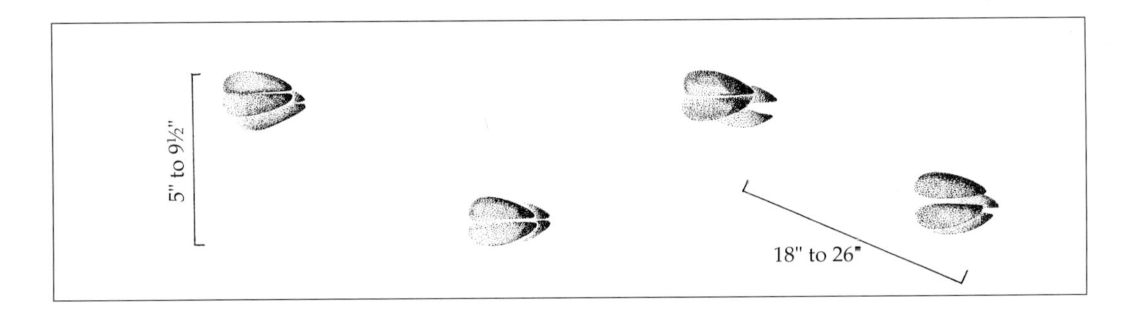

FIGURE 9.10
White-tailed deer running in deep snow may leave deep, rectangular impressions where all four feet land in the same hole. Strides of 9′ to 12′ are possible, and in some conditions, deer running full tilt may have strides up to 28′.

SIGN: ***Deer Runs.*** A deer run is one way deer deal with deep or crusted snow. A typical winter run is a path used by many deer on which the snow is packed down, making travel much easier between bedding or feeding areas.

In summer, a close examination of a deer run will reveal a choppy surface, the soil churned up by the passage of many hoofs. Leaf litter may obscure the soil's surface; try feeling the ground with your hands for the lumpy effect. Trails of animals without hoofs will be smooth by comparison and may take sharp turns to avoid objects such as rocks and small stumps. Deer runs do not usually turn sharply. Look for fresh sign, as some runs may be seasonal. Be aware, too, that many other animals will use a deer run.

SIGN: ***Deer Yards.*** In severe northern climates, where snow is deep, deer will seek shelter in coniferous areas, where there is less snow and the deer expend less energy moving around. An area where a number of deer go to seek shelter is called a deer yard, and it can cover a huge amount of territory. The largest deer yard in Ontario, for example, known as the Loring Deer Yard, is a region of about three hundred square miles between Trout Creek and Parry Sound. According to a survey conducted in 1987, between twelve thousand and fifteen thousand deer concentrate in the yard for eight weeks every winter. Farther south, winter deer yards are smaller and scattered and may be found in remnant bogs and forested river floodplains with coniferous cover. If you come across a deer yard, look for high browse lines in hemlocks and other conifer species and for well-used runs. Yards are usually located on south-facing slopes.

SIGN: ***Deer Beds.*** Deer make beds year-round, but their bedding areas are most easily found in snow. They are usually on high ground, on knolls or the south side of a hill, where they are exposed to the most sunlight. Often when deer bed down together, they have the tendency to face in different directions, as though they were jointly keeping watch. Sometimes you can actually see the whole outline of the deer (Figure 9.11). Deer also will bed down where a hill levels off and then continues down. They choose such areas because they can pick up sounds and

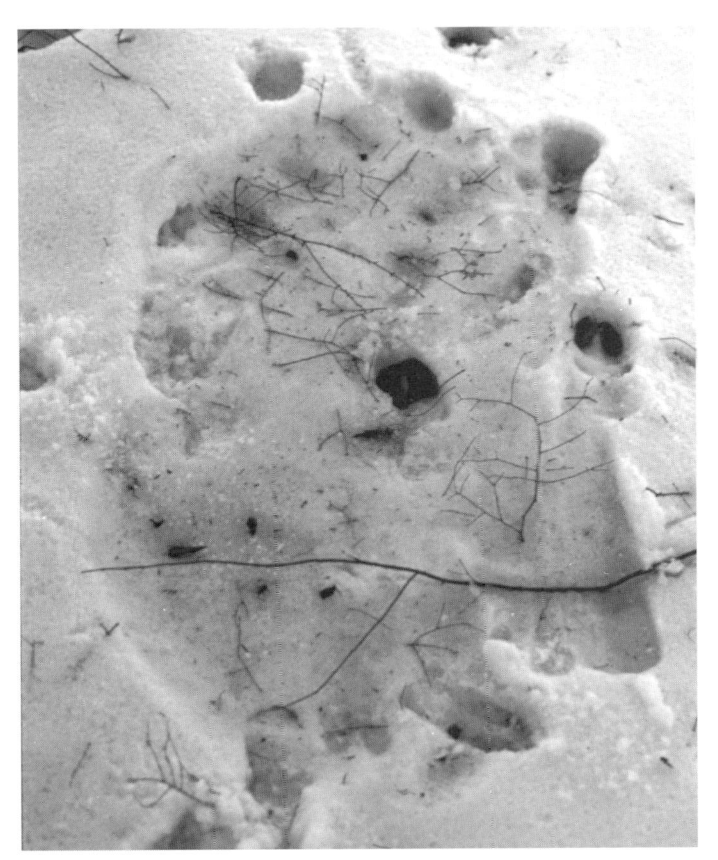

FIGURE 9.11
White-tailed deer beds may show a lot of detail. In this one, the impression of the deer's rump is to the lower left, the hind leg is to the lower right, and the two folded front legs are to the upper right. You can determine the size of the deer by measuring the bed from the center of the lower folded front leg diagonally across to the rump. A large deer's bed measures 41", a small deer's 25".

scents that travel up the hill. In winter, deer may bed down under conifers, especially hemlocks, whose branches bend down to form a tent that will help keep the animal's body heat from dissipating too rapidly. Contrary to popular belief, I have found substantial evidence of deer beds receiving multiple use, although I have not been able to determine whether the same animal used it repeatedly or a different animal moved in.

SIGN: *Browse.* Deer have incisors only on the bottom jaw, so they have to tear at the branches rather than cut through them cleanly. Figure 9.12 shows an oak sapling that has been browsed by a deer. Note that the ends are broken off and very rough-looking. Often, when the deer scrapes along the bark and then finally bites the branch off, about one-half inch at the tip will be brown, and the browse will have a frayed, squared-off end. Browsed leaves also will be squared off. When you compare deer browse with that of a snowshoe hare (Figure 2.16 on page 105), you can see that deer leave a very rough, chewed-up branch,

FIGURE 9.12
Since white-tailed deer have only bottom incisors, they leave rough, torn, or squared-off cuts when browsing.

FIGURE 9.13
This demonstrates the impact of forty deer per square mile on forest regeneration. The electric fence keeps deer out of the left side, with its lush growth of mostly poplar. On the right side, deer have eaten back all the new growth.

whereas rabbits and hares make a forty-five-degree cut that looks as though it was done with a knife.

Sometimes it's easier to tell what kind of animals are frequenting an area by looking at what is *not* there. For example, Figure 9.13 is a photograph of an electric fence used to keep deer out of a certain area. Notice that on the left side of the fence, where there are no deer, there is a lot of tree regeneration, mostly poplar. On the right side of the fence, where there are deer, there is no regeneration. In this area, the deer population is known to be forty per square mile, large enough to have a major impact on tree species preferred by deer. Similarly, Figure 9.14 shows a stump

FIGURE 9.14
This is the same scene from a different perspective. The stump in the foreground continues to shoot up new growth, but the deer browse it as quickly as it appears.

FIGURE 9.15
This browse line, made by a dense population of white-tailed deer, is typical of that made by any large concentration of the deer family or other hoofed browsers.

that has made many gallant efforts to regenerate, sending up tender shoots from the energy reserves in its deep root system, but the deer have been eating them faster than it can replace them. As a rule, you can use preferred species regeneration to estimate approximate deer populations in a particular area. Hemlock regeneration, for example, will be somewhat limited at deer populations of twelve per square mile. At fifteen per square mile, the regeneration should show some signs of being inhibited. At around twenty per square mile, hemlock regeneration will probably not occur at all.

Deer prefer to browse between human knee and hip height. When that range has been stripped, the deer will browse lower to the ground. Eventually, they'll stand on their hind legs to get food, but only as a last resort. Figure 9.15 shows a browse line in a hardwood forest; you can see that the vegetation is cut off almost in a straight line. Figure 9.16 shows a deer reaching for viburnum and the browse line formed by repeated browsing.

FIGURE 9.16
This white-tailed deer stretches high for viburnum. A browse line is evident, with most forage beyond easy reach.

SIGN: ***Incisor Scrapes and Antler Rubs.*** White-tailed deer will eat the bark of several species of trees—hemlock (Figure 9.17), witch hazel (Figure 9.18), sumac (Figure 9.19), fir, elderberry, alder, apple, cherry, striped maple, mountain ash, willow, shadbush, and others, depending on locale. Using their bottom incisors, they scrape at the bark with upward movements of their heads, leaving the tiny branches intact. This sign is often mistaken for an antler rub, but in antler rubs, the smaller branches are usually broken, and the scrape looks much smoother.

FIGURES 9.17 *(left)*, 9.18 *(right)*, *and* 9.19 *(bot.)* *Members of the deer family (in this case, white-tails) will scrape the bark of different trees for food, using an upward motion of their incisors. (Left is hemlock, right is witch hazel, and bottom is sumac.) As is apparent here, the appearance of the incisor marks may change with the type of tree scraped.*

Antlers are made of living tissue, which contains nerves and blood vessels, and they grow at an amazing rate. They begin to sprout in the spring, reaching full size sometime in September, and drop off in late December or January, depending on the locale. (Horns also begin as living tissue, but, unlike antlers, they are not shed annually. Once formed, horns are there for life.) When antlers are still growing, they are covered with what is usually referred to as "velvet," a soft, skinlike casing that looks exactly like velvet. This casing falls off sometime between September and October, just before or at the start of mating, or rutting, season. At this time, the bucks also start to rub their antlers against trees. Although at one time it was thought that this rubbing was intended simply to remove the velvet, recent studies suggest that the action has more significance. As the buck rubs his antlers up and down against a sapling, sometimes rubbing very slowly in an almost mesmerized state, he is actually scenting the tree. It's thought that these scented trees are used as territorial markers that tell other deer of their presence and perhaps dominance.

Figure 9.20 shows a typical deer rub. It is very smooth in appearance and shows no incisor marks. The larger the diameter of the sapling, the larger the buck doing the rubbing. I have measured trunks with rubs on them that have been four to five inches in diameter, which would indicate a *really* large buck. (I recently found a rubbing tree that was an amazing seven inches in diameter.) Rubs are usually found on one-and-a-half- to two-and-a-half-inch saplings and usually, but not always, on aromatic trees such as pines, spruces, and hemlocks. The height of the rubs from the ground ranges from ten inches to a maximum of about forty-eight inches. Bucks will return to the same tree year after year. The tree in Figure 9.21 has actually grown in a bowed shape from all the rubbing it has received over the years. Also note the frayed bark at the top and bottom of the rub, another identifying feature. Incisor scrapes are made with an upward movement and will not result in frayed bark at the bottom.

SIGN: ***Scrapes.*** During the rutting season, bucks will use their hoofs to make scrapes in the ground under an overhanging branch. If you follow a deer run in October or November, look for a spot where the run passes un-

FIGURE 9.20 *(left)* *White-tailed deer antler rubs are usually smooth in appearance.*

FIGURE 9.21 *(right)* *Rubs often have frayed bark at the top and bottom. This tree is bowed from many years of rubbing.*

NOTE: *New data collected indicates that overhanging branches that deer scrape under usually measure from 43" to 66" above the ground. Litter from the scrape may be thrown back 36" to 180".*

der a branch, and check the ground for signs of a scrape. The buck will often scrape the ground with its front hoofs, making a triangular disturbance 24" to 41" long by 19" to 30" wide, with the triangle pointing in the direction in which the buck is facing. Then, bringing its rear legs up near its forelegs, it will urinate on its own hind legs. The urine will run down over the hock glands, located at the joint of the leg on the inside, and carry the buck's scent down to the ground. After scenting the ground in this way, the deer will bite and pull at the overhanging branch, then scent the branch with its forehead glands. The branch will often be broken and/or twisted. Bucks will come back to these areas to scrape them again. They have been recorded following the scent of a doe, grunting as they go, after the doe has visited one of these elaborate scrapes.

SIGN: *Digs.* Deer will dig through snow and leaf litter under oaks for acorns. You'll be able to see the scrape marks on the ground from the deer's hoofs, and there will usually be bits of chewed-up acorn shells lying around. For some reason, when deer chew acorns, they usually drop some. If you examine the acorns, you can actually see the marks of the molars on the outside of the shell. Wild turkeys also dig for acorns, but they swallow the acorns whole.

Deer also dig for fern roots, especially when other preferred food sources are not available.

SIGN: *Scat.* Deer scat (Figure 9.22) is extremely variable, changing from season to season and food source to food

FIGURE 9.22
White-tailed deer scat is found in many forms. The smaller pellets to the left are from a winter diet of dry foods. The pellets bunch up into larger masses (right) when the diet consists of moist foods available in summer.

FIGURE 9.23
Hard deer pellets are often defecated in small piles about 4" to 6" in diameter. Individual pellet size is usually less than 1/2" in diameter. On rare occasions, individual pellets have been measured at 5/8" in diameter and 1 3/4" long.

source. Its shape is determined by the moisture content of the food. If the food is very dry, as in winter, when deer browse mainly on twigs, the pellets will be very hard and fibrous (Figure 9.23); if it is succulent, as in spring and summer, when the deer feed on herbaceous material, the pellets will be very soft, sometimes forming a single clump of pellets stuck together (Figure 9.22, upper right) or a solid mass.

Figure 9.22 shows the range of pellet shapes and stages of compressibility. Figure 9.24 is a scat age comparison. Looking from left to right, in the upper left corner are some very fresh pellets, glistening with an outer covering of muculike material that soon begins to dry or wash off as it is exposed to weather. As the scat ages, it is easier to see the wood fiber. Hard pellets (lower right) may stay around for years.

FIGURE 9.24
When deer scat is fresh, it has a shiny, wet appearance (upper left), which is lost in a matter of days as the scat weathers (upper right). After a month or two, the scat becomes lighter in color and plant fibers begin to show (lower left). When the scat is several months old or older (it may last more than a year), you can see the fibers clearly and the pellets are about ready to break up (lower right).

Mountain Sheep: Bighorn

Ovis canadensis

Thinhorn

Ovis dalli

MOUNTAIN SHEEP ARE Old World sheep that migrated into North America across the Bering land bridge during the Pleistocene. According to Valerius Geist, in his book *Mountain Sheep: A Study in Behavior and Evolution,* mountain sheep in Eurasia were "one of man's first domestic animals if not the first." The North American species, however, have never been domesticated.

Mountain sheep are divided sharply into two major groups—bighorn and thinhorn—the chief difference between them, apart from the size of their horns, being their geographic distribution. Bighorns (the two major subspecies are the Rocky Mountain bighorn and the California bighorn) are found high in the Cascade, Coastal, Sierra Nevada, and Rocky mountains south of the Peace River in British Columbia to Mexico. They are grayish brown in color, with short, thick underhair covered by long, whitish guard hairs with brown tips. Their back is darker brown, and they have a white rump patch. Males are up to six feet long and occasionally weigh up to 325 pounds; females are

FIGURE 9.25
If you have ever watched mountain sheep, such as this thinhorn ewe, skillfully negotiate their craggy domain, you cannot help being in awe of them.

about four and a half feet long and 150 pounds. They are the largest sheep in the world.

As their name suggests, male bighorns (rams) have massive, curled horns that can weigh more than twenty pounds and are useful for head butting during the autumn rut. Horns are never shed, and those broken off during butting sessions do not grow back. During the rut (November to early December), males vie for control of the entire female band, resulting in spectacular bouts in which the animals ram one another from distances of thirty feet and at speeds of more than 20 miles per hour. The males have double-layered skulls interconnected with bone to absorb these tremendous shocks, and few encounters result in serious injury, although some studies have suggested that males with the largest horns have the shortest life spans.

Thinhorn sheep (including Dall's, Stone's, and Kenai sheep) are more northerly, inhabiting alpine meadows along the highest mountain ranges north of the Peace River, including the Brooks and Alaska ranges in Alaska, the Kenai Peninsula, and the Mackenzie Mountains in Canada's Yukon Territory and Northwest Territories. Unlike bighorns, which are not found on the prairies, thinhorns will sometimes venture out onto tundra in search of food. Thinhorns are whiter (except for Stone's, which are brown or black and look, according to Geist, like "Dall's sheep in evening dress") and smaller. Males are five and a half feet long and weigh up to 200 pounds; females are four and a half feet long and 125 pounds, with a smaller face and horns. In comparison to the bighorn's horns, the thinhorn's have a more corrugated keel on the outer curl, and the tips are spread farther apart.

Mountain sheep are probably best known for their ability to negotiate steep, rocky terrain. They are excellent climbers and jumpers, capable of leaping off ledges twenty feet high. Whether ascending or descending tallus slopes, their agility and surefootedness are consummate skills.

According to Geist, "mountain sheep are some of the most specialized grazers, for they can live on hard, abrasive, dry plants." They have a larger rumen than deer of the same size, so digestion of less luxuriant grasses takes longer and is therefore more efficient. They eat all available grasses and forbs. After the first snowfall, bighorns participate in short migrations to winter feeding grounds, usually twenty-five to forty miles away, and their lambing range is generally found between the summer and winter grounds, in areas with dry southern exposures and precipitous terrain. Migration routes are used for generations, with the young bighorns learning the routes from their elders.

Years of heavy predation, chiefly by human hunters (Francis Parkman recorded that mountain sheep provided the chief source of meat for wagon trains along the Oregon Trail), have reduced mountain sheep in all their ranges, but there have been signs of a revival. For example, the California bighorn, reduced by 1961 in both Canada and the United States to fewer than seventeen hundred animals, now numbers about five thousand, thanks to strict wildlife protection programs. Apart from man, mountain sheep have the ungulates' usual problems with coyotes, grizzlies, wolverines, and golden eagles. Wolves are its chief predators within their range. Although in some areas cougars were once thought to be decimating sheep herds and were hunted for that reason, a recent study conducted in British Columbia's Fraser Plateau found that cougars almost always tend to take males, usually subdominant males, and so have little effect on reproduction patterns, since only dominant males mate. Also, cougars prey more on elk and mule deer, and therefore actually help sheep by reducing their competitors for food. Perhaps the biggest threat to wild sheep comes from habitat loss to domestic sheep, since alpine grasslands revert to sagebrush expanses when overgrazed. "Market hunting and diseases introduced by domestic livestock took their toll," writes biologist James K. Morgan, "but the deterioration of the grasslands, I feel, was the major factor in reducing the bighorns to scattered remnant herds."

TRACKS. Since all mountain sheep subspecies are more or less the same size, their tracks are similar. I will use those of some Dall's sheep (Figure 9.26) I photographed in Alaska to illustrate mountain sheep tracks in general. Note their similarity to mule or white-tailed deer tracks. The hoofs are cloven, with rough foot pads for better traction on rocky terrain. They are, however, somewhat less heart-shaped, with more rounded tips and straighter, more

FIGURE 9.26
A Dall's sheep made this hind track, which is superimposed over a front track. Mountain sheep tracks are not as heart-shaped as those of white-tailed deer, and they tend to look boxier, with straight sides.

rectangular sides. Ewe tracks are 1¾″ to 2⅛″ long by 1½″ to 1⅞″ wide. Male tracks may reach 3½″ long by 2½″ wide.

TRAIL PATTERN. Sheep walk in a typical alternating pattern, with the hind foot falling slightly off the track made by the forefoot, leaving a somewhat double-registering trail. Strides are 16½″ to 25″, and trail widths vary from 6″ to 12″.

SIGN: *Beds.* Mountain sheep bed down on rocky promontories with loose surface material and a good view, often with a secure cliff in back. They make their beds by scraping the ground with their forefeet to make a slight depression. Since they are creatures of habit, they will use the same bed year after year, but they will rescrape it, so the depression will become quite deep over the years. The beds are oval, measuring approximately twenty-four inches long (sometimes up to forty-eight inches long), sixteen inches wide, and one inch to ten inches deep. Geist says a mountain sheep bed may be mistaken for a grizzly dig, but if it's that deep, there will almost always be a large accumulation of sheep dung nearby. There may be a strong smell of sheep urine about the place as well.

FIGURE 9.27
Mountain sheep scat is highly variable, depending on whether the diet is of woody or succulent materials. Some pellets can be very smooth-looking, ranging in color from black to gray.

Sheep also have been known to bed down in caves. Frederick Wooding, in *Wild Mammals of Canada,* reports that in a deep cave high up on the walls of the South Nahanni River's First Canyon in the Tlogotsho Mountains, scientists discovered the bones of ninety Dall's sheep that, when carbon-dated, were found to be two thousand years old.

SIGN: *Scat.* Mountain sheep droppings (Figure 9.27) are very similar to those of white-tailed deer. They can be dark black to gray. The gray color and smooth, polished appearance seem to set sheep scat apart from that of other hoofed animals.

Caribou

Rangifer tarandus

THE WORD *CARIBOU* comes from the Micmac *halibu*, meaning "pawer" or "scratcher," a reference to the huge holes this large deer makes in the snow as it forages for lichen. Caribou were once found as far south as Maine but now live almost exclusively in the Arctic, except for small herds in Quebec and Newfoundland. A total of some 1.5 million animals are divided into three groups: woodland caribou, the largest of the three, range from the Mackenzie Mountains in the Yukon Territory to the sixty-fifth parallel; Peary caribou, the smallest and nonmigrating subspecies, are found on the Queen Elizabeth Islands and along parts of the Greenland coast; and barren ground caribou, the most numerous group, are found along the arctic tundra from Hudson Bay to the Mackenzie River and extending into most of Alaska.

Caribou are extremely gregarious animals. The more than thirty established caribou herds are from 10,000 to more than 100,000 strong. The herds winter in the southern extreme of their range, near the tree line where the sparse taiga vegetation offers some protection from the weather. In spring, just before calving season, the herds begin their long trek north, sometimes traveling vast distances to the northern calving grounds. One herd, the barren ground Porcupine herd, comprising about 110,000

FIGURE 9.28
Whether it is a lone animal or a herd of thousands, there is something truly magical about watching caribou move across the tundra.

animals, winters on the Alaska-Yukon border and every year migrates eight hundred miles north to the Beaufort Sea coast. As Barry Lopez notes in *Arctic Dreams*, these well-defined calving grounds offer distinct advantages: "The number of predators is low, wolves having dropped away from the herds at more suitable locations for denning to the south. Food plants are plentiful. And these grounds either offer better protection from spring snowstorms or experience fewer storms overall than adjacent regions."

Despite the huge numbers, caribou are in decline in some parts of their range. Peter Matthiessen reports that caribou in Alaska have been "in serious decline" since the turn of the century, when a vast fire on the Kenai Peninsula destroyed all the edible lichen and wiped out many of the native populations of caribou. Domestic reindeer, *Rangifer tarandus tarandus*, were imported from Siberia in an attempt to restore the stock, and these had spread into Canada by 1929, but in general the introduction has not been successful. Mixtures of reindeer and caribou seem to have accentuated the weaknesses of both subspecies.

Caribou coloration is variable, but the most common color is a soft brown with a creamy neck and mane and creamy-white on the belly. The face is darker brown, and the snout and underside of the short tail are whitish. Peary caribou are gray to nearly white, and the more southerly barren ground and woodland caribou are more brownish. The average height at the shoulders for the barren ground caribou is three and a half to four feet; adult males can measure about six and a half feet long and weigh up to 600 pounds, with females weighing only up to 350 pounds.

Unlike any other deer, both male and female caribou have antlers, although those of the male are much bigger. (Females in southern ranges, such as in Ontario, Quebec, and Newfoundland, are often found without antlers.) Male antlers are semipalmated, especially the single, flat brow tine that extends down almost, but not quite, past the nose. The main beams can be twenty to sixty inches long, with racks up to fifty-eight inches wide. Females' antlers are much shorter than males' and lack the brow tine. Antlers begin to sprout in early spring. The velvet drops or is rubbed off as the herds move north to their summer range. Males shed their antlers after the autumn rut, but females retain theirs until calving time in the spring.

Calves are born from late May to June, about one month before mosquito, blackfly, warble fly, and botfly season. After calving, females and calves join the main herd and begin their southerly migration.

In the summer, caribou graze on grasses, forbs, sedges, moss, and lichen (they have special gut juices that enable them to digest the complex chemicals contained in lichen). In winter, the animals mainly eat lichen, if available, and will browse willow and birch shrubs. They also will eat dropped antlers and have been known to eat lemmings when those small mammals' populations are at their peak. The dominant predators of caribou are human beings (for centuries, caribou have provided food, clothing, and utensils for Inuit populations), wolves, grizzly bears, lynx, and wolverines. Golden eagles have been known to snatch newborn calves.

TRACKS. Caribou have the widest and roundest feet of all deer, and their tracks appear as widely separated crescents 3″ to 4¾″ long by 4½″ to 5¾″wide. If the dewclaws are registering, they will increase the track length considerably, to 6½″ to 8″. Caribou dewclaws, reduced to remnants in most ungulates, are still well developed, as they help keep the animal atop deep snow. When the an-

FIGURE 9.29
Caribou tracks look like two widely separated crescents facing each other. The dewclaws of this track are registering fairly parallel to the direction of travel, indicating a walking gait.

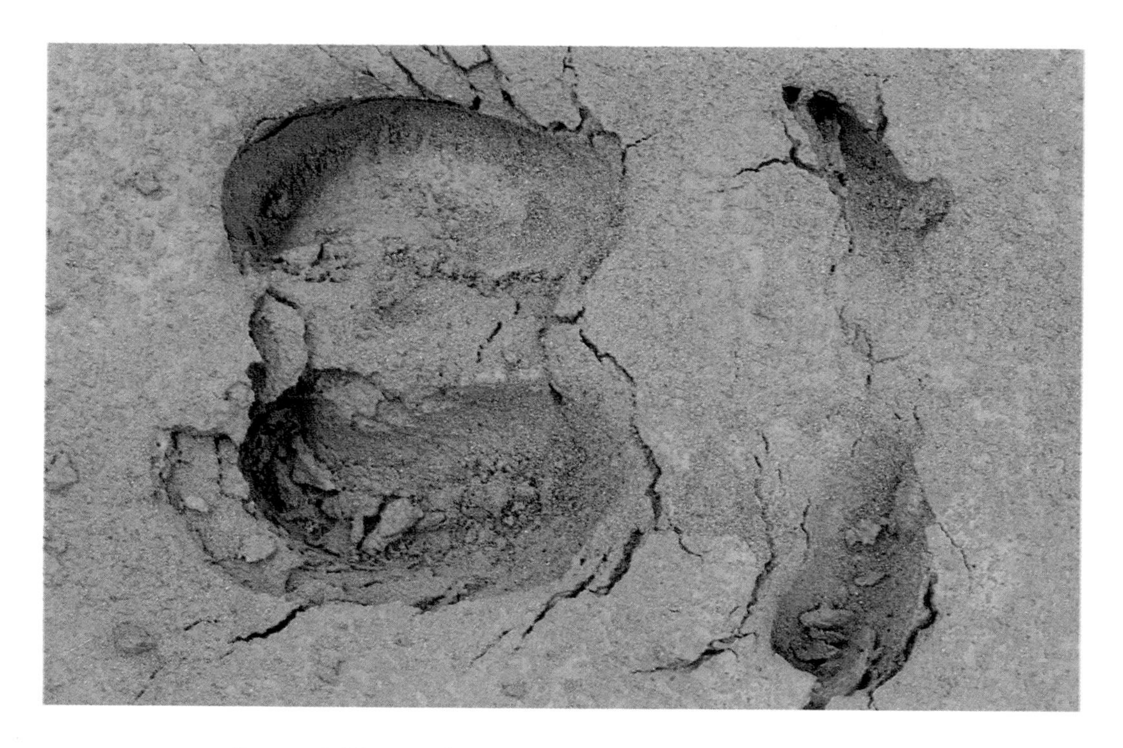

FIGURE 9.30
The dewclaws in this caribou's track are much more perpendicular to the line of travel than those in the previous photo. This indicates a faster gait.

FIGURE 9.31
The caribou's walking pattern has fairly long strides and a wide trail width. Tracks will both direct- and double-register.

imal is walking, the dewclaws fall parallel to the trail (Figure 9.29); as the animal picks up speed, the dewclaws spread out and mark the trail perpendicular to the direction of travel (Figure 9.30). Caribou hoofs change seasonally. In winter, the soft inner part of the hoof hardens and shrinks, leaving the outer portion to become an edge that bites into the snow for extra traction. At this time, dense hair also forms between the toes to protect the foot from the cold.

TRAIL PATTERNS. Caribou move in a variety of gaits: walk, fast walk, trot (an alarm response to predators), pace, and gallop. While grazing, they walk in an alternating pattern that seems leisurely but often moves

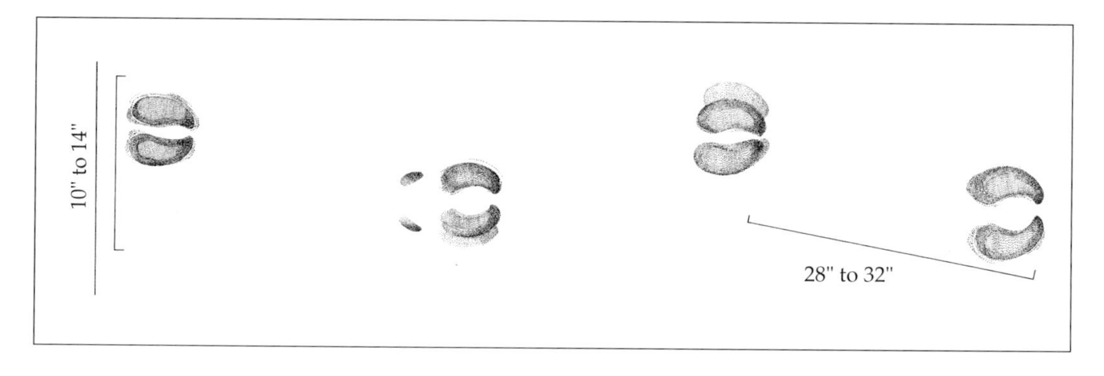

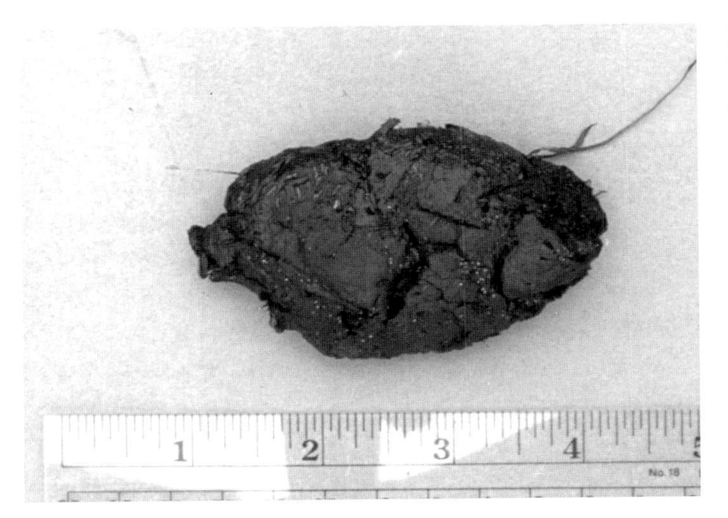

FIGURE 9.32
This fresh caribou summer scat is from a diet of moist vegetation. When food sources lack moisture, scat will be in the form of pellets resembling those of white-tailed deer and elk.

them at a pace of 4 miles per hour. The walking pattern (Figure 9.31) has a stride of 28″ to 32″, with a trail width of 10″ to 14″. Caribou can gallop at speeds over 40 miles per hour, although they cannot maintain such speeds for long. One gallop pattern I found on some glacial silt in Alaska had groups of four that measured an astounding 9+′ and strides of almost 5′.

Since caribou migrate along the same routes year after year, they leave deeply scoured trails in the tundra. They are good climbers, and their trails often go up steep cliffs or snowbanks even though more easily negotiated routes are available.

SIGN: *Antler Rubs.* Males begin to shed their velvet while still in the summer range, among low brush. Although these small trees offer little resistance to the rubbing animals, signs of their passage will be evident in worn bark and torn branches.

SIGN: *Scat.* Summer and winter scat resembles that of white-tailed deer and elk: a soft plop in summer (Figure 9.32) when the vegetation is more lush, small pellets in winter when their diet contains less moisture and a lot of fiber. Pellets can be very similar to those of white-tailed deer—flat or concave at one end, pointed at the other.

Elk
Cervus elaphus

EARLY SCANDINAVIAN settlers misnamed this animal "elk," thinking that it was a Scandinavian moose. The native name for the animal is *wapiti*, a Shawnee word meaning "white deer," probably in reference to the elk's large, buff-colored rump patch ringed by a distinctive dark brown. Except for a dark brown mane on the male's head and neck, the elk has a tawny, glossy, almost cream-colored coat, short in summer but with longer guard hairs in winter. Taxonomically, the wapiti is the same species as the European red deer, which has recently been introduced in North America, but the elk is much larger. Its average length is six and a half to ten feet, its tail is three to eight inches long, and its height at the base of its muscular neck is about five feet. Males weigh six hundred to a little over a thousand pounds; females are about 25% smaller.

Although the animals have been driven out of most of their former range due to loss of habitat and hunting, there are still an estimated 400,000 elk on the continent, mostly in the West. In Canada, they are found from Vancouver Island east almost to Manitoba; in the United States, smaller herds are found in isolated pockets throughout the West from Montana to the Pacific Coast, and some have been introduced to reserves in Pennsylvania. Elk summer in high alpine meadows, grazing on forbs and grasses,

FIGURE 9.33
A bull elk is pictured here with a small harem. Some harems may contain up to sixty cows. One of the bull's predominant features is its large antlers, which can weigh up to thirty pounds and reach 5′ in length.

Figure 9.34
Elk tracks are larger and rounder than those of white-tailed deer and mule deer. The direction of travel is to the right.

and come down to the forest's edge in winter, when the high grasslands are covered with snow. During the winter, their diet consists mostly of available grasses and fibrous browse. In various parts of their winter range, they'll eat species such as trailing blackberry, oak brush, aspen, serviceberry, willow, big sage, and snowberry. They've also been known to eat the cambium layer of trees.

The elk's haunting call is a buglelike sound that starts on a low note, rises gradually, and then flattens out to a sort of grunt or hee-haw. The call is made during the rutting season (September to November). Biologists believe that the bugling, which requires large amounts of energy, warns off other males and attracts females, who respond to the loudest and most frequently bugling males. Antlers are another mating tool. Males without them, even if they are large and strong, cannot keep females in their harem. The antlers on an older male will weigh up to thirty pounds and grow five feet long with a forty-seven-inch span. They rise in two beams upward and backward from the forehead, with tines, or points, growing off them. Tines increase in number in mature animals. An animal with twelve tines (six per side) is called a royal stag, while those with fourteen or more are imperial stags. The antlers begin to grow in April or May, reach maximum size in late July, and begin to lose their velvet in August or September. Woodland elk drop their antlers in February; those in open ranges seem to retain theirs longer.

The elk's chief predator is the mountain lion, but wolves and black bears also prey on stragglers, and wolverines have been known to kill calves.

FIGURE 9.35
This track is splayed more than the one in the previous photo. Here also, the direction of travel is to the right.

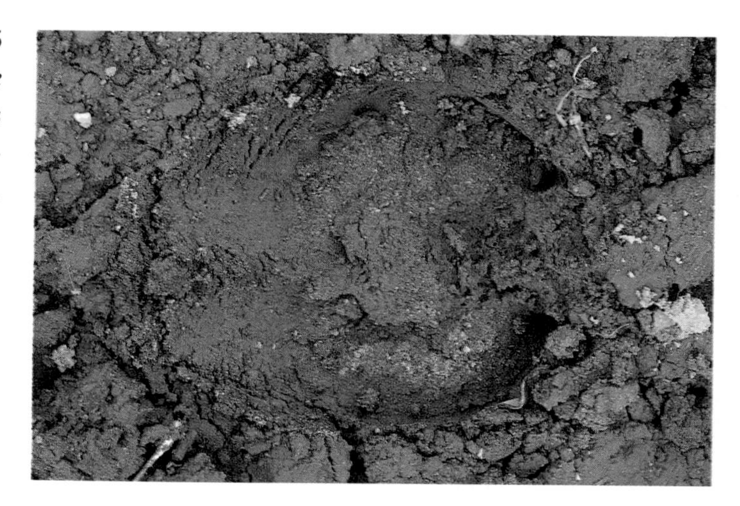

TRACKS. The elk's hoofs are both larger and rounder than those of the white-tailed deer and the mule deer. The hoofs of the Roosevelt elk (named after Theodore Roosevelt in 1897 and the largest subspecies in the United States) measure 3⅛″ to 4⅞″ long by 3⅛″ to 4⅝″ wide (Figures 9.34 and 9.35). Elk tracks are smaller and rounder than moose tracks, but there is some overlap with those of a yearling moose, especially if substrate conditions tend to round out the appearance of the yearling's tracks.

TRAIL PATTERN. The elk's walking pattern is an alternating, mostly double-registering trail with the hind track superimposed on and partly obliterating the front track. Often the hind foot will register slightly ahead of the front foot. The walking stride is 22″ to 34″, and the trail width is 7″ to 11″ (Figure 9.36).

FIGURE 9.36
The elk's walking pattern is usually a double register. The hind track often has a slight lead in the overlap of the front track.

SIGN: *Wallows.* Elk dig shallow depressions in the ground with their hoofs and antlers, urinate or defecate in the depressions, and then roll or wallow around in them to cover their bodies with the mixture. Female elk (and many trackers) become quite excited when they come upon one of these wallows.

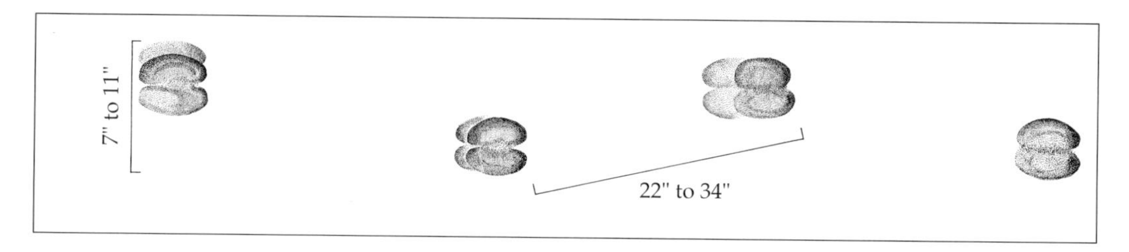

SIGN: ***Antler Rubs.*** The velvet from an elk's antlers begins to fall off in August or September. At this time, males will rub their antlers against small saplings, scraping the bark off the trunks. Some people believe they also leave the scent from their facial glands on the tree as a territorial marker.

SIGN: ***Scat.*** Elk scat in winter (Figure 9.37), when the animals are feeding mostly on browse and dry grasses, resembles moose scat, but the pellets are much smaller and there aren't as many of them. The pellets are slightly elongated or bell-shaped, at times having little points. When more succulent foods are eaten, the pellets begin to lose their shape (Figure 9.38). Summer scat may be more like cow patties but smaller—flat chips or plops 5" to 6" in diameter.

FIGURE 9.37 *(top)*
These winter elk pellets were collected from several different defecations to demonstrate their various shapes.

FIGURE 9.38 *(bot.)*
The scat samples here are from a diet with more moisture than the diet resulting in those pictured above. The pellets are starting to mass together. In summer, scat will look more like cow patties.

Moose

Alces alces

NEXT TO THE BISON, the moose is the largest land animal in North America, weighing up to fourteen hundred pounds (a bull Kenai moose from Alaska, *Alces alces gigas,* can reach eighteen hundred pounds) and measuring up to eight and a half feet in length. The second biggest moose after the Kenai are thought to be found in Maine. A hundred years ago, Thoreau saw a cow moose in Maine that measured, "from the tips of the hoofs of the forefeet, stretched out, to the top of the back between the shoulders . . . , seven feet and five inches. I can hardly believe my own measure, for this is about two feet greater than the height of a tall horse."

Moose are found from Alaska through most of Canada to the Atlantic Coast. In the East, they range south into Maine, New Hampshire, Vermont, Minnesota, Isle Royale in Lake Superior, and recently Massachusetts and New York. In the West, they extend south through the Rockies into northeast Utah and northwest Colorado.

Studies of moose populations in Canada and the United States have turned up some interesting statistics. One such study of moose and wolf populations has been conducted over the past thirty-four years on Isle Royale. Although I discuss this study elsewhere (see pages 19-20), it is worth repeating here that, despite the conventional assumption that wolf predation is responsible for the relatively low numbers of moose in some areas, the moose on Isle Royale are healthy despite, or perhaps because of, the wolf packs.

Wolves will prey on moose, but they tend to take the very young and the very old, possibly because they are less able to defend themselves. (Other factors, such as food supply, severity of winters, and availability of other prey, also play a part in wolf predation on moose.) According to David Mech, perhaps the world's leading authority on wolves, a healthy moose is capable of fending off an entire wolf pack. One winter, Mech observed 131 confrontations between wolves and single moose on Isle Royale. Of that number, about half simply outran the wolves (moose can run up to 35 miles per hour), and the other half stopped to defend themselves with their hoofs and drove the wolves off; only six moose were killed. Although wolves can kill healthy animals, the most recent studies have confirmed that they feed largely on calves (34% of the wolf kill)

FIGURE 9.39
Well equipped for living in the north country, moose are the largest deer in the world. They have long, powerful legs and can measure up to 7½′ high at the shoulders. The spread of a bull's huge palmate antlers can be over 5′ across.

and adults over seven years old. The older moose tend to be either sick or injured. Out of twenty moose killed by wolves over two winters, ten had arthritic joints and six others had periodontal infections that prevented them from eating effectively. Removing old or defenseless animals from a population leaves only healthy animals to pass on their genes to future generations, ensuring that, in the long run, the vigor of the species is maintained.

Moose populations seem to be more severely affected by deer than by wolves. Deer carry a parasite called *Parelaphostrongylus tenuis,* a nematode worm (also called a brain worm) that does not affect the host deer but is harmful to moose, sometimes causing a rapid progression of symptoms from blindness and disorientation to paralysis and death. Studies in Ontario's Algonquin Park have correlated fluctuations in the moose population far more closely with the northern movement of white-tailed deer than with changes in wolf densities.

When bulls are in rut, usually from late August to October, they will utter a series of deep grunts that, although they don't seem to be loud, are so deep that they can be heard from quite a distance. In most areas, you can get pretty close to bulls and cows, but you have to be very careful of them. The bulls can be a little disgruntled at this time of year, especially if they've just been rejected by a cow. And although bulls, with their big racks, may look more threatening than cows, don't let the cow fool you. Every year I meet some wildlife photographer friends in Baxter State Park, in Maine, to photograph moose. One evening in camp, we noticed that Bill Fournier was looking a little peaked, and his clothes were torn in places. When we asked him what had happened, he said that a cow moose had charged him and he had had to climb a tree to get out of her way.

My own close call came late one afternoon at a pond in Baxter. I was sitting at the entrance to a well-used moose run leading to the pond. Three hundred yards away stood a massive bull moose that had been courting a cow accompanied by her calf. The cow was paying no attention to the bull, and I could see that the bull was getting pretty frustrated. He turned and stared at me for a time, then started walking toward me—not fast, but steady, and straight at me. I unloaded two rolls of film on him before I realized I could no longer fit his rack into the frame. I looked up and saw that he was less than fifteen feet from me and still coming. I'm not ashamed to say that I ran. I don't really remember how I got behind a tree, but later one of my friends who had been watching and laughing from across the pond told me that I had hopped to safety like a rabbit. As soon as the bull saw my ignoble retreat, he seemed satisfied and moved off into the forest.

TRACKS. Moose tracks are quite large. One track (Figure 9.40) from Maine was just over 6" long. When the

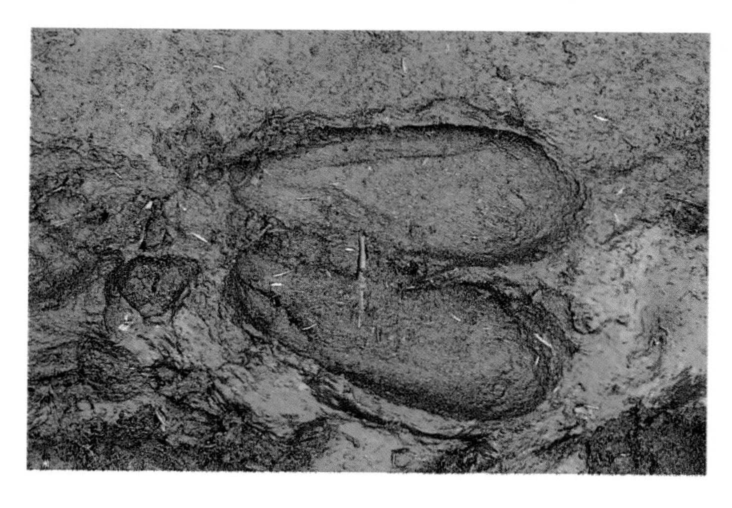

FIGURE 9.40
This adult moose track measures 6⅛" long without the dewclaws showing. Moose tracks are often heart-shaped like white-tailed deer tracks, although they are not usually as pointed.

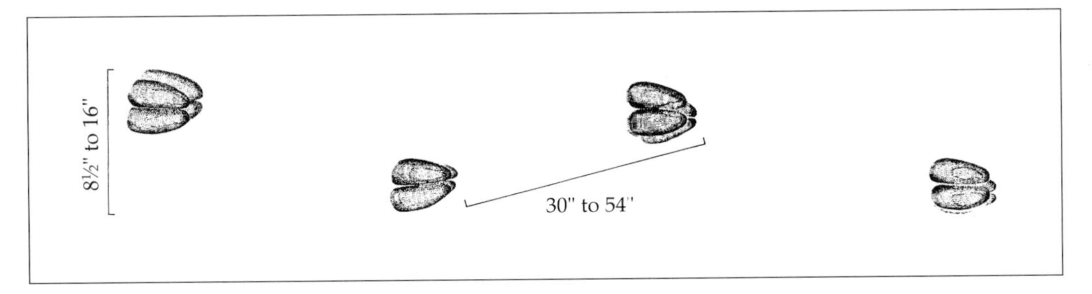

FIGURE 9.41
The alternating pattern of the moose is generally a double register with a wide trail and long strides.

dewclaws register, as they often do in soft mud or snow, the tracks can be up to 10½″ long. They resemble deer tracks in structure but are much larger and, in large adult bulls, blunter. Cows and young bulls leave pointier tracks. As with deer, the sharper end of the heart points in the direction of travel.

Most moose tracks range from 4″ to 6⅞″ long by 3½″ to 5¾″ wide. Yearlings have smaller tracks, of course. Tracks of calves I measured in Alaska and Maine (a small but wide sample) ranged from 2¾″ to 3½″ long by 2¼″ to 3″ wide in the fall. Wyoming moose are the smallest moose in North America; Alaska moose are the largest. In the northeastern states, you're probably safe assuming that any track over 4¾″ wide is that of a mature bull, as long as the width has not been exaggerated by the substrate or speed.

TRAIL PATTERN. Moose walk in an alternating pattern similar to that of the white-tailed deer, although a moose will probably double-register more often. Use the walking pattern for determining species: Moose strides are usually 30″ to 54″ long, and their trail width is 8½″ to 16″ (Figure 9.41).

SIGN: *Browse.* Moose browse (Figure 9.42) is similar to deer browse, and sometimes it's hard to distinguish between the two. There are, however, some plant species—such as balsam fir—that moose eat and deer usually leave alone. Moose will avoid barbed plants, such as raspberry, blackberry, barberry, and juniper, which deer will eat. Deer will also eat sweet fern (in Maine, Pennsylvania, and Massachusetts and sparsely in Wisconsin and Minnesota), hemlock (in the Northeast), mountain laurel (in the Northeast, the North Carolina mountains, and Pennsylvania and lightly in Wisconsin and Minnesota), and elderberry, but

FIGURE 9.42
Moose browse looks the same as that of the white-tailed deer and most other cervids, with twigs torn and squared off. The browse in this photo is on hobblebush.

FIGURE 9.43
This fresh rut pit was made by a bull moose during the "rut," or mating season. Moose dig these pits by scraping the ground with their front hoofs.

moose will avoid these plants. Moose browse is very high, often up to seven feet above the ground; deer browse will usually be from about four feet down, as long as the animals are not pressed for food. Favored browse species for moose are willow, aspen, white birch, and mountain ash. If moose browse is hard to find, focus on balsam firs, where the evidence is more apparent.

SIGN: *Rut Pits.* Bulls will make rut pits, or wallows (Figure 9.43), by digging with their front hoofs and then urinating into the pits. This is done during the rut and plays an important role in the mating ritual. Wildlife photographer Bill Byrne was a fortunate (and extremely rare) witness to one of these rituals. In September 1991, he was photographing moose in Baxter State Park when he came upon a bull and a cow in the woods near a stream. The cow was browsing on birch and mountain ash trees, and the bull started making a pit, digging at the dirt with his front hoofs. The cow made a soft, mooing sound but made no move toward the bull. The bull moved several feet and began to make a second pit (Figure 9.44). Squatting over this pit, he urinated into it. At that point, the cow made another mooing sound and rushed to the bull's side, circling him and mooing softly. She then nudged the bull out of the pit and, as he stood by, dropped down into it, wallowing with her neck and head close to the ground and vocalizing. She got up, dropped down again, and wallowed in the scrape, still vocalizing. She then seemed to settle down. The bull moved eight to ten feet away and lay down facing her. Ten

FIGURE 9.44
A cow moose stood nearby watching the bull dig and then urinate in this fresh rut pit. When he had finished, the cow pushed the bull out of the pit to roll in it herself.

to fifteen minutes later, the cow got up, went over to the bull, and rubbed her muzzle on his antlers. They touched muzzles, and she returned to the pit.

These pits may be twenty to forty inches long, thirteen to thirty-two inches wide, and three and a half to six inches deep. If they have urine in them, you'll certainly be able to smell it. Bulls also have been known to wallow in their own urinated pits. Biologists believe that bulls use the pits to tell other bulls of their presence.

SIGN: ***Incisor Scrapes and Antler Rubs.*** Just as deer scrape the bark of hemlocks with their incisors, moose scrape red maples (Figure 9.45), especially when a tree is young and has smooth bark, as well as aspens and other trees. Incisor scrapes are often about shoulder height—much higher than deer scrapes. Look for the individual teeth marks of the moose as it scraped in an upward motion, or sometimes at a slight angle. If the scrape shows evidence of a downward motion, it may be an antler rub.

Moose, like other cervids, will rub their antlers against trees and low shrubs. Moose rubs will have the same smooth appearance as white-tailed deer rubs, with one

FIGURE 9.45 *(left)* *Like other members of the deer family, moose scrape the bark of different trees with their incisors (the example here is on red maple). Although often at shoulder height, incisor scrapes have been measured as high as 90" above the ground.*

FIGURE 9.46 *(right)* *Moose antler rubs look like those of deer but are usually much higher. Rubs may range in height from 18½" to 100" above the ground. This bull rub was almost 7½' high.*

big difference: height. Deer rubs usually are not more than forty-eight inches from the ground, while moose rubs may be up to one hundred inches high, with most of the rubbing found between thirty-nine and eighty-one inches. A tree I examined in Algonquin Park (Figure 9.46) was three inches in diameter, and the rub marks were eighty-nine inches above the ground.

SIGN: ***Scat.*** Like the scat of most members of the deer family, moose droppings vary according to the amount of moisture and fiber the animals have been eating. In summer, when moose enjoy an herbaceous diet, they will leave scat in the form of a plop (Figures 9.47 and 9.48) that may be 7½″ to 11″ in diameter. In winter, when their diet is drier and more fibrous, they will excrete pel-

FIGURES 9.47 *(top)* and 9.48 *(bot.)* *These are examples of summer moose scat, the results of feeding on succulent vegetation. The scat in the top photograph is 11″ in diameter.*

FIGURE 9.49 *(top)*
In winter, moose browse mostly on dry, woody materials, resulting in the pelletized scat shown here.

FIGURE 9.50 *(bot.)*
This scat shows a variation that occurs when the moose's diet is changing from winter to summer forage.

lets that are sometimes oval and 1″ to 1¾″ long (Figure 9.49) and sometimes blocky. In spring and fall, the pellets may look like they've been squeezed together (Figure 9.50).

According to *The Moose in Ontario,* an excellent book edited by Mike Buss and Ron Truman and published by the Ontario Ministry of Natural Resources, calf pellets are 1″ long, a little larger than deer pellets. Adult bull pellets are larger and tend to be fatter and blockier. Cow pellets are generally longer and narrower. Pellets provide a rough method for determining the sex and age of the animals in a given area.

Bison
Bison bison

THE BISON OR BUFFALO (from the Latin *bubalus*, meaning "wild ox") is the largest terrestrial animal in North America. There are two subspecies—plains bison and wood bison—but the wood bison is now almost extinct, with fewer than one thousand animals left in two Canadian national parks. The wood bison is somewhat larger than the plains: seven feet tall at the shoulder, compared to six and a half feet for the plains (females are about a foot shorter). The males of both subspecies weigh nine hundred to two thousand pounds, and the females weigh seven hundred to eleven hundred pounds. A full-grown male bison can be ten to twelve and a half feet long, from its low head to the tuft at the end of its eighteen-inch tail. It has a shape familiar to anyone who has ever seen an old American nickel: massive, shaggy head (for head butting during the rut); short, powerful neck; and humped shoulders. Its foreparts are covered with a long, dark, woolly mane and beard (more on the plains than the wood), and its hind quarters are lighter and seemingly naked. The bison has short legs with large hoofs and short black horns that curve out, up, and in at their pointed tips, which are about three feet apart. The wood bison's horns are longer and straighter than the plains bison's, with less hair around the base.

These days, the bison's only significant predators are wolves and grizzlies, which usually have to content themselves with stray calves or old, weakened adults. Coyotes sometimes associate with bison, possibly hunting small animals in their midst or waiting for scavenging opportunities. On two occasions at the National Bison Range in Montana, however, coyotes that wandered into a bison herd were horned and trampled to death. The bison's primary defense is stampeding, as it can run up to 32 miles per hour—faster than a wolf.

In general, the bison's breeding season, or rut, occurs between July and October, peaking around mid-August, when the animal's long, grunting calls can be heard miles away. Males compete for females by standing side by side to assess each other's size. If there is an obvious difference, the smaller male backs off. If the two animals are of similar size, they square off and butt heads, and will sometimes take advantage of an exposed flank to inflict injuries. The strongest males in a herd may acquire harems of up to sev-

FIGURE 9.51
Bison often conjure up images of Native Americans and the Old West. They are also reminders of people's insensitivity to other species. Throughout the nineteenth century, hunters reduced the herds of bison from fifty million to fewer than fifteen hundred individuals.

enty females. Females give birth to a single calf that is able to stand one hour after birth and run within four hours.

For ten thousand years, from the last Ice Age until the coming of Europeans to North America, bison blanketed the prairies. The first white men to see what is now Kansas didn't really see much of Kansas at all, for the land was covered from horizon to horizon with bison. In 1839, John Bidwell, an early traveler on his way to California, wrote en route, "I think I can truthfully say that I saw more buffaloes in one day than I have ever seen of cattle in my life. I have seen the plains black with them for several days' journey as far as the eye could reach." Frank Gilbert Roe, in his book *The North American Buffalo,* cites a traveler in the 1850s who reported that his westward train trip was held up for three days while a single migrating herd crossed the tracks. Ernest Thompson Seton estimated that as many as 75 million bison roamed the prairies before the seventeenth century. Native Americans hunted them for meat and hides and, until the coming of white people, didn't seem to affect the size of the herds. Stories that Native Americans would kill thousands of bison only for their tongues should be tempered with Bidwell's eyewitness account: "We . . . followed the practice of Rocky Mountain white hunters, killing them just to get the tongues and marrow bones, leaving all the rest of the meat for the wolves to eat. But the Cheyenne, who traveled ahead of us for two or three days, set us a better example. At their camps we noticed that when they killed buffaloes they took all the meat, everything but the bones."

Throughout the nineteenth century, white hunters reduced the great herds from 50 million to fewer than 1,500. Many were killed to provide meat for railway crews. In the 1880s, William "Buffalo Bill" Cody was hired to provide one Kansas Pacific Railroad crew with 12 fresh bison carcasses every day for 18 months. He earned his nickname by killing 6,400 animals. Millions more were slaughtered to clear the land for homesteading and to subdue the Native American tribes that depended on free-ranging bison herds to continue their nomadic tradition. "Never before in all history were so many large wild animals of one species slain in so short a space of time," wrote Theodore Roosevelt, himself one of America's great hunters.

The last bison in the East was killed in Pennsylvania in 1799. "The last Canadian buffalo herd was dispatched in 1878," writes Canadian science historian Don Gayton, "but the American herds hung on a few years longer. The final curtain fell about 1891, in central Montana, when the last free-ranging herd was destroyed." By the end of the century, there were a thousand bison left in Canada and five hundred in the United States. Laws to protect the vanishing species were finally passed in both countries in the 1890s, and captive breeding programs since then have resulted in a present population of roughly 100,000 animals, most of them free-ranging once again in refuges such as Wood Buffalo National Park in Manitoba, Canada, and Yellowstone National Park in the United States. Many of these animals are crosses between plains and wood bison.

One of the side effects of removing the bison from the prairies has been a change in the prairie habitat. Bison thrived on the shortgrasses that covered the plains—the fescues, stipas, and native wheatgrasses that grew in various parts of the West. Evidence collected by Gayton suggests that although vast bison herds could trash the ground in an area where they stopped to graze, the prairies were so huge that it might be years before that particular area was grazed again—time for the sensitive shortgrasses to reestablish themselves. Those native grasses did not stand

up well to continuous cattle grazing, however, and in the early decades of this century, they began to die out, creating the dustbowl environment that culminated in the droughts of the thirties. Botanists had to introduce the faster-germinating crested wheatgrass from Europe to provide fodder for cattle and roots to hold down the soil.

FIGURES 9.52 *(top)* and 9.53 *(bot.)*
Bison tracks may be round (top right) and resemble those of cattle, though some will have a slightly more elongated shape (top left). On a hard surface, they may reveal only the outer edge of the hoof. Note how round the track below appears.

TRACKS AND TRAIL PATTERN. Bison tracks are similar to those of domestic cattle, except they are larger and, in most cases, slightly rounder (Figure 9.52). Adult bison tracks usually measure 4¼″ to 5¾″ long by 4″ to 5¾″ wide, with the hind track being slightly smaller than the front. On a hard substrate, the space between the two sides of the double hoof may not register, and the tracks can be confused with those of an unshod horse (Figure 9.53). In snow, the bison's dewclaws usually show up in the tracks, and a heavy foot drag often is evident.

Walking strides are 24″ to 36″, with trail widths 10″ to 19″.

SIGN: *Wallows.* All ages and both sexes of bison dig shallow depressions in the ground, usually eight to fourteen feet across and a foot deep (Figure 9.54). They roll in the dust of these depressions to rid themselves of molt-

FIGURE 9.54
Bison of all ages and both sexes roll in wallows to help remove molting hair and rid themselves of biting insects. Wallows may measure 8′ to 14′ across.

ing hair and biting insects. These wallows form catch basins for rain. The bison will roll in them again to cover themselves with mud as further protection from insects. During dry spells, male bison will urinate in these wallows to form mud and then roll in them. Some biologists believe that this is ritual behavior, possibly connected to establishing dominance.

SIGN: *Rubs.* "Buffalo have a great need to rub," Gayton writes, "and on treeless prairie, the only objects tall and rigid enough for them were the rare boulder erratics, huge stones that somehow missed being pulverized by glaciers." These boulders, worn smooth and even shaped by ten thousand years of bison, remain on the plains even though the animals are gone. They are usually at the bottom of a slight depression formed by countless hoofs circling the stone as the bison rubbed against it or at the confluence of many nearly vanished bison trails, most of which can now be discerned only from the air.

In areas where there are trees (in the higher winter ranges), both sexes of bison rub against them, although during the rut, males will rub more often than females. The bark will be worn off many of these trees in a ring all the way around the trunk, and limbs will be trampled into the ground. Whole stands of timber also may be affected by bison during the rut. One study found that 51% of large pines in parts of Yellowstone National Park have been horned by bison.

FIGURE 9.55
This bison scat from a moist diet resembles that of a domestic cow, though bison scat is usually bigger.

SIGN: ***Scat.*** Bison scat is similar to that of the domestic cow, but it can be larger, usually measuring 10″ to 12″ in diameter. Droppings range in consistency from a soft patty (Figure 9.55) to a more segmented form that readily breaks up into wafers, sometimes called buffalo chips (Figure 9.56). The result of a drier diet, chips were used by Native Americans and pioneers for fuel.

FIGURE 9.56
Bison scat from a drier diet forms into wafers, also called buffalo chips.

CHAPTER 10: BIRDS

Aves

Wild turkey, bathed in dawn's light, Martha's Vineyard, Massachusetts.

MOST OF MY ACQUAINTANCES know I spend a lot of time tracking mammals, so when I was hired to monitor loons for the state of Massachusetts, a few friends teased me: "How are you going to find the loons, Paul? You won't be able track them, you know!" It was obvious I couldn't track the loons on the waters of Quabbin Reservoir. However, any bird that touches ground potentially leaves sign of its passage. These signs range from the distinctive large track of a great blue heron, to the detailed wing marks of a raven landing in the snow, to the somewhat rectangular holes bored into mud by the American woodcock as it searches for food. Or you might come across the shattered remains of clams, mussels, and crabs dropped by gulls onto rocks along the seashore, or a collection of white-coated scat in a ruffed grouse's roost.

Great blue heron

There are about 650 species of birds known to nest in North America. Over the last century, methods of identification have evolved from using actual specimens to a fairly reliable system based on field observations. Some species, however, are so similar that they cannot be distinguished by sightings and must be identified by breeding song. Other birds are extremely arboreal and rarely touch the earth, so positive field identification is not always possible, even for the astute bird-watcher.

Although it can be difficult to identify birds–which spend much of their time airborne–strictly through their tracks, scat, and other sign, there *are* many birds that can be identified this way. Some birds leave very distinctive "signatures," like the holes left by the pecking of a yellow-bellied sapsucker or the cavities made by the pileated woodpecker. Other sign is more ambiguous, like small indentations in dirt called "dust bowls"–wallowing places where birds cover themselves with dirt to rid themselves of lice, mites, or other parasites. You'll often find feathers in and around the bowl that were loosened in the process.

Learning to identify bird tracks and sign is a valuable tool in confirming the presence of birds that you are not able to see or hear. Walking up to the remains of a small animal in the snow and being able to determine that it was killed by an avian rather than mammal predator, and to further distinguish whether a hawk or an owl killed it, is not only exciting but may be critical information in a wildlife study.

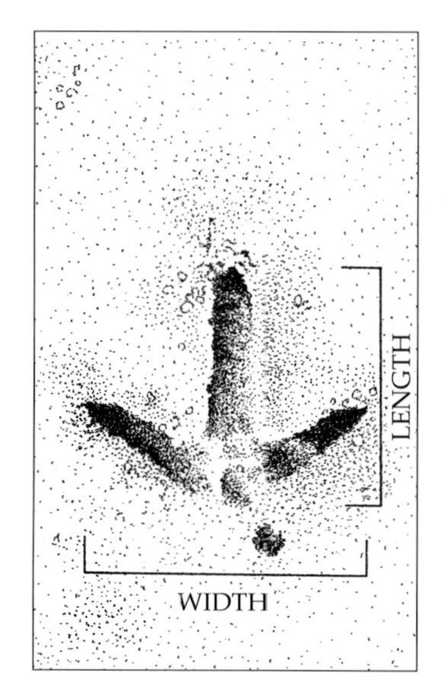

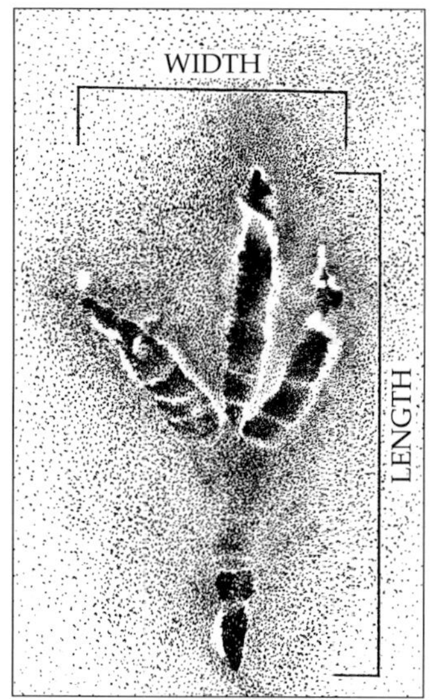

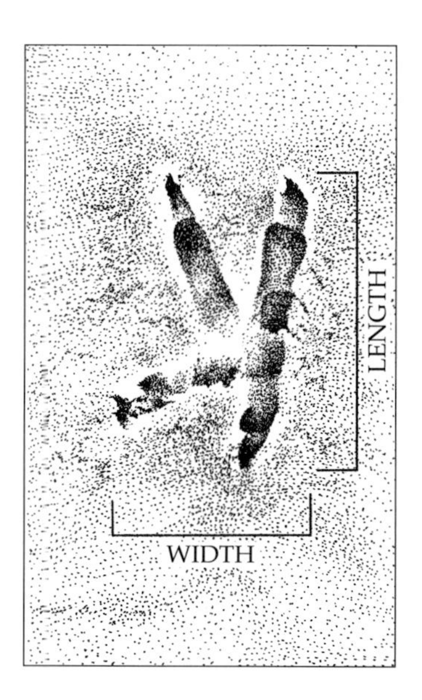

FIGURES 10.1, 10.2, 10.3
These illustrations depict the categories of birds defined according to track types: left, Category 1; middle, Category 2; right, Category 3. Each illustration demonstrates how to measure track lengths and widths for each category.

TRACKS. Given the number of bird species that exist, it would be impossible to cover them all in this book. I have grouped the birds into three major categories (with subgroups) based on certain shared characteristics in their tracks. Category 1 is comprised of ground birds, some shorebirds and waders, and waterfowl; Category 2 is perching or songbirds and long-legged waders; Category 3 is woodpeckers, roadrunners, and owls. These categories do not follow any other system of classification usually found in bird identification guides. They are strictly my own method of organizing the tracking information for ease in identification and differentiation between the tracks of different species. Figures 10.1, 10.2, and 10.3 depict representative tracks from each of the three categories of birds and show how to measure the tracks. See pages 320-321 for measurements for specific birds.

Other important factors to keep in mind when bird tracking are the season and habitat. Some species will be found in your area only when migrating, or may breed in your area but winter elsewhere. Where did you find the tracks–at the seashore or in an inland forest? If you're in a wetland habitat, is it a freshwater marsh, the shore of a pond, a wooded swamp, or a salt marsh? You'll need to consult a good bird guide for habitat use and range maps if you aren't familiar enough with the birds, their habitat use, or with the area in which you are tracking.

Killdeer

Ground Birds, Shorebirds, Waders, and Waterfowl. This category consists of three subgroups: ground birds, such as wild turkey, ruffed grouse, American woodcock, ptarmigan, ring-necked pheasant and quails; some wading and shorebirds, including sandhill cranes, greater yellowlegs, sandpipers, killdeers, and gulls; and waterfowl, such as geese, ducks, and swans. All of these birds have three toes pointing forward and one short toe pointing backward. In some species, especially ducks, geese, and gulls, the short toe often does not register in the track. In others—wild turkey and grouse—it usually registers as a nail hole to the rear of the three forward-pointing toes. See Figures 10.4, 10.5, 10.6, 10.7 and 10.8 for sample tracks from this category.

Gulls, geese, ducks, and swans have webbing connecting the three forward-pointing toes. This assists them in swimming and when it shows in the tracks as in Figures 10.6 and 10.7, it also helps to distinguish these species from the others in this category. On a hard substrate, however, the webbing may not be evident. A closer look at the tracks of these web-footed birds shows that the two outside toes curve in toward the center toe (Figures 10.6, 10.7, and 10.8). However, in the tracks of turkey, pheasant, cranes, and others in this category *without* webbed feet, the two outside toes tend to be straight or, in some cases, curve away from the center toe (Figures 10.9 and 10.10). The curvature of the outer toes is an important difference when distinguishing between wild turkey and Canada goose tracks, since they are often the same size and can be found in the same habitat (lakeshores, for example). Also, tracks of these two species look very similar on hard surfaces where the goose's tracks do not show webbing. Note, too, that the toes of ground birds often appear more robust than those of wading or shorebirds.

FIGURE 10.4 *(bot. left) Wild turkey tracks in mixed fine and coarse sand. Wild turkey tracks measure from 4″ to 5″ long by 4¼″ to 5¼″ wide.*

FIGURE 10.5 *(bot. right) Two ring-necked pheasant tracks in sand. The hind toe registered in the track on the left but not in the track on the right. Ring-necked pheasant tracks measure 2″ to 3″ long by 1¾″ to 2¾″ wide.*

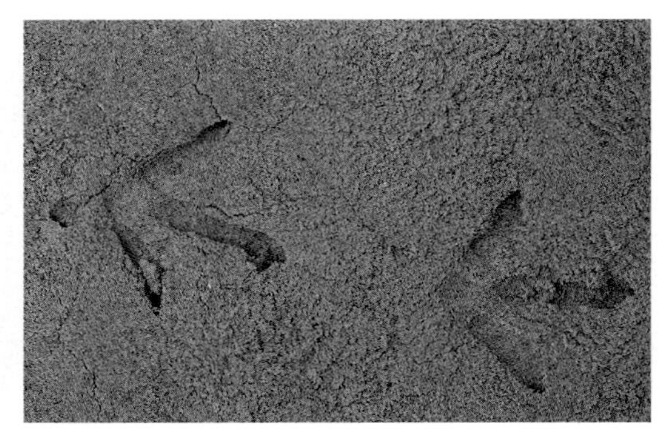

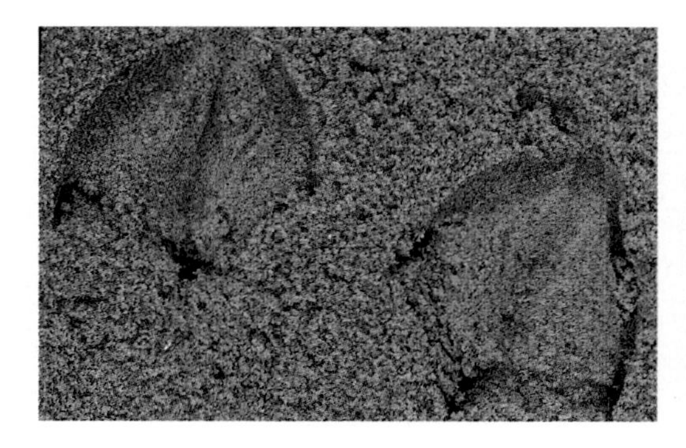

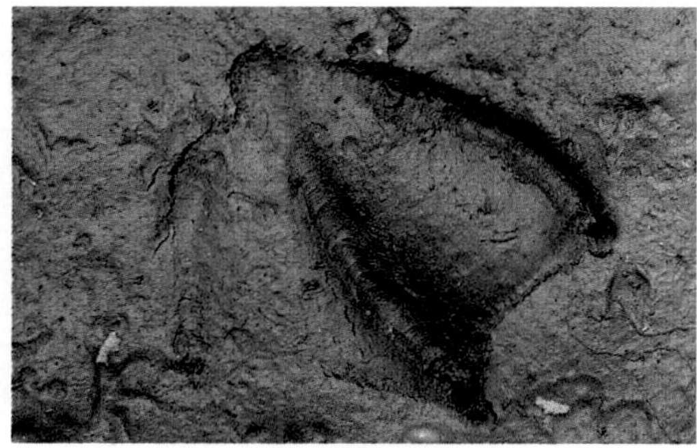

FIGURE 10.6 *(top left)*
These wood duck tracks in sand show webbing. The hind toe is apparent in the track on the right at the top of the track. Wood duck tracks measure 1⅞" to 2¼" long by 1⅞" to 2⅛" wide.

FIGURE 10.7 *(top right)*
A Canada goose track in mud shows webbing. Track measurements for Canada geese range from 4" to 4⅞" long by 4" to 5" wide.

FIGURE 10.8 *(middle right)*
The three large tracks from left to right are: great black-backed gull (track sizes 3⅛" to 3¾" long by 3⅜" to 4" wide), herring gull (track sizes 2½" to 2⅞" long by 2¾" to 3¼" wide), and ring-billed gull (track sizes 1⅞" to 2⅜" long by 2" to 2⅜" wide). The smaller tracks are killdeer (track sizes 1" to 1⅛" long by 1 3/16" to 1½" wide).

FIGURE 10.9 *(bot. left)*
This woodcock track in soft mud has thin toes compared to those of ruffed grouse. Woodcock track sizes are 1½" to 2" long by 1⅝" to 1⅞" wide.

FIGURE 10.10 *(bot. right)*
A very detailed ruffed grouse track in mud shows the indentation made by the hind toe to the rear of the track (right side of photo). Ruffed grouse tracks measure 1¾" to 2¼" long by 2" to 2½" wide.

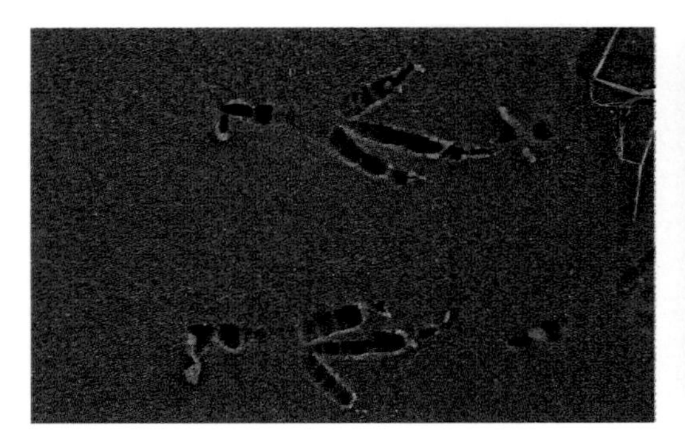

FIGURE 10.11 *(left)*
Crow tracks side by side in wet sand indicate that the bird was hopping along. Crow tracks measure 2⅞" to 3½" long by 1⅜" by 1⅞" wide.

FIGURE 10.12 *(right)*
These song sparrow tracks in very fine sand are quite small. Track measurements for sparrows are 1¼" to 1½" long by 7/16" to ⅝" wide.

Perching or Songbirds and Long-legged Waders. Perching or songbirds, such as robins, rock doves (domestic pigeons), crows, ravens, and magpies, and some long-legged wading birds, such as great blue and other herons, egrets and ibises, form the second category. In these birds, the hind toe is well developed and at times is as long (or almost as long) as the front center toe (Figures 10.11, 10.12, and 10.13). In some species, the hind toe, as it points straight back, may occasionally form a straight line with the front center toe. The three forward-pointing toes are often asymmetrical; they are of different lengths and are arranged at different distances or angles from each other. The tracks show three asymmetrical toes pointing forward with one long toe pointing backward. Please note that because of the great diversity of species in this group, the degree of asymmetry is quite varied. The rock dove, blue jay, and great blue heron tracks in Figures 10.14, 10.15, and 10.16 are other representative examples of this group. Hawks and eagles have similar characteristics but have very robust toes (Figures 10.17 and 10.18).

Northern saw-whet owl

Woodpeckers, Roadrunners, and Owls. The third category consists of flickers and most other woodpeckers, roadrunners, and sometimes owls, where two toes point forward and two point to the rear, forming a rough K- or X-like pattern (Figures 10.19 and 10.20, page 310). Owl tracks may also show two toes pointing forward, one to the side and a very short toe pointing to the rear or off to the side.

Nails are usually evident in all three categories of bird tracks, from the small, very sharp nails of perching birds to the large, dull nails of geese and other waterfowl. Note

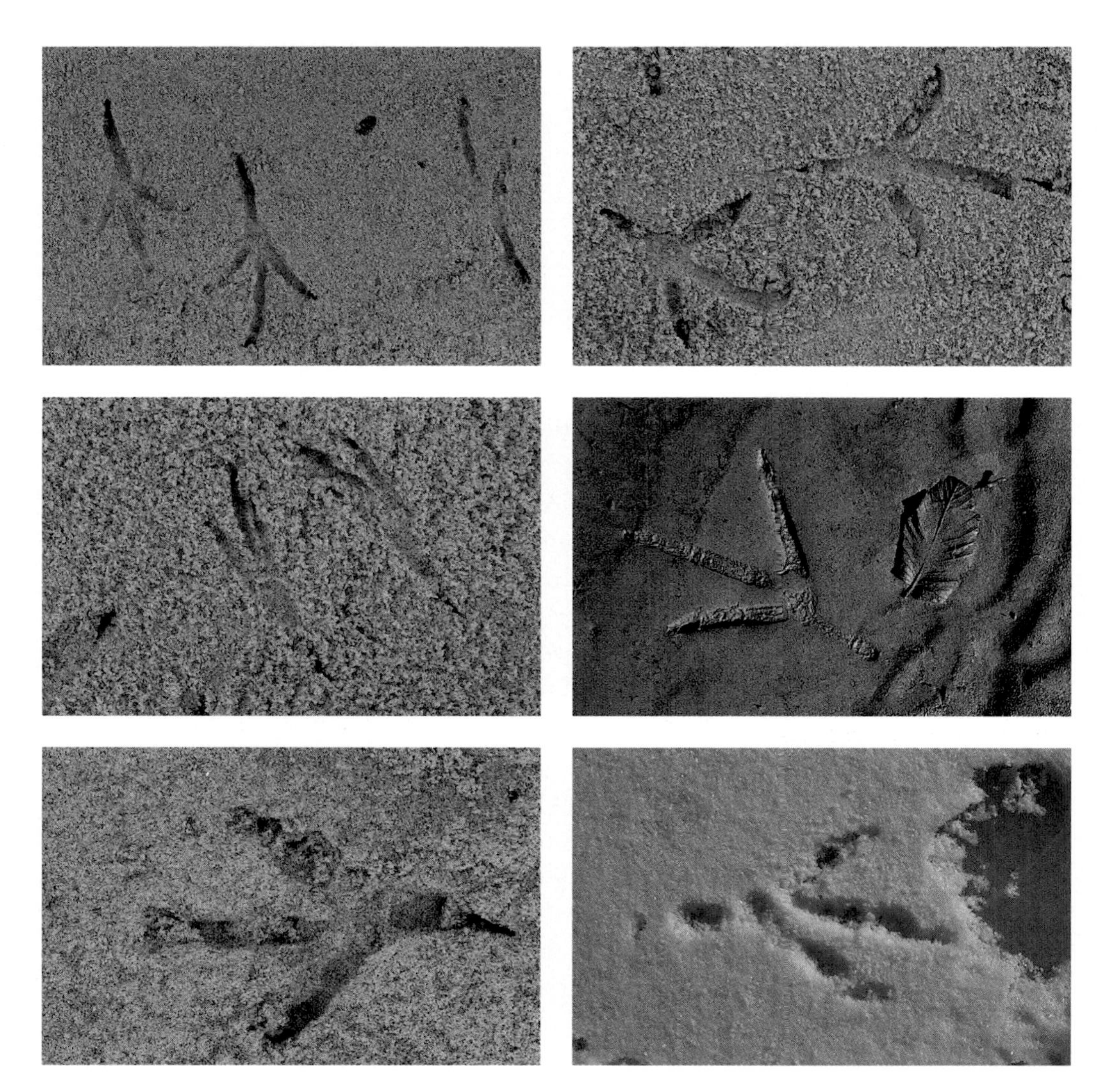

FIGURE 10.13 *(top left)* *Three robin tracks in sand show movement in different directions. Robins spend time both on the ground and in trees. They leave an alternating walking pattern and a hopping pattern. Robin tracks measure 1 9/16" to 2" long by 1 3/16" to 1 1/4" wide.*

FIGURE 10.14 *(top right)* *Soft sand has perfectly recorded these rock dove (domestic pigeon) tracks. Track measurements are 2" to 2 5/8" long by 1 3/4" to 2 7/8" wide.*

FIGURE 10.15 *(middle left)* *Blue jay tracks in sand show how closely the outer toes fall to the center toe in this species. Tracks measure 1 5/8" to 2 3/8" long by 1/4" to 5/8" wide.*

FIGURE 10.16 *(middle right)* *River silt provided the perfect substrate to record this great blue heron track. Track measurements are 6 3/8" to 8" long by 4" to 6" wide.*

FIGURE 10.17 *(bot. left)* *A broad-winged hawk track in very soft sand shows the toes looking very robust. Broad-winged hawk tracks measure 2 3/4" to 3 1/4" long by 2 1/8" to 2 1/4" wide.*

FIGURE 10.18 *(bot. right)* *A bald eagle left this track in the snow as it walked over to feed on a white-tailed deer carcass. Bald eagle track measurements are 6" to 7 1/2" long by 3" to 5" wide.*

FIGURE 10.19 *(left)*
This barred owl track forms a rough K in soft sand. Like other large birds of prey, its toes look thick and powerful. Its tracks measure 2⅞" to 3⅝" long by 1⅝" to 2⅝" wide.

FIGURE 10.20 *(right)*
These yellow-bellied sapsucker tracks in sand are typical of most woodpeckers. Yellow-bellied sapsucker tracks measure 1⅜" to 1¾" long by ⅜" to ⅝" wide.

that some birds, such as woodcocks, sandpipers, and yellowlegs, may not readily show nails.

The size of individual tracks is not usually a good indication of gender in bird species. However, several studies of wild turkeys have proved useful in this regard. The tracks of male wild turkeys (toms) in the eastern United States can easily be 5" long, not including the hind nails. You can be fairly sure that tracks 4¼" or longer are indicative of a tom; those under 4⅛" belong to females, small toms, or jakes (males born the previous season.) Other track measurements can also be helpful in determining gender: tracks 4¾" or wider and middle toes ½" or wider would also indicate a tom.

TRAIL PATTERNS. When it comes to trail patterns, the three categories we established for individual tracks no longer hold true. Except for ground dwellers, such as wild turkey and grouse, there is no set rule. Generally speaking, birds that spend a lot of time on the ground–wild turkeys, sandpipers, pheasants, geese, and herons–will leave an alternating pattern of tracks similar to the alternating pattern left by a person walking. Birds that spend a lot of time in trees, such as sparrows, crows, finches, chickadees, and juncos, will leave a hopping pattern–two tracks side by side as the bird hops along–as well as the alternating walking pattern. Eagles, hawks, and owls will usually walk, not hop, when on the ground.

SIGN: *Tree Cavities.* Like mammals, birds leave sign as they search for food. The pileated woodpecker creates some of the most conspicuous bird sign in the forest. The woodpecker locates its preferred food–carpenter ant colonies–by listening for them in live or dead trees. This large woodpecker "drills" into the tree with its beak to get at the ants. It is able to dislodge hefty pieces of wood, sometimes creating extensive cavities well over a foot in length or a foot deep (Figure 10.21). At other times, holes may not be more than an inch long. Typically these holes are rectangular in shape, but may also be irregular. They may be arranged two or more in a group. A pileated woodpecker will peck wherever the ants are, whether it's at the very base of the tree or high in its limbs. In its search for food, no other woodpecker in North America makes holes anywhere comparable in size to the pileated's.

Another woodpecker that leaves a lot of tree holes is the yellow-bellied sapsucker. These holes, however, are small, often less than ¼" in length, round or oval in shape, sometimes squared at the ends. The holes are often arranged in neat parallel rows, working their way partly around trunks and limbs (Figure 10.22). The sapsucker bores its way through to the cambium (inner) layer of bark, causing the sap to flow. The bird will wipe up or suck the oozing sap with its brush-like tongue. It may also eat insects that have been attracted by the sap. I've seen the sapsucker's work on hemlocks, tulip poplars, hickories, fruit trees, and mountain ash.

The pileated and other woodpeckers will also make holes in trees for nesting, but these cavities are usually round. Wood ducks, hooded mergansers, bluebirds, nuthatches, tree swallows and some owls also use tree cavities for nesting.

FIGURE 10.21 *(top)* *Excavated pieces of tree next to the trunk of this balsam fir, along with the large hole at its base, offer clear evidence of a pileated woodpecker's work.*

FIGURE 10.22 *(bot.)* *In search of sap and insects, the yellow-bellied sapsucker makes small holes in trees, often less than ¼" in length. The holes on this tree expanded as the tree grew.*

SIGN: ***Ground Holes.*** Besides holes in trees, look carefully for small holes at muddy spots in moist bottomland forests, especially among alder thickets. American woodcocks often visit these muddy places in search of earthworms. They have long bills specially adapted for extracting these worms. The tip of the upper mandible is flexible, allowing the bird to grasp the worm while probing, without even opening its bill. This probing results in groups of somewhat rectangular holes in the mud from ⅛" to 5⁄16" in diameter (Figure 10.23). Look for the woodcock's tracks (Figure 10.9), which are 1½" to 2" long (excluding the hind toe), with more slender toes than those of ruffed grouse . The woodcock doesn't weigh much, though, so the mud has to be soft for its tracks to register. The American woodcock's

FIGURE 10.23
These tracks and small holes were made by an American woodcock as it probed into the mud for earthworms. The holes may often appear rectangular and measure 1/8" to 5/16" wide.

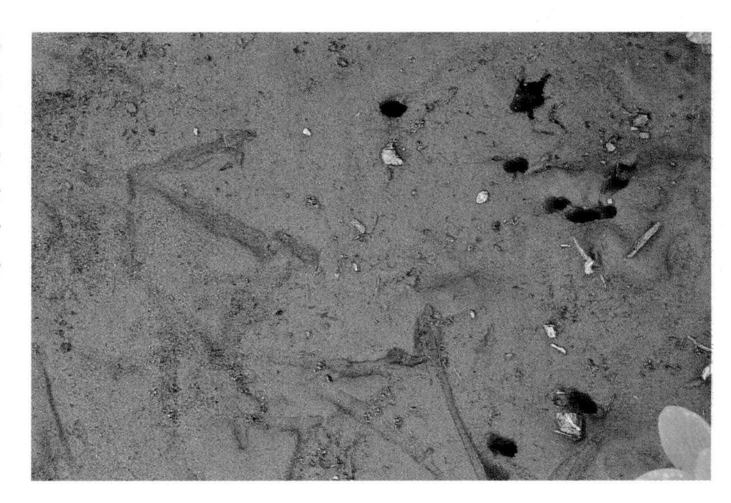

close relative, the common snipe, often shares the same habitat, has similar feeding habits, and makes similar probing holes. Many shorebirds also make probing holes but in different, more exposed habitats.

Also look for holes in banks or sandy cliffs along rivers and lakeshores where belted kingfishers, bank swallows and northern rough-winged swallows burrow into the earth for their nests. Burrowing owls also make their nests in holes in the ground in prairie dog habitats in the central and western United States, in the prairies of southwestern Canada, and in Florida.

SIGN: *Scat and Pellets.* The easiest way to tell the droppings of birds from most mammals is to look for the white coating of uric acid on the bird scat. Because mammals urinate and defecate through different pathways, their urine is not defecated with the scat. Many birds, including songbirds, ground birds, ducks, geese, swans, and woodpeckers, on the other hand, defecate their uric acid with their undigested solid wastes. It most often appears

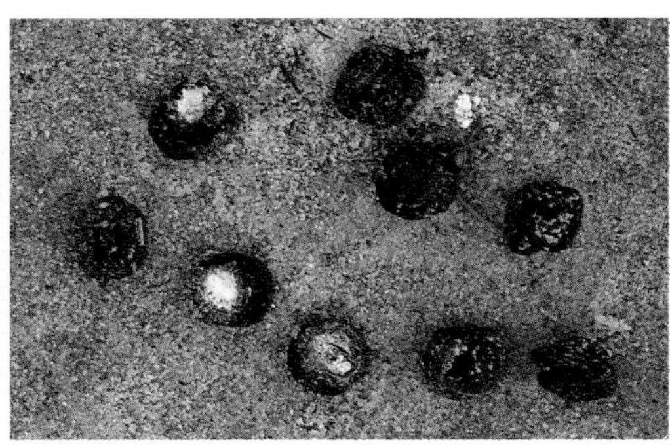

FIGURE 10.24 *(left)* *Ring-necked pheasant scat is about 3/16" to 7/16" in diameter and is irregular and bulbous in appearance.*

FIGURE 10.25 *(right)* *These bobwhite droppings are 1/4" to 9/16" in diameter. The coils that form the circle are around 1/16" in diameter. Note the white uric acid.*

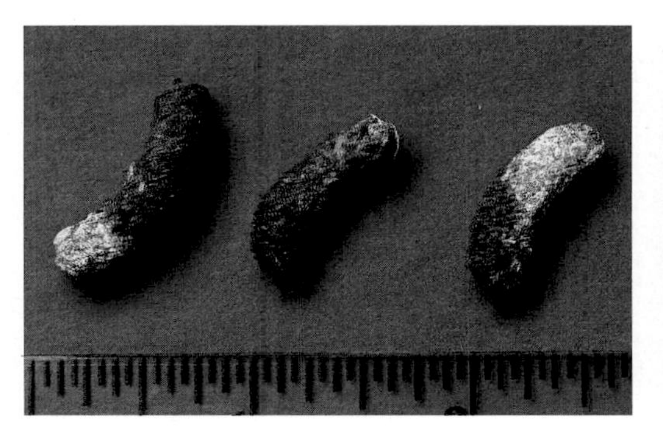

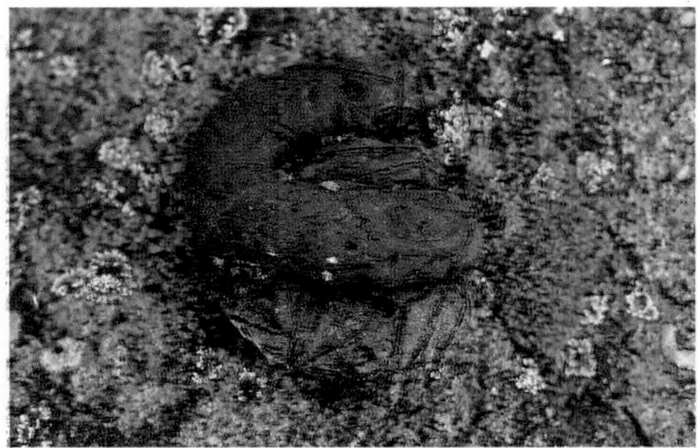

as a semi-liquid white substance coating part of the scat (Figures 10.24, 10.25, and 10.26). Be aware, however, that not all bird scat will have a coating of white. A good example of this is the ruffed grouse soft dropping (Figure 10.27). Also, as scat ages and weathers, the white coating wears off. The uric acid coating on the goose droppings in Figure 10.28 has mostly disappeared because of aging. A general rule of thumb is most scat containing a white liquid coating is that of a bird, but the absence of uric acid does not necessarily indicate a mammal scat.

Birds of prey and herons, gulls, crows, and ravens do not leave white-coated droppings. Their uric acid is defecated separately from their solid waste and is sometimes referred to as whitewash. It is most prominent where the birds eat or roost. These birds do not defecate their solid wastes. Instead they regurgitate undigested food substances (usually hair, bones, feathers, etc.) in oval balls called pellets (Figures 10.29 and 10.30).

Since many different birds, varying greatly in size, void in this manner, it is not always clear which pellet was cast by a particular species. Most pellets that are found are usually those of owls or hawks.There are a few distinctions between these two types of birds to aid in the identification of their pellets. Hawks have stronger digestive juices than owls and usually manage to digest all the small bones they have consumed (Figure 10.29). Sometimes hawk pellets are nothing but balls of animal hair, the only remnants of the hawk's prey. Owls, however, cast pellets that contain many small delicate bones (Figure 10.30). Entire skulls of small animals such as voles, mice, or moles are often discovered intact. (See Figure 10.31 for a comparison of hawk and owl pellets.)There is nothing more fascinating than taking an owl pellet apart and examining its contents. I've ob-

FIGURES 10.26, 10.27
Ruffed grouse scat comes in two forms: the more common hard scat (left) and the less common soft scat (right). This soft scat is variable in shape and color and usually consists of a very fine, putty-like material with little or no smell. This sample measured about 1" across.

FIGURE 10.28
These old, dried-up Canada goose droppings demonstrate a variety of sizes and colors in the scat. This bird's diet consists mostly of marsh plants, especially root stocks.

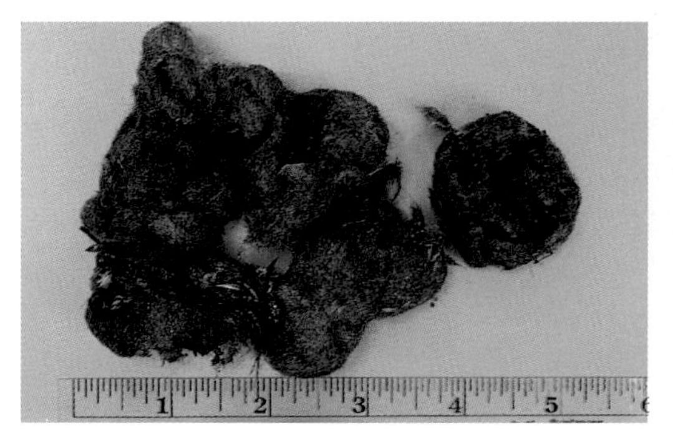

FIGURE 10.29 *(left)*
This large hawk pellet consists mostly of hair. Hawk pellets do not normally contain the delicate bones found in owl pellets.

FIGURE 10.30 *(right)*
These owl pellets (sometimes called castings) are from small to medium-sized owls. The contents are mostly rodent hair and well-preserved rodent bones and skulls.

served many people become spellbound when they first witness an owl pellet being dissected.

I have seen owl and hawk pellets ranging in size from $1\frac{3}{8}$" long to over 4". These larger pellets can be mistaken for red fox or coyote scat. You can often tell the pellets from the scat, however, through smell. When fresh, red fox or coyote scat has a musky odor to it, while the bird pellet is fairly odorless. In most cases, scat will also have fecal material mixed in with hair and bones, but bird pellets lack this fecal material. The multiple small, well-preserved bones found in owl pellets offer another clue that the specimen is not a canine scat.

Another valuable identification tool in distinguishing between hawks and owls is the different ways in which they expel uric acid. When hawks expel their uric acid they usually lean forward and eject it with some force, projecting it behind them. Owls normally drop their uric acid straight down. Wildlife rehabilitators I know who tend these birds have told me how difficult it is to clean the hawk's enclosure compared to the owl's. The hawk's

FIGURE 10.31
A comparison of the contents of a hawk pellet (left) and an owl pellet (right) shows all the small bones in the owl pellet, while the hawk pellet contains only one large bone. Because hawks have much stronger digestive juices than owls, their systems dissolve many of the smaller bones.

uric acid is all over the walls and bottom of the enclosure while the owl's waste is just on the floor.

How all this applies to tracking is easily demonstrated by the kill site in Figure 10.32. A gray squirrel was discovered here, killed and half eaten by its predator. It became evident upon examining how the uric acid was projected away from the kill site that the squirrel was killed by a hawk. An owl would not have thrown its uric acid back in this manner.

FIGURE 10.32
Streaks of white uric acid ejected away from the kill sight clearly denote the presence of a hawk. Owls drop their uric acid waste straight down. The kill is a gray squirrel.

Size of scat is not often a reliable factor in determining species or gender of birds, though it can be useful in some cases. Studies of wild turkey droppings have demonstrated that the sex of turkeys can be determined by the size and shape of their scat. Toms (male turkeys) usually have a fairly straight or J-shaped tubular dropping (Figure 10.33). Hens' droppings are curled, corkscrewed, or bulbous shaped (Figure 10.34). Biologists believe the reason for this distinction in shape is physiological and not a difference in diet or some other behavior. These differences work best when comparing adult birds, but are not infallible: turkeys will also leave soft droppings that have no relevance to gender. These soft droppings are brown to dark brown or almost black, and are similar to the soft grouse dropping in Figure 10.27.

Researcher Wayne Bailey working at the Patuxent Research Refuge in Laurel, Maryland, found that an adult tom's dropping may reach almost ⅝″ in diameter, while a young turkey's scat seldom reached a diameter of ⅜″. Hens' scat, on the other hand, measured between 3⁄16″ and 5⁄16″ in diameter. From this and other research, it is safe to conclude that a dropping over ⅜″ in diameter is most likely that of an adult tom. Droppings that measure less than ⅜″ in diameter could be those of an immature male, or a mature or immature female turkey.

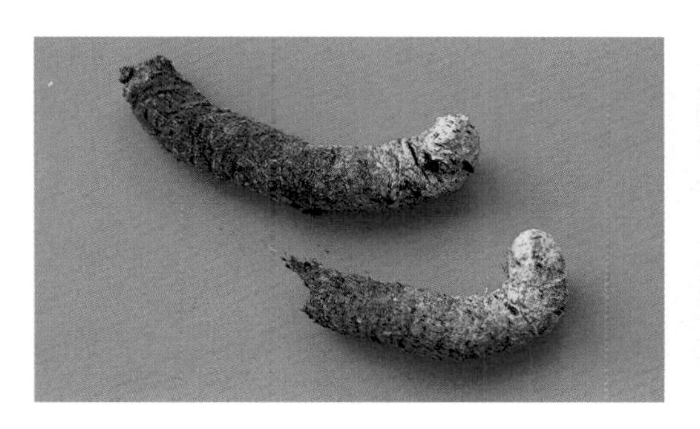

FIGURES 10.33, 10.34 *(bot.)*
Male and female wild turkey droppings can look quite different. On the left, the J-shaped scat of a tom; on the right, typical female scat.

Track and Trail Data
QUICK REFERENCE CHARTS
ALL DIMENSIONS ARE IN INCHES

Small Rodents

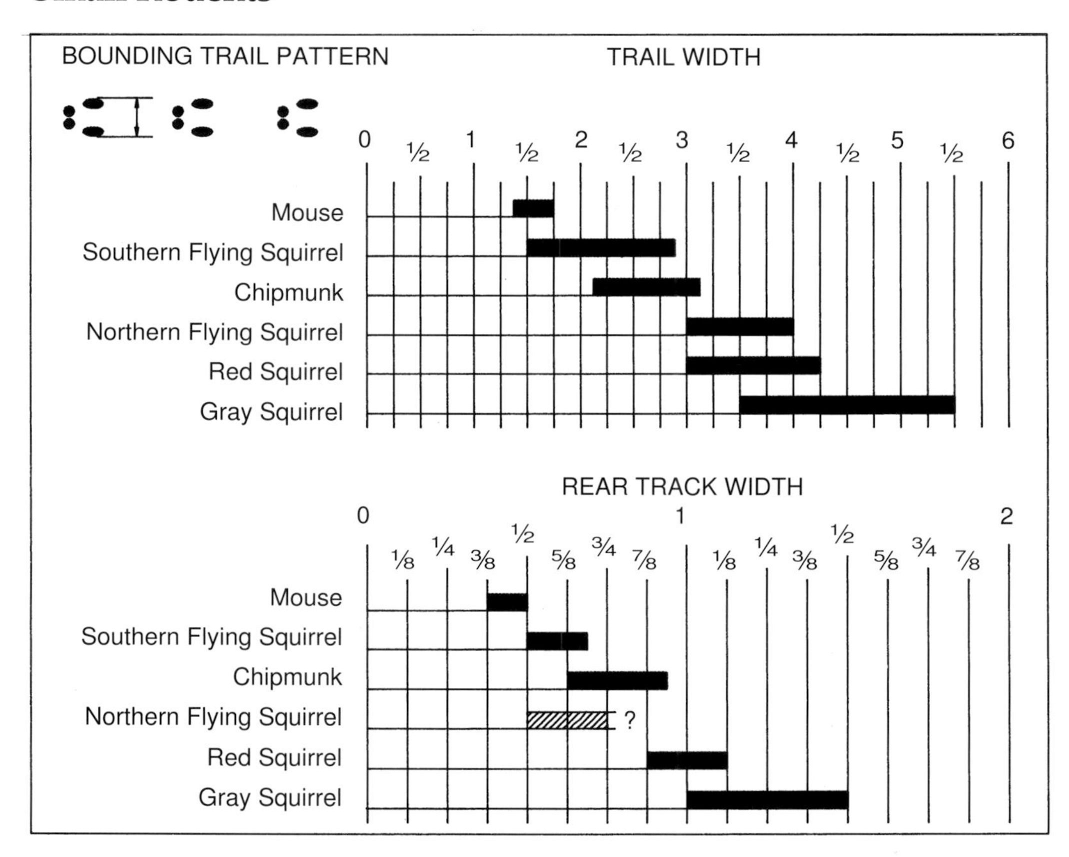

Weasel Family

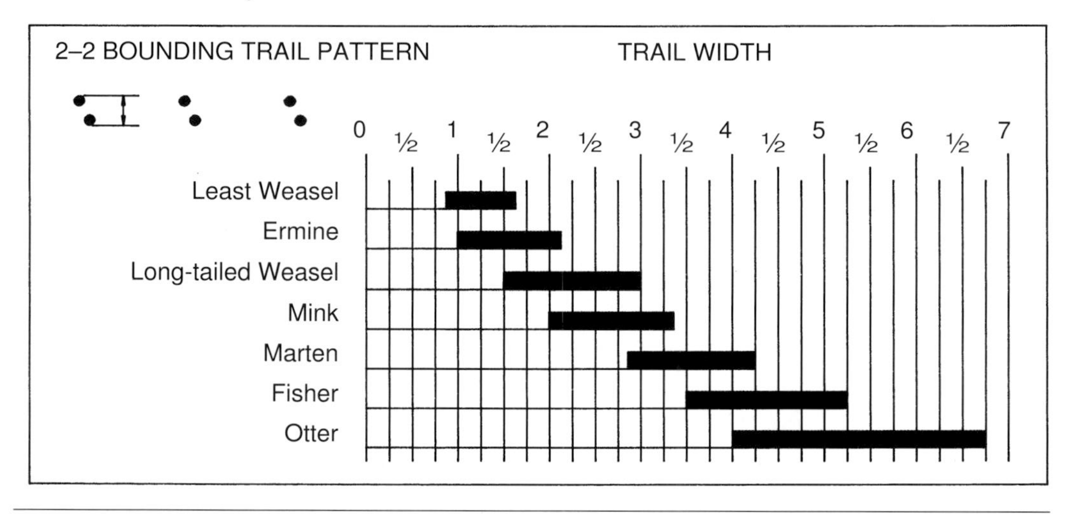

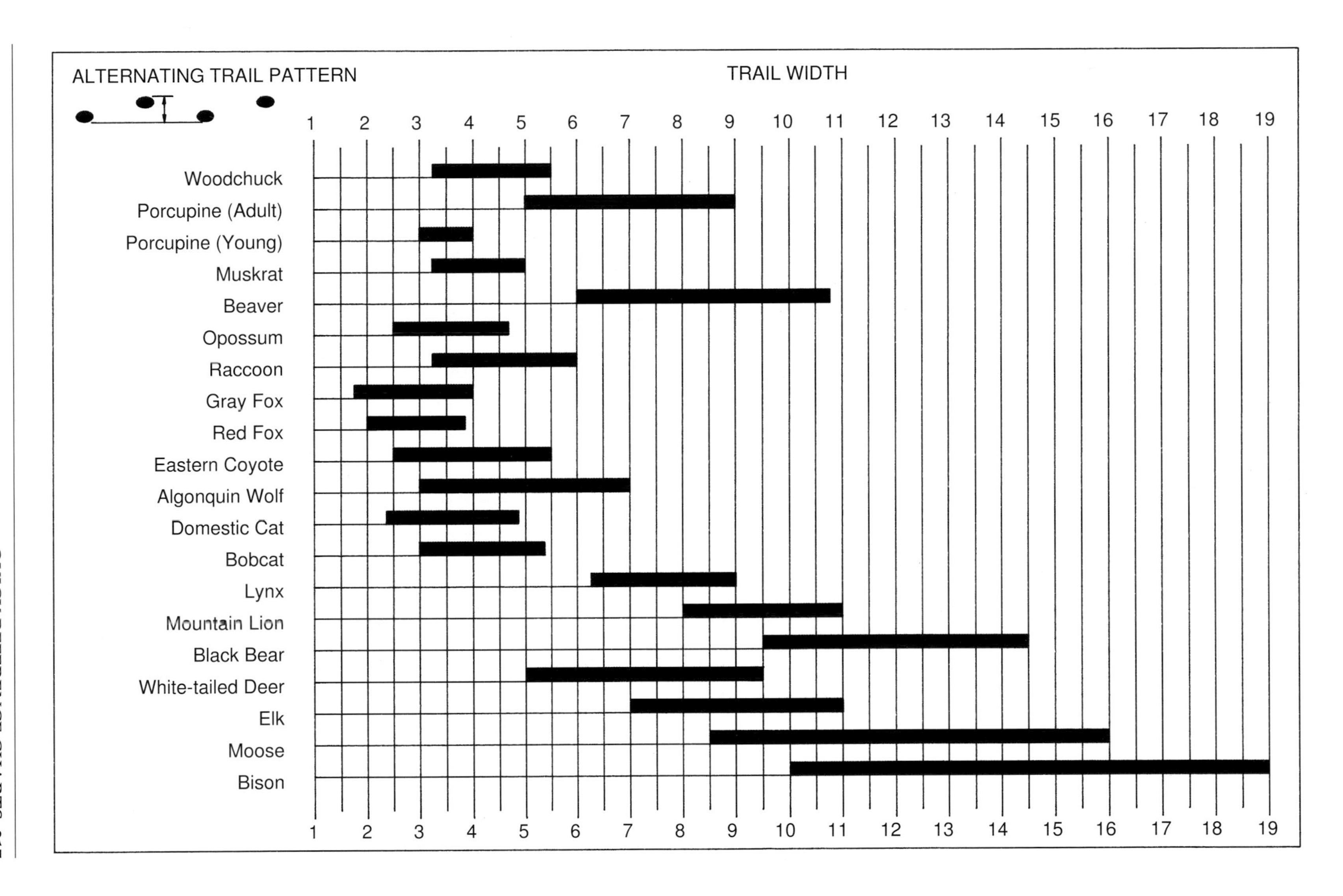
ALTERNATING TRAIL PATTERN
TRAIL WIDTH
1
2
3
4
5
6
7
8
9
10
11
12
13
14
15
16
17
18
19
Woodchuck
Porcupine (Adult)
Porcupine (Young)
Muskrat
Beaver
Opossum
Raccoon
Gray Fox
Red Fox
Eastern Coyote
Algonquin Wolf
Domestic Cat
Bobcat
Lynx
Mountain Lion
Black Bear
White-tailed Deer
Elk
Moose
Bison

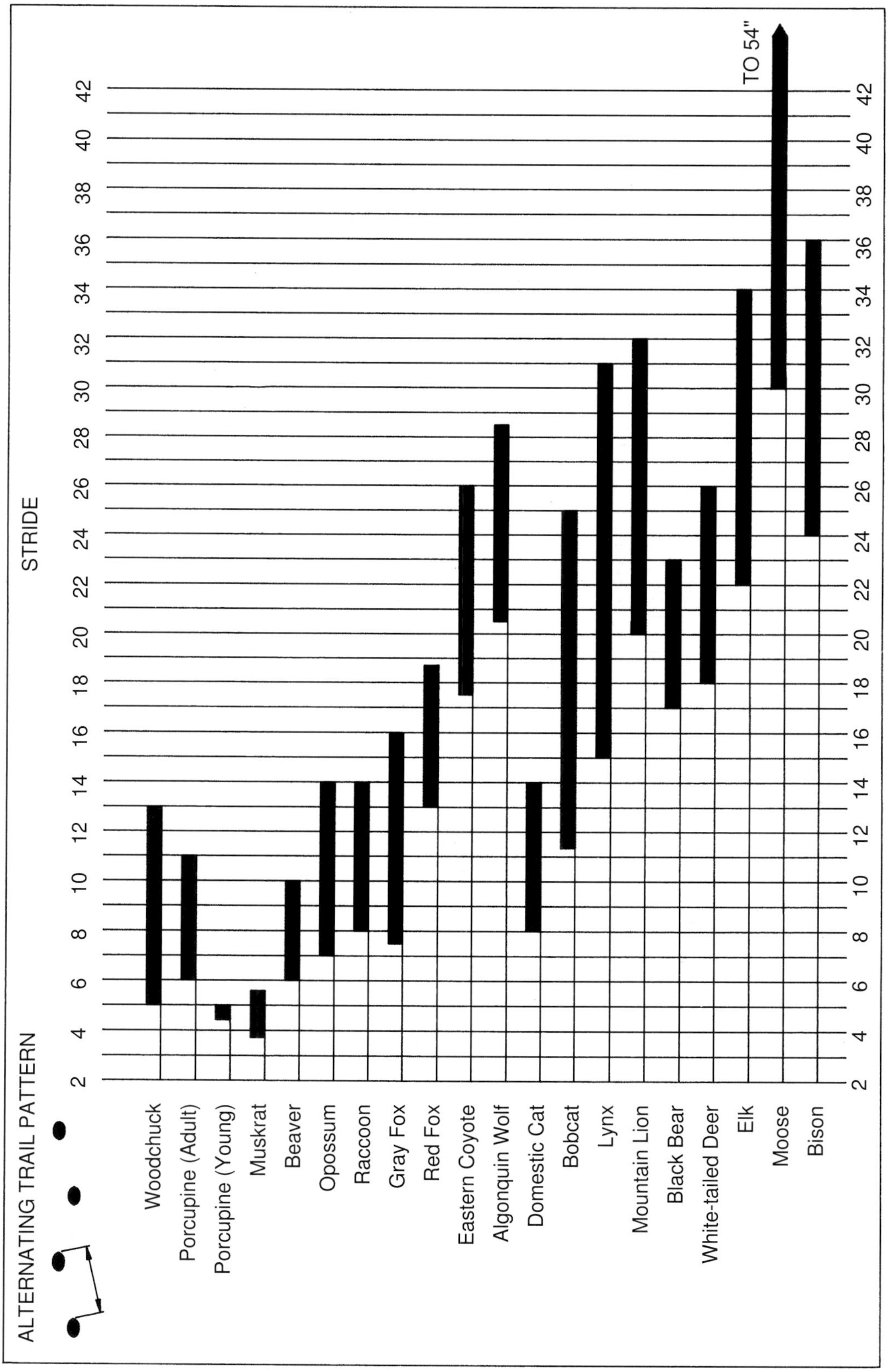

ALTERNATING TRAIL PATTERN

STRIDE

2 4 6 8 10 12 14 16 18 20 22 24 26 28 30 32 34 36 38 40 42

TO 54"

Woodchuck
Porcupine (Adult)
Porcupine (Young)
Muskrat
Beaver
Opossum
Raccoon
Gray Fox
Red Fox
Eastern Coyote
Algonquin Wolf
Domestic Cat
Bobcat
Lynx
Mountain Lion
Black Bear
White-tailed Deer
Elk
Moose
Bison

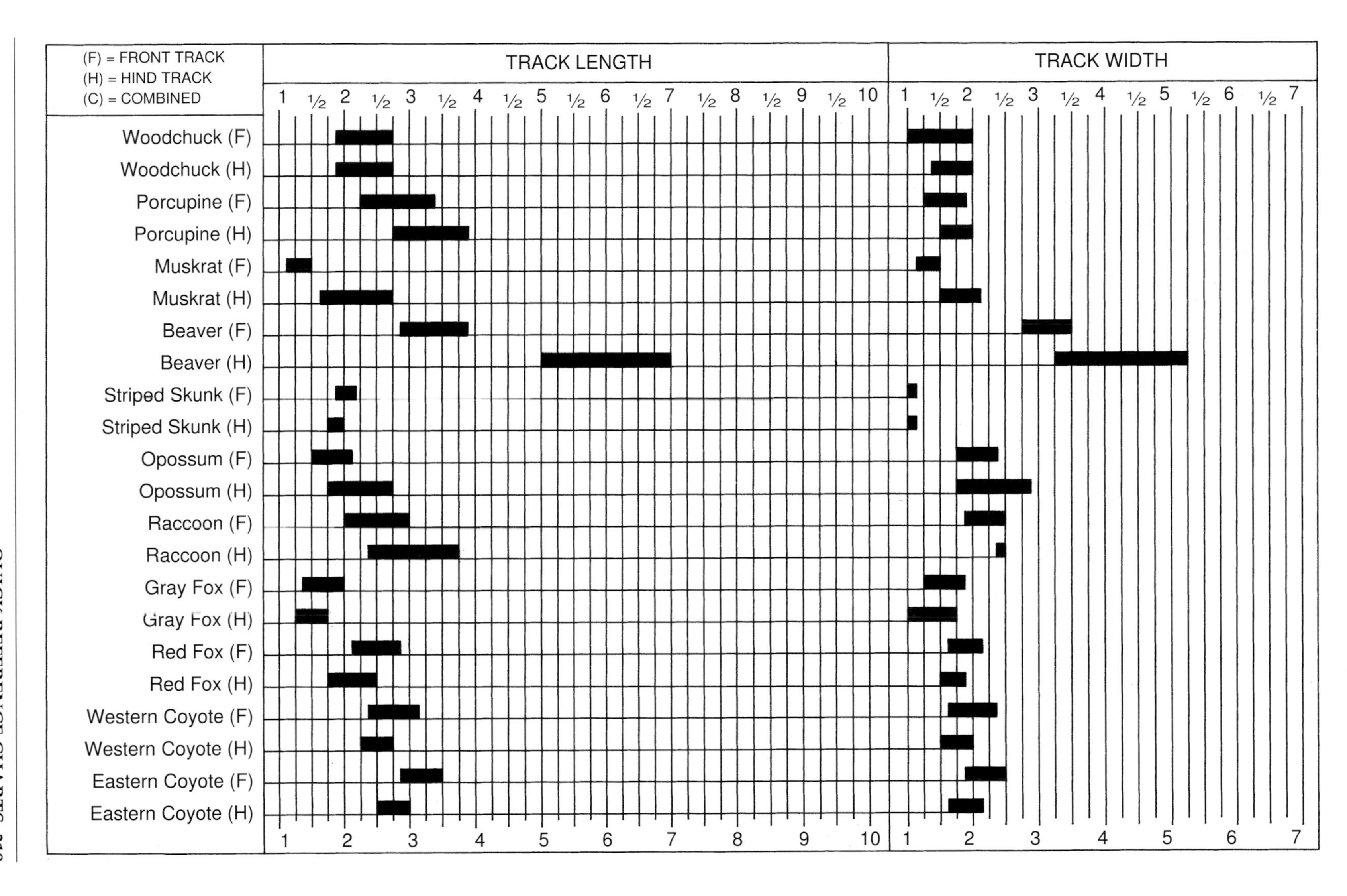
(F) = FRONT TRACK
(H) = HIND TRACK
(C) = COMBINED
TRACK LENGTH
TRACK WIDTH
Woodchuck (F)
Woodchuck (H)
Porcupine (F)
Porcupine (H)
Muskrat (F)
Muskrat (H)
Beaver (F)
Beaver (H)
Striped Skunk (F)
Striped Skunk (H)
Opossum (F)
Opossum (H)
Raccoon (F)
Raccoon (H)
Gray Fox (F)
Gray Fox (H)
Red Fox (F)
Red Fox (H)
Western Coyote (F)
Western Coyote (H)
Eastern Coyote (F)
Eastern Coyote (H)
1 1/2 2 1/2 3 1/2 4 1/2 5 1/2 6 1/2 7 1/2 8 1/2 9 1/2 10
1 1/2 2 1/2 3 1/2 4 1/2 5 1/2 6 1/2 7

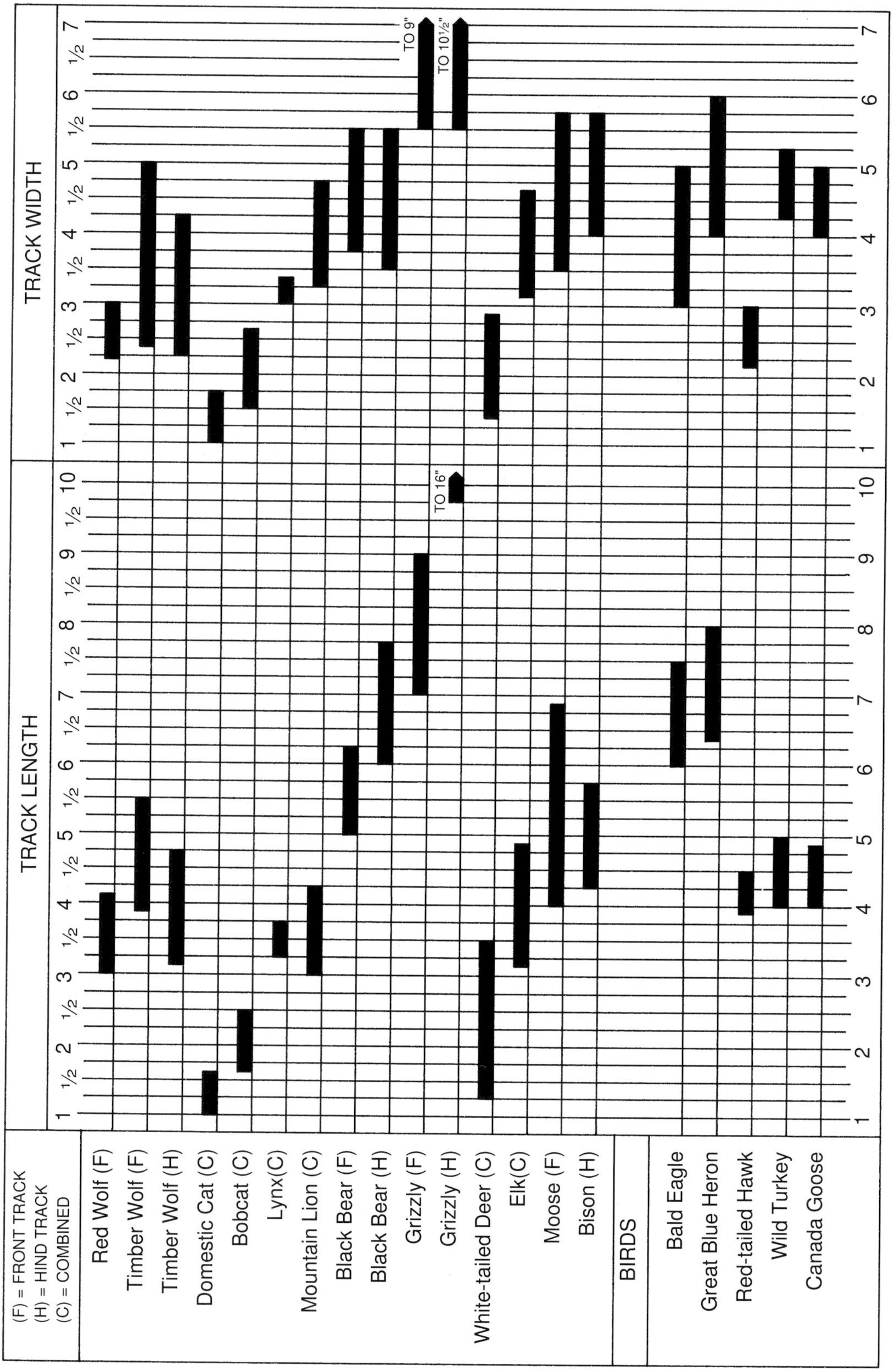

(F) = FRONT TRACK
(H) = HIND TRACK
(C) = COMBINED

TRACK LENGTH

1 ½ 2 ½ 3 ½ 4 ½ 5 ½ 6 ½ 7 ½ 8 ½ 9 ½ 10

TRACK WIDTH

1 ½ 2 ½ 3 ½ 4 ½ 5 ½ 6 ½ 7

Red Wolf (F)
Timber Wolf (F)
Timber Wolf (H)
Domestic Cat (C)
Bobcat (C)
Lynx(C)
Mountain Lion (C)
Black Bear (F)
Black Bear (H)
Grizzly (F) — TO 9"
Grizzly (H) — TO 16" — TO 10½"
White-tailed Deer (C)
Elk(C)
Moose (F)
Bison (H)

BIRDS

Bald Eagle
Great Blue Heron
Red-tailed Hawk
Wild Turkey
Canada Goose

1 2 3 4 5 6 7 8 9 10

1 2 3 4 5 6 7

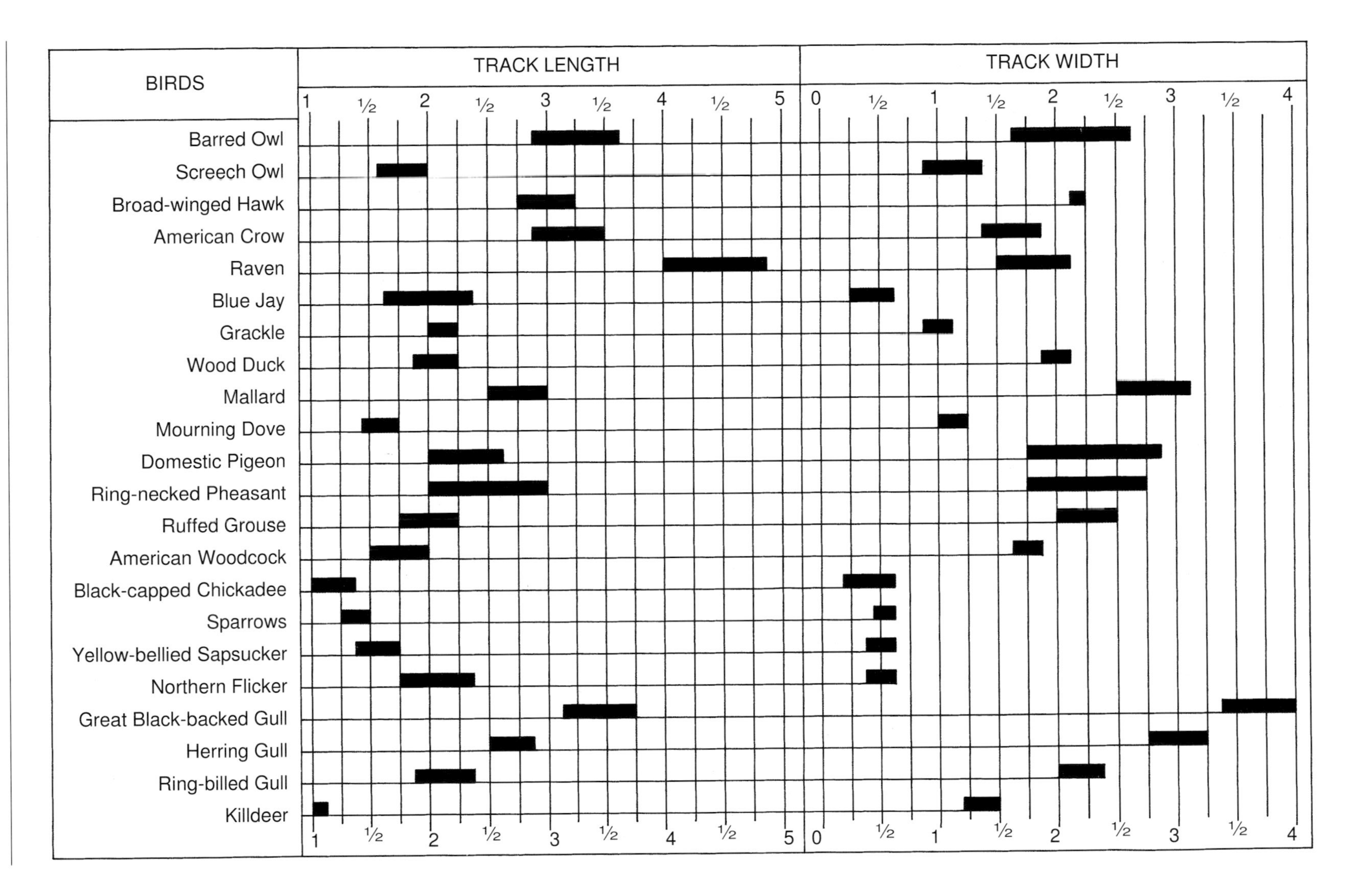
BIRDS
TRACK LENGTH
TRACK WIDTH
1
1/2
2
1/2
3
1/2
4
1/2
5
0
1/2
1
1/2
2
1/2
3
1/2
4
Barred Owl
Screech Owl
Broad-winged Hawk
American Crow
Raven
Blue Jay
Grackle
Wood Duck
Mallard
Mourning Dove
Domestic Pigeon
Ring-necked Pheasant
Ruffed Grouse
American Woodcock
Black-capped Chickadee
Sparrows
Yellow-bellied Sapsucker
Northern Flicker
Great Black-backed Gull
Herring Gull
Ring-billed Gull
Killdeer

BIBLIOGRAPHY

Anderson, Peter. *In Search of the New England Coyote.* Chester, Conn.: Globe Pequot Press, 1982.

Audubon, John James. *Journey of John J. Audubon.* Cambridge, Mass.: The Business Historical Society, 1929.

Bailey, R. Wayne. "Sex Determination of Adult Wild Turkeys by Means of Dropping Configuration." *Journal of Wildlife Management* 20, no. 3 (1956): 220.

Banfield, A.W.F. *The Mammals of Canada.* Toronto: University of Toronto Press, 1974.

Barker, Will. *Familiar Animals of America.* New York: Harper & Brothers, 1965.

Bekoff, Marc, ed. *Coyotes: Biology, Behavior and Management.* New York: Academic Press, 1978.

Bekoff, Marc, and Michael C. Wells, "The Social Ecology of Coyotes." *Scientific American,* April 1980, 130–148.

Burt, William H., and Richard P. Grossenheider. *A Field Guide to the Mammals.* Boston: Houghton Mifflin, 1976.

Buss, Mike, and Ron Truman, eds. *The Moose in Ontario.* Book 1: *Moose Biology, Ecology and Management.* Toronto: Ontario Ministry of Natural Resources, 1990.

Cardoza, James E. "A Summary of Information on the Phylogeny, Distribution, Life History and Management of the Coyotes, with Emphasis on the Northeast." Appendix to Massachusetts Federal Aid in Wildlife Restoration Report W-35-R-23, Job I-8, 1981.

———. "Coyote Concerns" (draft). Massachusetts Federal Aid in Wildlife Restoration Project W-35-R, Massachusetts Audubon Public Service Information Sheet, 1989.

Chanin, Paul. *The Natural History of Otters.* New York: Facts on File, 1985.

Chapman, Joseph A., and George A. Feldhamer, eds. *Wild Mammals of North America. Biology, Management, and Economics.* Baltimore: Johns Hopkins University Press, 1982.

Danner, Dennis, and Norris Dodd. "Comparison of Coyote and Gray Fox Scat Diameters." *Journal of Wildlife Management* 46, no. 1 (1982): 240–241.

DeGraaf, Richard, and Deborah Rudis. "New England Wildlife: Habitat, Natural History, and Distribution." Gen. Tech. Report NE-108. U.S. Department of Agriculture, Northeastern Forest Experiment Station, 1987.

Dickson, James G., ed. *The Wild Turkey, Biology and Management.* A National Wild Turkey Federation and USDA Forest Service Book. Harrisburg, PA.: Stackpole Books, 1992.

Einarsen, Arthur. *Determination of Some Predator Species by Field Signs.* Corvallis: Oregon State College Press, 1956.

Forrest, Louise. *Field Guide to Tracking Animals in the Snow.* Harrisburg, Pa.: Stackpole Books, 1988.

Forsyth, Adrian. *Mammals of the Canadian Wild.* Camden East, Ontario: Camden House Publishing, 1985.

Geist, Valerius. *Mountain Sheep: A Study in Behavior and Evolution*. Chicago: University of Chicago Press, 1971.

Godin, Alfred. *Wild Mammals of New England*. Baltimore: Johns Hopkins University Press, 1977.

Goff, Gary, Joseph C. Okoniewski, Shari L. McCarty, and Daniel J. Decker. "Eastern Coyote *(Canis latrans)*." Number 16. Department of Natural Resources, New York State College of Agriculture and Life Sciences, at Cornell University, Ithaca, 1984.

Gosling, Nancy Wells. *Flying Squirrels: Gliders in the Dark*. Washington: Smithsonian Institution Press, 1985.

Grainger, David. *Animals in Peril*. Toronto: Pagurian Press, 1978.

Green, Jeffrey, and Jerran Flinders. "Diameter and pH Comparison of Coyote and Red Fox Scats." *Journal of Wildlife Management* 45, no. 3 (1981): 765–767.

Halfpenny, James. *A Field Guide to Mammal Tracking in Western America*. Boulder, Colo.: Johnson Books, 1986.

Halls, Lowell K., ed. *White-tailed Deer, Ecology and Management. A Wildlife Management Institute Book*. Harrisburg, Pa.: Stackpole Books, 1984.

Halpern, Daniel, ed. *On Nature: Nature, Landscape, and Natural History*. San Francisco: North Point Press, 1987.

Hay, John. *In Defense of Nature*. Boston: Little, Brown & Co., 1969.

Herrero, Stephen. *Bear Attacks: Their Causes and Avoidance*. New York: Nick Lyons Books, 1985.

Hoagland, Edward. *Red Wolves and Black Bears*. New York: Random House, 1976.

Hummel, Monte, and Sherry Pettigrew. *Wild Hunters: Predators in Peril*. A World Wildlife Fund Publication. Toronto: Key Porter Books, 1991.

Jackson, Harley H.T. *Mammals of Wisconsin*. River Falls: University of Wisconsin Press, 1961.

Lagler, Karl, and Burton Ostenson. "Early Spring Food of the Otter in Michigan." *Journal of Wildlife Management* 6, no. 3 (1942): 244–254.

Lawrence, R.D *The Ghost Walker*. Toronto: McClelland and Stewart Ltd., 1983.

———. *The Natural History of Canada*. Toronto: Key Porter Books, 1988.

———. *In Praise of Wolves*. Toronto: Collins Publishers, 1987.

Lehman, Niles, et al. "Introgression of Coyote Mitochondrial DNA into the Sympatric North American Gray Wolf Populations." *Evolution* 45, no. 1 (1991): 104–119.

Lopez, Barry. *Arctic Dreams: Imagination and Desire in a Northern Landscape*. New York: Charles Scribner's Sons, 1986.

———. *Of Wolves and Men*. New York: Charles Scribner's Sons, 1978.

Lyons, Paul J., and Paul Rezendes. "The Ecology of Eastern Coyotes on Quabbin Reservation." A report submitted to Friends of Quabbin, September 1988.

Martin, Alexander, H.S. Zim, and A.L. Nelson. *American Wildlife and Plants. A Guide to Wildlife Food Habits*. New York: Dover Publications, 1951.

Maser, Chris. *The Redesigned Forest*. San Pedro, Calif.: R & E Miles, 1988.

Mason, C.F., and S.M. Macdonald. *Otters, Ecology and Conservation*. Cambridge, England: Cambridge University Press, 1986.

Matthiessen, Peter. *Wildlife in America.* New York: Viking, 1959, revised 1987.

McNamee, Thomas. *The Grizzly Bear.* New York: Alfred Knopf Ltd., 1984.

Mech, L. David. *The Wolf: The Ecology and Behavior of an Endangered Species.* Minneapolis: The University of Minnesota Press, 1970.

Melquist, Wayne E., and M.G. Gornocher. *Ecology of River Otters in West Central Idaho.* Wildlife Monographs, no. 83. Washington, D.C.: The Wildlife Society, 1983.

Mumford, R.E., and J.O. Whitaker, Jr. *Mammals of Indiana.* Bloomington: Indiana State University Press, 1982.

Murie, Olaus J. *A Field Guide to Animal Tracks.* The Peterson Field Guide Series. Boston: Houghton Mifflin, 1954, revised 1974.

Muul, Illar. "Geographic Variation in the Nesting Habits of *Glaucomys volans.*" *Journal of Mammalogy* 55, no. 4 (1974): 840–844.

Palmer, Laurence. Revised by H. Seymour Fowler. *Fieldbook of Natural History.* New York: McGraw-Hill, 1975.

Ray, G. Carleton, and M.G. McCormick Ray. *Wildlife of the Polar Regions.* New York: Chanticleer Press, 1981.

Robbins, Chandler S., Bertel Brunn, and Herbert S. Zim. *Birds of North America, A Guide to Field Identification.* The Golden Field Guide Series. New York: Golden Press, 1966, revised 1983.

Roze, Uldis. "How to Select, Climb, and Eat a Tree." *Natural History,* May 1985, 63–69.

———. *The North American Porcupine.* Washington: Smithsonian Institution Press, 1989.

Rue, Leonard Lee III. *Sportsman's Guide to Game Animals.* New York: Outdoor Life Books, Harper & Row, 1968.

Sadler, Doug. *Reading Nature's Clues: A Guide to the Wild.* Peterborough, Ontario: Broadview Press, 1987.

Sale, Kirkpatrick. *The Conquest of Paradise: Christopher Columbus and the Columbian Legacy.* New York: Penguin, 1990.

Schwartz, Charles, and Elizabeth Schwartz. *The Wild Mammals of Missouri.* Columbia: University of Missouri Press, 1959.

Seton, Ernest Thompson. *Animal Tracks and Hunter Signs.* New York: Doubleday & Co., 1925.

Shaw, Harley G. "Mountain Lion Field Guide." Arizona Special Report No. 9, Arizona Game and Fish Department, Phoenix, November 1983.

Theberge, John B., ed. *Legacy: The Natural History of Ontario.* Toronto: McClelland & Stewart, 1989.

Verts, B.J. *The Biology of the Striped Skunk.* Urbana: University of Illinois Press, 1967.

Weaver, John, and Steven Fritts. "Comparison of Coyote and Wolf Scat Diameters." *Journal of Wildlife Management* 43, no. 3 (1979): 786–788.

Whitaker, John, Jr. *The Audubon Society Field Guide to North American Mammals.* New York: Alfred A. Knopf, 1980.

Wishner, Lawrence. "Chipmunks: Lively Lords and Ladies of Our Woodlands." *Smithsonian,* October 1982, 76–85.

Wooding, Frederick H. *Wild Mammals of Canada.* Toronto: McGraw-Hill Ryerson Ltd., 1982.

INDEX

Boldface indicates an illustration.

C

D

E

F

G

H

N

O

P

Q

R

S

U

About the Author

WITH NEARLY three decades of experience in the wilderness, Paul Rezendes teaches the language of the forest to thousands of students through his year-round seminars and outdoor workshops. A professional wildlife consultant and animal tracker, he is one of the leading experts in the field.

Rezendes is the author of *The Wild Within: Adventures in Nature and Animal Teachings* (Tarcher/Putnam 1999), and a stunning photography book coauthored with his wife, Paulette Roy, *Wetlands: The Web of Life* (Sierra Club Books/Random House, 1996). An internationally published photographer, his work has appeared in numerous books, calendars, and magazines, including National Geographic Books, *The New York Times Magazine*, *Sierra*, *New Age Journal*, and *Backpacker*. He lives in Athol, Massachusetts.

FOR INFORMATION regarding Paul Rezendes' Nature Programs, please contact:

Paul Rezendes
3833 Bearsden Road
Royalston, MA 01368-9400
e-mail: reztrack@tiac.net

or visit our website at www.mossbrook.com/rez/